FIRE SERVICE ETHICS

H. SCOTT WALKER

JONES & BARTLETT LEARNING

World Headquarters
Jones & Bartlett Learning
5 Wall Street
Burlington, MA 01803
978-443-5000
info@jblearning.com
www.psglearning.com

Jones & Bartlett Learning books and products are available through most bookstores and online booksellers. To contact the Jones & Bartlett Learning Public Safety Group directly, call 800-832-0034, fax 978-443-8000, or visit our website, www.psgearning.com.

Substantial discounts on bulk quantities of Jones & Bartlett Learning publications are available to corporations, professional associations, and other qualified organizations. For details and specific discount information, contact the special sales department at Jones & Bartlett Learning via the above contact information or send an email to specialsales@jblearning.com.

18944-5

Production Credits

General Manager and Executive Publisher: Kimberly Brophy
VP, Product Development: Christine Emerton
Executive Editor: Bill Larkin
Vendor Manager: Molly Hogue
Associate Marketing Manager: Hayley Lorge
VP, Manufacturing and Inventory Control: Therese Connell
Composition and Project Management: S4Carlisle Publishing Services
Cover Design: Kristin E. Parker
Text Design: Scott Moden
Rights & Media Specialist: Thais Miller
Media Development Editor: Shannon Sheehan
Cover Image (Title Page): © donskarpo/Getty Images; © FXQuadro/Getty Images
Printing and Binding: LSC Communications
Cover Printing: LSC Communications

Library of Congress Cataloging-in-Publication Data
Names: Walker, H. Scott, author.
Title: Fire service ethics / H. Scott Walker.
Description: Burlington, Massachusetts : Jones & Bartlett Learning, [2020] | Includes bibliographical references and index.
Identifiers: LCCN 2018028555 | ISBN 9781284171655 (pbk. : alk. paper)
Subjects: LCSH: Fire departments--Management. | Fire fighters--Professional ethics. | Public safety--Moral and ethical aspects.
Classification: LCC TH9158 .W36 2019 | DDC 174/.9363--dc23 LC record available at https://lccn.loc.gov/2018028555

6048

Printed in the United States of America
22 21 20 19 18 10 9 8 7 6 5 4 3 2 1

BRIEF CONTENTS

TABLE OF CONTENTS

SECTION 2 Ethics for Frontline Personnel 76

Chapter 5

Chapter 6

Chapter 7

Chapter 8

SECTION 3 Ethics for Administrative Personnel 172

Chapter 9

Chapter 10

Chapter 11

SECTION 4 Applied Ethics for the Fire Service. 248

Chapter 12

Chapter 13

Chapter 14

Flames: Drx/Dreamstime.com;
Steel texture: © Sharpshot/Dreamstime.com

DEDICATION AND ACKNOWLEDGMENTS

This book is dedicated to those fire fighters who embody all the virtues that are best about the fire service, who bravely stand tall in crisis, who serve faithfully, who act compassionately, and who shoulder great responsibility with dignity and integrity. Any cynic needing proof of humanity's worth need look no further than the local fire fighter.

This work would not have been possible without the assistance and unwavering support of my wife Dr. Karen Sears. She was my personal editor, sounding board, critic, and counselor. Most important, she provided encouragement and boundless patience as I plodded through its writing.

ABOUT THE AUTHOR

Chief H. Scott Walker (retired) is a 30-year veteran of the fire service. He has held the rank of fire fighter/paramedic, lieutenant, assistant chief, and fire chief.

Chief Walker received his bachelor's degree from Western Illinois University and completed his master's degree at Benedictine University. Additionally, he holds several state and national certifications including Executive Fire Officer from the National Fire Academy.

He currently serves on the faculty at Western Illinois University's Fire Studies Program, where he specializes in fire service ethics, management, and administration.

Chief Walker also operates a fire service ethics resource and blog at www.fireethics.com.

PREFACE

About This Book

Why Another Ethics Book?

If you were to do some research, you would find that there is no shortage of texts on the subject of ethics. Additionally, a quick visit to your favorite search engine would indicate that there are literally thousands of Internet pages dedicated to the topic of ethics. So it is a fair question to ask, "Does the world need yet another book on ethics?" I believe the answer to that question is yes! I wrote this book for three reasons:

1. **There is a literature gap.** As a professor within a university fire program, I was amazed to find that there were few, if any, textbooks specifically addressing the issue of fire service ethics. Even more surprising was the comparative lack of academic papers addressing the subject. When you consider that there are over 1 million fire fighters in the United States who are responsible for the protection of lives and property and who safeguard almost all the country's infrastructure, it would seem that the ethics of such an important endeavor would be relevant. Yet the issue of fire service ethics remains largely ignored.
2. **The fire service has unique ethical issues.** Consider that fire fighters have free access to people's homes and businesses. They perform emergency medical procedures, and they enforce fire safety codes. The ethical implications of those responsibilities and privileges are significant.
3. **As a profession, the fire service is unique.** Members of the profession have a unique culture and a unique relationship with the public they serve. The fire service exists for one purpose only: to assist others in need. Their mission literally affects people's quality of life, and often even has life-and-death implications. Very few professions are depended on, and trusted, more than fire fighters. Because the services provided and the conditions under which they are provided are unique, the fire service has unique ethical issues. This text identifies and explores those issues.

The Author's Intentions

This book is a good-faith attempt to raise awareness regarding fire ethics and to contribute something back to the profession that has been so very good to me. Specifically, the text is intended to accomplish the following three goals.

To Make Ethics More Accessible

Ethics has been a subject of study since the founding of civilization. The principles of ethics are well explored. There are over 3,000 years of known ethics-related literature. Every generation within every culture has in some way contributed to the body of ethical philosophy. Unfortunately, much of it is abstract and often unapproachable. There is a frustrating lack of consensus regarding ethical philosophy. There are numerous competing, and even contradictory, ethical systems.

Some will (and have) argued that ethics is ethics, and that ethics for the fire service is no different from ethics within any other profession. There is some truth to this viewpoint. However, it is in the application of ethical principles to the unique circumstances of the fire service when challenges appear.

There is no new great ethical insight within this text. What this book has attempted to do is to gather existing applicable ethical principles and place them in a context that is relevant to the 21st-century fire service.

To Initiate Discussion

That the fire service provides unique services under unique circumstances is self-apparent. That we would have resulting unique ethical issues is logical. Unfortunately, there is no consensus regarding ethical issues, and there is precious little discussion regarding the ethics issues facing the fire service. As women enter the fire service in greater numbers, and as gay fire fighters are increasingly "coming out," fire chiefs are having to deal with questions regarding fire fighters dating fire fighters. As more and more fire fighters are carrying personally owned body cameras, there are increasing ethics issues regarding ownership and distribution of material. An ever-increasing number of fire fighters own blogs, which opens the door for an ethics debate regarding free speech and the need to protect fire department interests. To date, fire chiefs are facing all these issues with little or no guidance from the profession itself and with precious few resources to help them make informed decisions regarding policy.

No other profession places such a discretionary burden on its practitioners as the fire service. The military, law enforcement, the medical profession, and the legal profession all provide a robust set of ethical policies and guidelines to their members. If the fire service is to develop as a profession, it needs to begin having frank discussions about what constitutes its ethical boundaries.

To Foster Ethical Intelligence

It is not the intention of this book to instruct individual fire fighters regarding what is right and what is wrong. A basic understanding of right and wrong, and of moral and immoral, are essential for daily life. However, simplistic understanding of these principles is often not enough to deal with complicated professional issues. Ethical dilemmas exist, and problems present themselves that have no easy solutions. Fire fighters and fire service leaders will find benefit in having tools to aid them in dealing with complex issues.

I do not want to tell the reader what to think, but rather I hope to provide the reader with a basis for critical thinking. As such, the text is intended to encourage exploration of ethical philosophy in decision making.

Using the Text

The text was designed to be a general resource for the fire service. The content contained within it can be utilized in both the academic setting and within individual department training programs. I realize that the fire service has many job descriptions, and students have differing needs. Theoretically, this text can be used by high school and college programs teaching future fire fighters how to prepare for their first job in the fire service. Additionally, the text may be utilized by a fire department as part of its in-service training program for rank-and-file fire fighters. Finally, the text may be used by fire officers or aspiring officers responsible for administrative duties.

To accommodate those needs, the text has been divided into four unique sections, each of which addresses a particular subset of topics relevant to particular needs and responsibilities.

Section 1 of the book is entitled *Foundations*. Its four chapters are intended to provide the reader with the basic understanding of ethics necessary to master the material in the succeeding three sections. Chapter 1 is recommended for all readers, as it addresses the important questions of ethical relevance to the fire service. Chapters 2, 3, and 4 are grounded in academic approaches to ethics and are specifically intended to be used in college courses. The material within the sections encourages critical analysis of ethical systems and an understanding of the basics of human behavior.

Section 2 of the book is entitled *Ethics for Frontline Personnel*. Its chapters address issues directly related to rank-and-file fire fighters and company officers. The section also deals with diversity—first as an abstract concept and then as it applies specifically to the fire service. Section 2 is intended to serve dual purposes.

First, it is designed to be included in an academic curriculum aimed at future fire fighters. Second, Section 2, combined with Section 4, encompasses the majority of material appropriate for inclusion in fire service training programs.

Section 3 of the text is entitled *Ethics for Administrative Personnel*. The chapters within the section explore ethics issues faced by fire chiefs and senior administrative personnel. Included within the section are chapters on the building and maintenance of an ethical work culture, the ethical responsibilities associated with administration, and finally ethics laws. Section 3, combined with Sections 1 and 4, is especially appropriate for officer development programs and collegiate programs in fire administration.

Section 4 of the book is entitled *Applied Ethics for the Fire Service*. The three chapters within the section deal with the application of ethics on the personal and department level. Included within the section are chapters on ethical decision making, mechanisms by which unethical behavior is engaged, and a review of current ethical issues affecting local fire departments. Section 4 is universal in its application and should be used in all collegiate programs, as well as in fire fighter and fire officer training programs.

Features of the Text

In addition to the general material within the text, each chapter contains student resources intended to assist in the learning process.

Case Study

Each chapter begins with a case study. The reader is encouraged to review the facts of the scenario, and then write responses to the discussion questions. Near the end of the chapter, commentary on the case study is provided. Readers should reevaluate their responses based on both the material within the chapter and on commentary provided in the case study review.

Challenging Questions

Each chapter contains a section entitled "Challenging Questions," which includes questions to provoke critical analysis of the material covered within the chapter. Program instructors are encouraged to use the questions within the section as a basis for group discussion or written assignments.

Key Terms

Toward the end of each section is a list of key terms and phrases. The key terms tend to illuminate significant elements within the chapter and are an excellent resource for reinforcing learning objectives.

Chapter Quizzes

At the end of each chapter is a quiz that encourages you to self-evaluate your comprehension of the essential elements presented in the chapter.

Part Two: Ethics Axioms

Throughout the text, certain assumptions are made regarding ethical obligation, accountability, and responsibility. The term "ethical attachment" is frequently used in reference to ethical obligations. As the reader moves through the material, he or she may question the basis for the assumption of moral culpability, or ethical responsibility.

I have based my assumptions regarding ethical obligation and responsibilities on proposed fundamental "truths" or ethical *axioms*. For the reader to comprehend fully the assertions I make throughout the text, he or she must be familiar with these axioms. The reader should also be advised that I distinguish between personal ethics and professional ethics. As such, there are two sets of axioms.

The axioms regarding personal ethics rest heavily on the combined principles of all three philosophic approaches to ethics. These include the pursuit of virtue, the pursuit of best consequences, and the concept of personal responsibility to others. The amalgamation particularly rests on the works of Aristotle, John Stuart Mill, and Immanuel Kant (all are referenced later in the book). I have synthesized their work into what I believe to be simple principles that are easily understood and universally applicable.

The axioms regarding professional ethics are more or less my assertions based on an amalgamation of ethical principles relative to the fire service and rooted in common sense.

In both cases, the axioms do not represent groundbreaking ideas. In the best traditions of the fire service, there was no attempt to reinvent the wheel. Rather, proven solutions were applied in a new application.

Axioms for Personal Ethics

Axiom one: Do no harm or, by inaction, allow harm to come to others.

Rationale: The quality of human life and the progress of the human race depends on personal safety and cooperation; absent these are brutality and anarchy. Fundamental to that truth is the restriction from harming others for personal gain or having a complete disregard for the welfare of others with whom you have contact. We live in a society that provides us with safety, quality of life, and a potential for human development. By accepting those benefits, we have a social obligation to extend the same benefits to other members of society.

Implications:

- It is unethical to cause physical, emotional, or financial harm to another person.
- Living within a larger society infers a social contract for mutual well-being. As such, each member of a society is to some extent responsible for the welfare of others.
- Social obligation toward other human beings extends beyond actively avoiding hurting others; each member of the society has a duty to minimize unavoidable harm to others.

Axiom two: Seek justice.

Rationale: Fairness and, to a greater extent, justice are essential to the concept of ethics. Ethics is rooted in human relationships, and the acceptance that others have a right to thrive and pursue happiness is the defining nature of healthy relationships.

Implications:

- The fair allocation of resources is fundamental to ethical behavior. Greed is contradictory to ethical behavior.
- Give others that to which they are entitled. Depending on the individual's relationship with others, entitlement may include: services, acknowledgment of achievement, time, assistance, affection, intangible goods, or money.
- All individuals should be given the equal opportunity to succeed.
- With human relationships comes responsibilities and obligations. Fairness and justice require good-faith efforts to keep promises, meet obligations, and honor responsibilities.
- Authority and privilege have attendant responsibilities. It is inherently unfair and unjust to take without equal contribution. Authority and privilege must be exercised equitably. This includes both the fair application of punishment and fair distribution of reward.

Axiom three: Contribute to the common good.

Rationale: To take without contributing is to use others for personal gain. Usury is inherently unethical. Except for those living in isolation (where ethics is irrelevant), quality of life is dependent on mutual cooperation. Because a person benefits from membership in a community, that person is obligated to contribute to it.

Implications:

- Ethics decisions should consider the general welfare.

- Individual action that affects others should be consistent with society's needs for group behavior. Act as you would want everyone else to act.
- Quality of life is dependent on each member of society abiding by cooperative rules of behavior. Each member is obligated to act in a way that is essentially cooperative with others' needs.
- Each individual benefits from advances in culture made by previous generations. Each individual should contribute to culture for the benefit of future generations.

Axiom four: Respect the humanity of others.

Rationale: By benefit of birth, each person is endowed with the right to dignity, self-determination, and equality.

Implications:

- Each of us has the responsibility to respect the human rights of others.
- Lying and deception show both a lack of personal integrity and demonstrate a disregard for the dignity of others.
- Each person has a right to be treated as he or she deserves, based on how that person treats others. Any form of discrimination that negatively affects someone without cause is inherently unethical.
- Any action that degrades human dignity, or is intended to diminish the perceived value of another's life, is unethical.
- Every individual is obligated to treat others as he or she would want to be treated.

Axioms for Fire Service Professional Ethics

Axiom one: The fire service and its members have an ethical obligation to provide the highest quality public safety services as made possible by circumstance and funding.

Rationale: Public safety is one of the organizational principles behind human society. Civilization cannot grow absent safety of person and property. The welfare of community members is dependent on the fire service for the provision of public safety. The fire service's organizational principle is to save lives and protect property; this is the reason it exists, which represents an implicit promise to do both. There is an ethically imperative obligation to uphold a promise where the implications of dependency are of supreme importance.

Implications:

- Fire departments and, by extension, their members have an obligation to act in the public's interest.
- Fire departments and, by extension, its members have an ethical duty to maintain competency to fulfill their promise of service.
- By accepting a position as a fire fighter, an individual accepts the responsibilities inherent in the fire service's obligations to provide service.
- Fire fighters have an obligation to maintain health and physical fitness to fulfill their promise of service.
- Fire fighters have an ethical obligation to maintain cognitive skills necessary to fulfill their promise of service.
- Fire departments have an ethical obligation to provide the required resources for the provision of public safety, including adequate training, equipment, and staffing.

Axiom two: Fire fighters have an ethical obligation to conduct themselves in a manner consistent with the trust and faith placed in them by those they serve.

Rationale: The fire service exists for the stated purpose of assisting others in need. Because a local fire department is almost always the sole provider of its services, and because of the nature of our services, there is a significant degree of dependency on fire fighters by citizens. People in need have little choice but to call the fire service as a last and only resort. As a result, they are often vulnerable and forced to trust that fire fighters will act in a way consistent with honoring that trust.

Implications:

- Fire fighters should act consistently with the general welfare of those requesting their help.
- Fire fighters should provide services equally to all, regardless of race, religion, nationality, sexual orientation, social standing, or political affiliation.
- Fire fighters must always respect the dignity, humanity, and reputation of those they serve.
- Fire fighters make a promise to protect others. They have an ethical obligation to avoid harming others through action, inaction, or word.

Axiom three: As government agencies, fire departments control and utilize public resources, and have an ethical obligation to manage those resources judiciously and equitably.

Rationale: Whether career or volunteer, fire departments are funded by public money. Therefore, the public owns the fire department equipment. There is a fiduciary and ethical responsibility to spend public funds judiciously and to maintain public equipment.

Implications:

- Fire department administrators have an ethical obligation to conservatively budget and expend public money.
- As fire departments utilize public resources, the public has a reasonable expectation that fire department funds will be used for their benefit.
- As a public agency utilizing public funds, fire departments have an obligation to hire or admit members fairly.
- Fire fighters have an ethical obligation to treat public resources with the highest regard, including protection from loss and providing proper maintenance.
- By joining the fire department, a person makes a commitment to its policies and procedures. The department and its officers have a right to expect loyalty to the department's mission, to policy, and to legitimate authority.

Axiom four: Each fire fighter has a responsibility to the welfare of the department and to its members.

Rationale: A fire department provides a much-needed service on which many depend. Whether compensated or not, when an individual joins a fire department, that person creates a covenant with the department to assist in the delivery of those services. By joining the fire department, there is an implicit agreement to act in its welfare, and to do otherwise is deceitful.

Implications:

- A department's reputation is essential to its success and longevity. Damage to the department's reputation and standing within the community invariably has adverse effects on numerous stakeholders. A fire fighter does not have the right to damage the reputation of the department and, by extension, the reputation of other department members.
- Firefighting is dangerous work in which individual safety is often dependent on the skills and determination of others. Each fire fighter has an obligation to to meet competency requirements as they contribute to the safety of fellow fire fighters.
- Firefighting is a team-based exercise requiring competency among all its members. As such, each member has a responsibility to pass on knowledge, mentor, and promote professional development.
- Fire departments deliver services to people desperate for help. It is a reasonable assertion that the mission of the fire service is more important than individual self-interest. By joining the fire department, each member assumes responsibility for the continued improvement of the department.

Axiom five: Fire officers have an ethical responsibility to fulfill the duties bestowed on them as a condition of rank.

Rationale: By becoming fire officers, fire fighters are granted special authorities that attach special responsibilities unique to supervision. The acceptance of promotion is essentially a contract between the department and the officer to fulfill those duties.

Implications:

- Fire officers are ethically obligated to act in the public interest through the provision of emergency services, risk-reduction activities, and the responsible management of public resources.
- Fire officers are ethically responsible for the health and safety of the fire fighters they supervise. This is done by assuring core competency, compliance with industry standards and best practices, and the careful application of risk-benefit analysis.
- Fire officers have an ethical obligation to be loyal to their superior officers, department initiatives, and department policy.
- Successful delivery of public safety requires a well-trained and dedicated workforce. Fire officers have a responsibility to teach, mentor, provide professional development, and maintain discipline among those they supervise.

H. Scott Walker

FAIRFAX FIRE & RESCUE
PROTECTIVE
ADKINS
0701
0702
FAIRFAX COUNTY FIRE & RESC
FIRE & RESCUE DEPARTMENT
FAIRFAX COUNTY
PRO

SECTION 1

Foundations

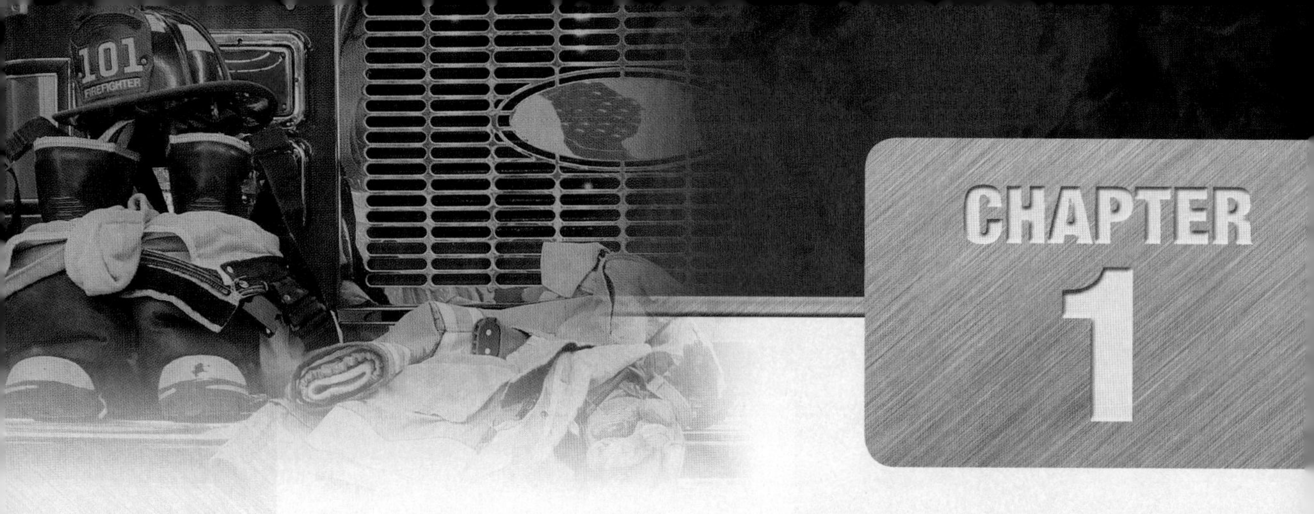

CHAPTER 1

Introduction

"Our very lives depend on the ethics of strangers, and most of us are always strangers to other people."

—Bill Moyer

OBJECTIVES

After studying this chapter, you should be able to:

- Define ethics and understand why it is important to the fire service.
- Evaluate the personal benefits achieved in the study of ethics.
- Assess the relationship between ethics and social order.
- Determine the roles and responsibilities of others in fire service, such as fire administrators, fire chiefs, and fire inspectors.
- Differentiate between personal and professional ethics.

Case Study

You're a battalion chief for the Anywhere Fire Department. Two of your crew, fire fighters, Smith and Jones, operate a janitorial service on their days off. One afternoon fire fighter Smith asks for a shift transfer because he cannot work with Jones any longer. You ask what the problem is and are told that Smith believes Jones has been stealing from the janitorial business. You explain to Smith that shift transfers are only arranged when there is an opening and then offered on a seniority basis. Smith then asks if he can file charges against Jones for stealing from him. You ask if they are both equal owners in the business and if they have any contractual obligations. Smith informs you that, yes, they are equal partners and that their arrangement is informal without any written contracts. You tell him that he does not have grounds to file charges because Jones does not appear to have violated any laws or department policies. Smith asks if he can appeal to the chief's office.

- If true, certainly Jones's theft of money from his partner is an ethical issue. Are there other ethics issues involved? What issues affect you as their supervising officer?
- Does Smith or Jones have an obligation to conduct themselves any more or less ethically than anyone else in the community? If so, how are they different?
- Should ethical violations conducted while off duty be under the purview of the fire department?
- How would you proceed?

What Is Ethics?

If you were to research the question, What is ethics? you would likely find as many definitions as sources, and they would be vaguely similar. If you were to ask your friends, their answers would most likely fall into one of five common themes:

- Ethics is about following your conscience.
- Ethics has to do with religion and morality.
- Ethics is being a good person.
- Ethics is about abiding by the law.
- Ethics is about doing what society says is right.

Each of these themes seems right, but none are complete. Religion seeks to provide a framework whereby you can identify "right and wrong." Yet if ethics were confined to religion, then ethics would apply to only religious people. Obviously, most non-religious people can and do act ethically. Conversely, history is full of examples of atrocities committed not only by self-professed religious people but in the name of religion. As such, ethics is not confined entirely to the concept of religion.

At one time or another we all have been advised to follow our conscience, yet so often our conscience can confuse and even betray us. Many of us have experienced an internal conflict arising from competing "goods" or, conversely, the necessity of choosing between the lesser of two evils. Our conscience seems to develop by the values we are taught and our life experiences. Some even argue that, to some degree, the conscience is **innate**. These factors all conspire to make the conscience essentially **subjective**. Can conscience be subjective and still have ethical relevance? If right and wrong are entirely subjective concepts, how do they have any real meaning in a world where we interact daily with others (Cooper, 2012)?

Many ethics definitions focus on the ideas of morality, good versus evil, and right versus wrong. *Merriam-Webster's Collegiate Dictionary* defines ethics as "rules of behavior based on ideas about what is morally good and bad: an area of study that deals with ideas about what is good and bad behavior: a branch of philosophy dealing with what is morally right or wrong: a belief that something is very important" (Merriam-Webster, 2003). At first glance, it would appear that **morality** and ethics are synonymous with each other, yet scholars are quick to point out that such is not the case. The concept of morality is usually tied to universally accepted principles of behavior, such as "thou shall not kill"; whereas ethics are essentially confined to human interaction. In the case of personal ethics these interactions are described in terms of fairness, justice, and the greater good. Professional ethics behavior boundaries are applied to relationships in the context of employment. Personal and professional ethics will be further discussed in Chapter 2, *Judging Behavior*.

The idea that ethics is about *being a good person* certainly sits well with our common understanding of ethics. Still, individuals who consider themselves to be good people frequently find themselves embroiled in ethics scandals. History is rife with tyrants, villains, and monsters convinced they were serving a greater good. If ethics is rooted in individual perception, is it not dependent on individual circumstance? Entire schools of ethics ideology wrestle with the issue of situational ethics. Are good and bad dependent on time, place, and circumstance? If killing is immoral, when is war justified? Who defines good and bad? Can almost any "sin" be justified somehow? In short, exactly what does being a good person mean, and who decides it?

Obeying the law is certainly consistent with most people's ideas concerning ethics. Much of our common law strives to enforce the values and ideals generally accepted by society. Yet the law once legitimized slavery and denied women the right to vote or even own property. Values change, laws change; does right and wrong change with the law? Does the law actually represent the collective agreement of the people on issues of right and wrong? Should it? Much of the U.S. Constitution is designed to protect the rights of the individual from the tyranny of the majority. Many argue that separation of church and state in concert with the doctrine of the protection of individual freedoms would suggest the state has little capacity to delineate right or wrong.

Does society have the most reliable insight into what is right and what is wrong? At one time or another, society has been deeply divided on almost all social and moral issues. While there are certain universal principles within ethical systems, moral imperatives vary from nation to nation, from group to group, and from time to time. In many ways, societies' views on right and wrong are written in pencil, in other words, subject to change. Does this somehow argue that right and wrong are subject to interpretation by the individual? How can society survive without consensus on acceptable versus unacceptable behavior?

The definition of ethics is not easily approachable. Not surprisingly, the study of ethics can be confusing. The fact that the concept of ethics seems to be apart from the ever-shifting sands of social norms, morals, the law, and personal values yet somehow tied to them has confounded students and philosophers for centuries. This much is absolute: Ethics is very much about social interaction. Your actions are judged as good or bad almost entirely on their relative impact on others. The study of ethics is complicated by the complexity and fluidity of social expectations.

To understand the text that follows, we should begin with a definition of the term **ethics**, however flawed that definition may be.

For our purposes, this text will consider the study of ethics as the pursuit of three general goals.

1. Ethics seeks to define behavior boundaries within social relationships.
2. Ethics seeks to establish generally accepted standards of right and wrong based on personal obligations, benefits to society, compliance to behavior standards, and specific virtues.
3. Ethics refers to the study and development of personal standards of behavior through continuous evaluation of one's own belief system and conduct.

Why Be Ethical?

The study of ethics is only relevant and worthwhile if you assume that ethical behavior is likewise relevant and worthwhile. The question, Why should anyone be ethical? may seem obvious, but it is at the center of a much larger issue. To be ethical, you must understand the rationale for **ethical restraint**.

From the time we are toddlers, our parents, teachers, and caregivers instruct us in the expected behaviors. Our first lessons are intended to ensure our physical safety as we explore our world; however, soon we are taught behaviors that are rooted in social expectations whose values are somewhat less evident. These lessons may include the importance of politeness, proper dress, etiquette, and, most importantly, obedience. The lessons of obedience vary in nuance and expediency. Conformity to the rules of the home and the law are relatively straightforward, whereas conformity to behaviors rooted in values, traditions, ideals, and morality are often taught subtly and subjectively. It is the subtle and subjective elements that shape our **personal ethics**.

As young children, we obey our parents in part to gain their praise and in part to avoid punishment. However, children quickly learn that sometimes good behavior goes unnoticed and unrewarded. We also learn that punishment is dependent on supervision. With these discoveries, the child begins a lifelong process evaluating the positives and negatives of following the rules of good behavior Figure 1-1.

And so we return to the question Why should we be ethical?

Figure 1-1 As human beings, we go through a lifelong process of evaluating the pros and cons of following the rules.

Avoiding Punishment

As stated earlier, from the time we are children we are instructed to follow the rules imposed on us by parents, the legal system, schools, and religion. Our society has an intricate and robust set of methodologies to monitor and enforce compliance to its prescribed rules. Certainly, avoidance of negative consequence shapes our decision process when evaluating a particular action. Compliance is then intrinsically connected to the likelihood of discovery and severity of punishment. This raises a question: If fear of punishment is a primary motivation for our behavior, then why do we refrain from unlawful or immoral actions even when we're relatively sure we will not be caught? Some would suggest that religion in general and particularly the concept of an all-seeing God who shall punish or reward our actions, mediates our behavior. This argument is sometimes referred to as **argumentum ad baculum**. Undoubtedly, the promise of heaven and the fear of eternal damnation motivate many of us to be our better selves, yet most nonbelievers demonstrate the same level of moral behavior as do their religious counterparts. Conversely, many religious people struggle with moral and ethical paradoxes to the point where, in good conscience, they can't act consistently with religious dogma. Avoidance of punishment only explains behavior to a point. Children behave even when unsupervised, and drivers obey traffic laws even when law enforcement is not around to observe them.

There is a more philosophical question regarding obedience as well. Is fear of punishment an actual expression of ethics, or merely self-serving behavior? From a purely religious perspective, is obedience to God a good within itself? Does compliance with the law make a person lawful? If I do not break the law or commit a sin only out of fear of punishment, am I a good person or potentially bad person who lacks nerve?

Obedience to the law also brings up the question of the law's value. Is it more accurate to say God commands it because it's good, or that it's good because God commands it? Can this same test be applied to laws or rules imposed by outside authority? Is obedience to law in itself good, or is obedience only as good as the law itself? Obviously, there must be other factors that mediate behavior beyond avoidance of negative consequences (Archie, n.d.).

Ethical People Are Happier

At one time or another, we are all subject to stress caused by self-doubt and indecision. At some time, we all worry about choices made or a path not taken. Having a clear understanding of who we are, what we value, and what we believe in can do much to assist us with internal struggles in times of difficulty. The Greek philosopher Socrates is credited with the famous statement, "An unexamined life is not worth living." This statement is a touchstone in ethical study and will be revisited on several occasions within this book. If we consider what Socrates said, we see that he is arguing for the value of self-reflection and self-understanding. His point is that self-understanding and, more importantly, self-respect, are keys to individual happiness. Individuals with high self-esteem tend to be more optimistic about their futures and feel as though they are in greater control of their own lives.

Ethics Leads to Greater Success

If you follow the news, it is easy to get the impression that the world is populated by people trying to get ahead by cheating. We may be tempted to lament that honesty is a disadvantage in competitive endeavors, but the truth is much more complex. Ethical people have the advantage of possessing traits that dishonest people must fake. Trust is hard earned but an incredibly valuable commodity. People with reputations for honesty and trustworthiness are almost always more successful than those with reputations for deceptive practices. Simply put, we all prefer to deal with people we trust and we tend to avoid people we do not. Popular wisdom warns that "you cannot fool all the people all the time," as such individuals with reputations for high ethical standards find it much easier to consistently build and maintain collaborative relationships with others. Individuals with strong principles tend to be more comfortable in interpersonal relationships and free from the insecurities associated with self-doubt. This sense of confidence is a valuable tool for life success. A highly developed sense of self-worth can free an individual to take on challenges that lead to success.

Ethics Creates a Stable Society

As important as ethics are to individual development, happiness, and success, the principles embodied in ethical behavior are also critical for the maintenance of a stable society. Large numbers of people cannot coexist together peacefully without cooperative agreement to abide by basic rules of conduct. Commerce, art, science, and quality of life cannot exist without a lot of ethical people working together. The ability of a judicial system to maintain public order is very limited. Lawmakers cannot anticipate all the situations and demands of daily interactions among the population. Even if a body of regulations could be created that encompassed all the needed limitations and controls associated with a complex society, it would certainly lack the resources to enforce them. A government has a very limited ability to control the actions of its individual citizens, a point that was clearly demonstrated by the U.S. experiment with the prohibition of alcohol in the early part of the 20th century. The attempt to eliminate alcohol failed because a significant portion of the population simply ignored the law.

An orderly society ultimately depends on mutual agreement by a significant majority of its population to behave in a manner consistent with its shared values. Absent voluntary adherence to accepted principles and behaviors by the majority of a society, anarchy will eventually ensue (Archie, n.d.).

Ethics Is at the Core of Humanity

A distinguishing feature of human beings is our complex social bonds based on mutual cooperation Figure 1-2. At our core, we are social beings. No other species comes close to the variety and complexity of our interactions with each other. Daily we follow official and unofficial rules that allow us

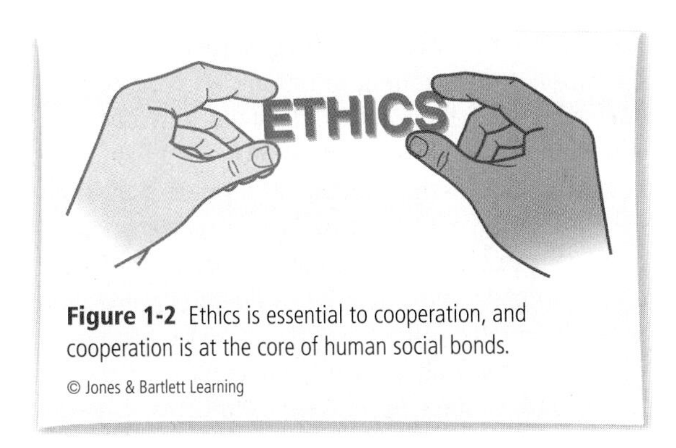

Figure 1-2 Ethics is essential to cooperation, and cooperation is at the core of human social bonds.

to live in harmony with our fellow citizens in all sorts of social situations. Etiquette, politeness, and self-restraint are all expressions of our cooperative efforts to coexist with others. While ethical restraint allows for the development of complex societies, those same sorts of restraints guide us in our interpersonal relationships. Cooperation and self-restraint allow a complex society to exist. As a result, our ability to live in relative harmony with others can be equated to our humanity. **Antisocial behavior** is demonstrated through a lack of ethics, morals, or social self-restraint and is often considered a sign of mental illness, and unprovoked violence is often labeled as "inhuman." Undoubtedly, many will hasten to argue that humanity is a condition of birth and not subject diminishment (because all men are created equal), yet it is clear that within the context of individual judgment, adherence to some universal principles tends to drive social acceptance (Ruggiero, 2009).

Why Study Ethics?

We would all like to believe that we know right from wrong. In our earliest childhood, we are taught what we should and should not do, what is expected, and what is right. Society provides numerous resources to guide us in our day-to-day lives. Our parents, churches, and teachers instill in us values, traditions, and principles. The justice system promulgates laws that mandate behavior. Most limits on our behavior are universally accepted and fairly clear-cut. We know we should not lie, steal, or commit unnecessary acts of violence. But social life is much more complicated than just following rules. We are also taught to value family, truth, fairness, social obligation, and responsibility. But what happens when these values conflict? What do we do when a family member commits an act that we consider morally corrupt? What happens when law and our sense of justice conflict?

Even with all the guidance provided by our society, each of us will often face **ethical dilemmas**. Books, movies, and television seem to constantly provide examples of "good" people faced with seemingly impossible decisions. Unlike fictional characters, our ethical dilemmas have real consequences. Our choices can damage relationships, place us in jail, cause us to lose our jobs, or, in extreme cases, jeopardize lives.

Each generation, in each culture, has held its behavioral imperatives, and that is true today as well. However, there has been a trend over the last 60 years toward a philosophy known as **moral relativism**. This increasingly accepted view of right and wrong is predicated on the belief that actions must be judged in context. That context involves an individual's experience, the time and place of the act, the situation's unique factors, and the intent of the action.

Moral relativism has the effect of creating a moving target when trying to decide what is ethical versus unethical, or moral versus nonmoral. The burden of subjectivity is clearly placed on the individual, yet ultimately our actions are judged by others. Consider that many people involved in scandal believe that their actions were justifiable. They are sometimes shocked to discover that what seemed like a reasonable action at the time appears reprehensible to dispassionate spectators. So often phrases such as "You had to be there" or "My intentions were good" are the best and only justification offered.

Shakespeare is famously quoted as saying "to thine own self be true." His point is that each of us must remain true to our own nature. Ultimately, it is our own conscience and our own values that will determine our actions. Conversely, while our actions are subjectively influenced by our conscience, they are **objectively** judged by others. We must study ethics because life is complicated and the consequences of our decisions can be significant.

Clarification

If nothing else, the study of ethics encourages introspection, and through that can come an understanding of why we choose a course of action (Ruggiero, 2009).

As mentioned earlier, a similar thought was expressed by the Greek philosopher Socrates, who expressed the idea that an unexamined life was not worth living Figure 1-3. He was advocating that each of us should embark on an inward journey of self-understanding. A key benefit in the study of ethics is self-analysis. We all have beliefs and principles we value, but how often do we give this serious consideration to the simple question, Why do we believe what we believe? To understand how our own experiences and beliefs shape our judgment is a powerful tool in the pursuit of happiness through the elimination of self-doubt. Having a clear sense of your own principles brings authenticity and integrity to your decision making. If Socrates were alive today, he might well advise us that to be happy we must be "comfortable in our own skin."

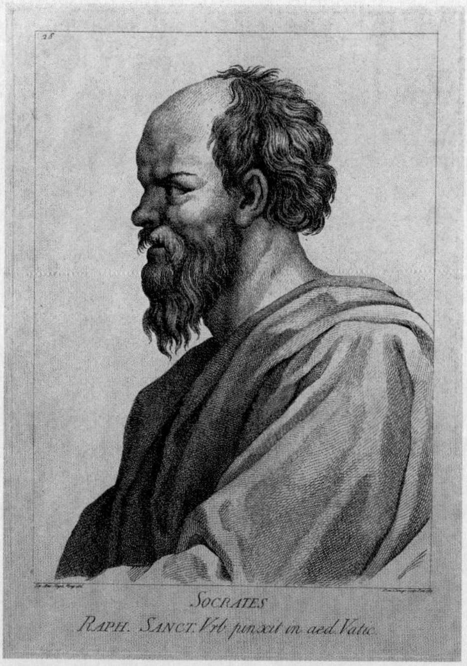

Figure 1-3 470 BC to 390 BC, Socrates was a classical Greek philosopher credited as one of the founders of Western philosophy.

Socrates. Line engraving by D. Cunego, 1783, after A. R. Mengs after Raphael by Raphael/ Wellcome Collection

Beyond gaining insight into ourselves, a study of ethics also does much to foster understanding of others. As we interact with others, we tend to judge them solely based on their actions or, at least, what we perceive to be their actions. In most cases, we do not have a full understanding of context, intention, or motivation. As the study of philosophy and ethics has diminished over the last several decades, Western culture has focused more on what is happening and less on why. The study of ethics tends to shed light on why people do what they do, and this can foster greater understanding and even empathy. Understanding the motivation that drives people's actions has several significant benefits. The first is a practical one: Understanding what drives behavior is to be able to better predict behavior. This has obvious benefits when dealing with people in both professional and social situations. In many cases, the difference between effective and impoverished leadership may well be an understanding of how a group's culture is affecting its values, traditions, and attitudes. As we shall learn later, these are critical elements that shape ethics.

The second benefit of studying ethics and how it affects others' actions is of a more personal nature. As social media have become more pervasive, society can more easily express opinions regarding current events. The value of this is that our society is much more transparent regarding moral and ethical diversity. Unfortunately, with transparency comes potential conflict. More than a few of our fellow citizens believe our society is in a state of moral decay. Many are offended and even fearful of what they see as a loss of traditional boundaries. With the study of ethics can come **tolerance**. Those who understand the relationship between personal values and experiences are likely to be less frustrated by the actions of others when those actions are unexpected or disappointing. Conversely, the focus on behavior absent understanding of contextual influence can lead to anger, confusion, and fear.

Is Ethics Study Important to the Fire Service?

In the preface of this text there is a discussion about why the author considered a book specifically dedicated to fire service ethics to be important. That discussion offers contextual information about this section. If you have not done so already, you are strongly urged to read the preface before continuing this section.

As members of their community, fire fighters certainly have as much need as any of their neighbors to be ethical. As such, all the arguments as to why any member of society will gain value in the study of ethics applies equally to fire fighters. A fair question might be, Do fire fighters, by nature of their career, have a greater need to study ethics than that of the general population? There are significant arguments to indicate that such is the case.

Fire fighters inherit tremendous responsibilities as a condition of employment Figure 1-4. We are entrusted to protect the lives of our fellow citizens as well as their most valued possessions and property. With those responsibilities comes significant and unique authority. Fire fighters routinely enter residential and commercial properties without the specific permission of the owner. In some rare cases, we do so against the express wishes of the property owner. In extreme situations, we may do this without a warrant if, in our opinion, it is necessary to protect lives or nearby property. Many fire departments provide emergency medical services. As a result, we routinely "lay hands" on persons. We frequently remove a patient's clothing, and, for their own safety, we sometimes restrain patients if we deem it necessary. Fire inspectors have the authority to interpret and enforce codes where compliance can cost property owners significant money. **Fire administrators** are responsible for millions of dollars of taxpayer money. They sign contracts, discipline employees, and write policies that have far-reaching effects on stakeholders both in the fire service and the community at large. Most importantly, fire officers make decisions on whether to risk the lives of their fire fighters in an aggressive attempt to save a building or to strategically sacrifice property. These decisions have an enormous impact on the economic and emotional well-being of home and business owners, balanced against the lives and health of their fire fighters. The ethical implications of such decisions are obvious, and also complex. With responsibility comes an ethical obligation to act responsibly. The complexity of our responsibilities coupled with the significant impact of our decisions makes the study of ethics very important for fire fighters.

Figure 1-4 Fire fighters inherit tremendous responsibilities as a condition of employment.

Clarification

The complexity of our responsibilities coupled with the significant impact of our decisions makes the study of ethics very important for fire fighters.

Are Fire Fighter Ethics Issues Unique?

Is the fire service profession in need of ethics standards beyond that of the general public? Do the professional standards of other service providers adequately instruct us? Certainly, fire fighters are not unique in that they have tremendous responsibilities and have been granted significant authority to aid them in meeting those responsibilities. The military, law enforcement, the medical profession, and the legal profession all have similar comparative weights of responsibility. Fire fighters do, however, operate within a relatively unique environment and with a relatively unique orientation.

The fire service, similar to law enforcement and the military, is organized to respond in defense of the public well-being. We are supported by public money to protect the public and our work, similar to theirs, is dangerous. The military and law enforcement have clearly defined ethics policies that are overseen by boards of review. Do the ethics principles of those organizations then adequately serve the fire service? Can the ethical guidelines of other emergency response organizations be applied to the fire service, or does the fire service have a need for a unique set of ethical guidelines?

The military's mission is the application of force against lives and property to maintain some goal perceived to be "worthy" of such force. Those duties are often performed in high-risk environments. This is, of course, an oversimplification of the military's mission. However, it is a fair representation of what makes it unique in its contribution to society at large. Essential ethical issues pertaining to the military typically center on the questions of when, how, and to what extent such force can and should be applied. This is balanced against relative risk to the combatants (Sandin, 2009).

Similar to the military, law enforcement frequently operates in dangerous conditions. Its mission is to maintain civil order, often by limiting personal liberties and at times using force. Also, similar to the military, policing ethics often focus on when, how, and to what extent its powers can be reasonably exercised. Law enforcement has the additional challenge of pursuing a mandate of apprehending criminals, but within the confines of specific restraints. This raises ethics issues regarding policy versus outcome. This issue has been repeatedly depicted in TV and movies. Characters are often confronted with the dilemma of following the rules even when they may impede an arrest or breaking the rules to ensure that some form of justice is meted out. Policing courtroom dramas often simplify and exaggerate reality, but often the ethical dilemmas are quite real. Police are empowered to take away liberty and to use force to do so if necessary. The associated ethics questions often balance a confrontation of desired outcome (justice), and the restriction of potential abuse of police powers. Certainly, both the military and law enforcement routinely engage in positive interactions, and their existence is both needed and beneficial. However, the core of their mission is essentially to protect through confrontation (Sandin, 2009) Figure 1-5.

The fire service also exists to protect, and we engage in high-risk activities. Where we differ from law enforcement or the military is that our core mission is not adversarial. We do not exist to apply force or to limit activity. We exist to combat a nonhuman opponent. It is, essentially, a servant-based mission. We exist solely to be a "Good Samaritan" force within the community. As a result, our ethical issues tend to be less defined and more elusive. Our core mission of saving life and property is essentially "good" without qualification, yet it is balanced against the self-serving desire of self-preservation. We commit ourselves to a noble cause, at great risk, but only up to a point. And that qualification can and does generate ethical issues. That we do this at taxpayer expense also creates fertile ground for ethical dilemmas (Sandin, 2009).

Someone may rightly point out that fire departments have the responsibility of enforcing fire codes, an act that is very similar to that of law enforcement. Certainly, the fire service can benefit from studying the literature addressing the ethics associated with law enforcement in this regard. Yet even in the responsibility of code enforcement, the fire service is unique in its methodology. While the codes are not

Figure 1-5 The ethics of firefighting differs from other professions.

© Jones & Bartlett Learning

open to interpretation, fire chiefs routinely exercise discretion in their application. The subjective analysis of a code's intent is very often the catalyst for fire chiefs and inspectors to allow "trade-offs" in a code or even to completely forgive certain requirements. This is often done on a case-by-case basis. This too can create ethical questions not faced by law enforcement who, at least in theory, enforce the law consistently and at face value.

Similar to the fire service, the mission of the medical and legal professions is to contribute to public welfare. Also similar to the fire service, they are essentially nonconfrontational in their relationship with their clients. They are ethically obligated to provide their best efforts without consideration of the recipient's worthiness. Their ethical issues tend to focus on process. Both professions have robust ethical guidelines that serve their practitioners in dealing with questions regarding their professional obligations and limitations. You can correctly infer that there are significant overlaps between these professions and the fire service, but do the ethical standards associated with these professions completely meet the needs of the fire service?

Figure 1-6 Fire fighters work in uncertain environments.

© Glen E. Ellman

Unlike doctors and lawyers, fire fighters usually work in rapidly evolving and dangerous environments Figure 1-6. Because fire fighters expose themselves to personal risk, they must deal with questions regarding risk versus benefit. Emergency scenes rarely give fire officers time for in-depth reflection. Further, fire fighters often deal with emergencies where need overwhelms resources. These are problems typically not faced by the legal profession and rarely faced by the medical profession outside emergency response. When medical personnel are required to perform triage, they operate on a specific set of criteria based on likely success. The relative value of the individual patient is considered the same. Fire fighters also make strategic and tactical decisions in which they balance resources against likely success; however, they have the added responsibility of also incorporating relative value into their "cost-benefit analysis." For instance, not all property is of objective and, more importantly, subjective equal value. A mansion may be objectively assessed as a high-value property, yet subjectively it may be less valued by the community than a 200-year-old church. The owners of any given property may place great emotional value on their property, yet fire fighters must at times choose which properties should be saved based on available fire suppression resources. Further complicating the analysis and its ethical ramifications is that these decisions also involve inherent risks to fire fighters—a condition not shared by doctors or lawyers in the provision of their services (Sandin, 2009).

Another factor that separates the fire service from other professions is reputation. Most professions place great value on their reputations. Yet with the exception of the military, no other profession enjoys the nearly universal admiration of the public as does the fire service. In America, the fire fighter has become a national symbol representing bravery, self-sacrifice, and competence. The hero image of the fire fighter has become ingrained in American culture and media. Likewise, it has completely permeated the fire service culture as well. Many fire fighters are drawn to the profession at least in part by its image. Fire fighters tend to express an unusually high level of job satisfaction, as witnessed by an extremely low amount of employment turnover. This is at least in part because of the nearly constant positive reinforcement they receive from the public. Fire fighters' unions have frequently used the depth of public respect to their advantage, both at the bargaining table and in the political arena. Perhaps no other profession embraces its image so completely, nor profits more from it, than does the American fire service Figure 1-7.

Figure 1-7 The relationship the fire service has with the public is a critical element in its culture.

With this great affection comes great expectation. Fire fighters are held to a higher standard of behavior than are most citizens. A fire fighter arrest or an internal disciplinary hearing is almost always considered newsworthy. Public response to bad behavior by fire fighters is rarely indifference. More typically there is great disappointment and outrage.

Ironically, an organization with such varied and important responsibilities and held to such a high standard of behavior as is the fire service has no universally codified standards of ethics. Nor does it have a professional board of review and oversight for its chief administrators.

The fire service is a unique profession, and few have so many varied and important responsibilities. Fire fighters work in such unique environments. Few workers are in such close and constant contact with each other as are fire fighters. Not many professions enjoy so much public respect and benefit from it as do fire fighters. It is evident that fire fighters must not only be aware of their ethical responsibilities but study them in depth.

The Growing Ethics Problem

The fire service has a growing ethics problem. Do a basic internet search utilizing the terms "fire fighter under investigation" and you will find numerous stories of fire fighter misbehavior. Visit any of the major fire service web portals and you will likewise find several weekly stories about fire fighters in trouble Figure 1-8. This is not to say that fire fighters today

Figure 1-8 The fire service has a growing ethical problem.

are any less ethical than they were a generation ago. There seems to be tangible evidence that millennials are somewhat more conservative in their life choices then were baby boomers at the same age. Yet there *seems* to be more incidences of fire fighter misbehavior than ever before. Whether accurate or not, the growing public perception that fire service leaders are potentially corrupt or that departments are staffed with unethical people is a threat that should not be ignored. The ethics problem facing the fire service is not generational, rather it is rooted in the culture in which the current generation works. Changes in culture have created several factors that conspire to reshape the relationship between public safety agencies and the populations they serve.

Rules Have Changed

One issue driving the growing fire service ethics problem is that public perception regarding behavior has changed **Figure 1-9**. America has become more conservative and more politically correct in its judgment of behavior by public officials. Foolish practical jokes, racist language, and pornography are simply no longer excused in the workplace. A generation ago the presence of strippers at a union function may have been dismissed as "boys will be boys." Today, such behavior is fodder for scandal. Fire departments can no longer be boys' clubs, nor fire stations man caves. Departments are taxpayer supported, and the public expects professional behavior.

News Media Has Changed

A few decades ago the media was more discrete in its reporting. Even if reporters were aware of certain types of bad behavior, they would not report it. The belief was that it was not newsworthy. As newsrooms have become profit centers, human interest stories have become more important, especially scandals. In particular, bad behavior by fire fighters "sells."

Just as important is the fundamental shift of how news is delivered and consumed. Even a decade ago most people got their news from television and newspapers. Today, most people acquire their news from the internet. This has two effects; first, no news is solely local news. Breach of conduct by fire fighters anywhere is likely to be reported everywhere. This gives the appearance of a greater occurrence rate of bad behavior than was experienced a generation ago. Second, the life span of a story has become infinite.

Figure 1-9 The public perception regarding behavior has changed.

Courtesy of Scott Walker

A generation ago, a local fire department scandal was reported locally, and it was newsworthy for perhaps a few weeks. Afterward, the department and the municipality moved on to other issues. Because of the Internet news delivery, stories are written and posted for a very long time, if not forever. Again, do an internet search of fire fighter misbehavior and you are likely to find stories that are several years old but still out there. This too gives the overall impression of a greater rate of scandal. The sheer number of negative stories to which the public is exposed has helped to make them cynical.

Social Media's Impact on the News

Perhaps the most significant cultural change since the invention of television is the rise of social media. Everyone with a cell phone and a camera is a potential news reporter. Just as importantly, pictures and video can be uploaded in real time. The average citizen has access to worldwide distribution of content via social media. "Catching" public officials in compromising positions is nearly a guaranteed **viral post**. The general public is not constrained by journalistic ethics and often will post video and photos out of context, and without hesitation present opinion and conjecture as fact. The unnerving truth is that fire fighters are constantly under public scrutiny and every word, every moment, and every action is liable to be posted worldwide. Because of constant scrutiny, bad behavior will be caught. And because of social media, it will be widely reported.

Our Approach to Ethics

The ethical environment in which the fire service operates has created a new reality with significant consequences. For too many years the fire service has approached ethics retroactively, focusing only on punishing bad behavior. We have assumed that ethical misbehavior is a result of a few bad apples and that scandals happen to other departments. This is a dangerous viewpoint. There is a lesson to be learned from law enforcement. While the general public still expresses support for local law enforcement, the constant barrage of stories of abuse of authority has had the effect of eroding public confidence in the profession. While still supporting local police, many of the public have come to believe that unethical behavior is inherent within the profession. The fire service cannot afford the same fate. The fire service absolutely depends on the public's goodwill. The fire service's reputation and status within the community helps us to recruit the very best people. Public support is critical to maintaining staffing during local budget fights. Public trust is essential for fire departments to perform their duties effectively. We must change our approach to ethics by becoming more proactive.

In today's social context, it is not enough to react to breaches of conduct by fire fighters. We must proactively teach ethics, and ethics education must be pervasive throughout every level of the fire service.

A Lesson Learned from Fire Fighter Safety

A comparison can be made between the fire service's attitude toward fire fighter safety in the 1980s and current attitudes toward fire service ethics. In both cases, there was a general awareness of a problem but no clear initiative to deal with it. In the 1980s there was a general acceptance of fire fighter injury as a foregone conclusion; it was considered part of the job. It was also believed that accident-producing injuries were unpredictable and unavoidable. There was also widespread denial that catastrophic injury happened within a local department; it was something that happened in other places. In the early part of the 21st century attitudes toward fire fighter safety changed. The profession adopted new attitudes toward safety. There was a call for a culture change regarding safety. Leadership in the fire service insisted that injury was predictable and avoidable. Fire fighter safety became a priority at all levels of the fire service and was infused in all activities and training.

We need the same approach for ethics. Ethics violations are destroying careers, derailing department initiatives, and ruining morale Figure 1-10. Ethics violations are predictable; almost every fire officer is

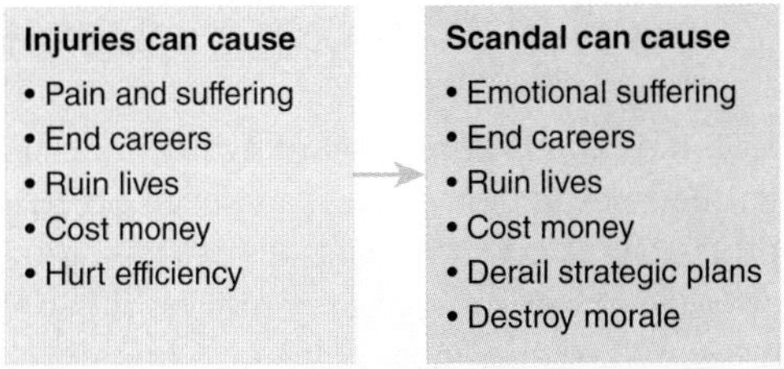

Injuries can cause	Scandal can cause
• Pain and suffering	• Emotional suffering
• End careers	• End careers
• Ruin lives	• Ruin lives
• Cost money	• Cost money
• Hurt efficiency	• Derail strategic plans
	• Destroy morale

Figure 1-10 Injury and scandal have many consequences.

aware of at least one member of his or her department who is most likely to get in trouble. Behavior problems usually start off as small and escalate; there are warning signs. There is a hard and fast rule in accident prevention that says, "That which can be predicted can be prevented." The same axiom applies to the fire service ethics. Through education, culture change, and vigilance, the fire service can proactively prevent ethics violations.

The fire service has too much at stake to continue to ignore ethics. As is often the case, the solution to the problem begins with education.

WRAP-UP

Chapter Summary

- The term "ethics" means many things to many people. For many, it is the set of rules that guide our interactions with each other. For others, ethics is an internal benchmark that we use to judge ourselves and others.
- Ethics is often considered the equivalent of morality, and while the terms are related, they are also unique.
- The study of ethics seeks to understand the relationship of the various factors that modify the behavior within a social group.
- For society to function, individual members of a community must agree on rules of conduct.
- The study of ethics seeks to define behavior boundaries within social relationships; it seeks to establish generally accepted standards of right and wrong based on personal obligations, benefits to society, compliance to behavior standards and specific virtues.
- Members of the fire service provide a critical public service in ensuring public safety. They have tremendous responsibilities in the face of significant risks. With those responsibilities comes significant obligation.
- Fire fighters are bestowed rights and privileges that have significant ethical implications.
- The fire service is a unique entity within a community; it exists solely to help others in need. The parameters of its mission create unique ethical challenges, and it is held to a very high standard of behavior.
- As a public institution, a local fire department is under considerable scrutiny, and any ethical breach is magnified in the public eye. It is imperative that fire fighters understand their ethical responsibilities and their communities' expectations.
- The fire service needs to invest itself in the subject of ethics at the local level and as a profession.
- The individual actions of fire fighters reflect on their entire department, and even on the profession. Our ethical issues are complicated, and we must proactively teach our fire fighters expected behavior limits. In short, ethics is important, and we cannot ignore it.

Key Terms

antisocial behavior aggressive, impulsive, and often violent actions that violate protective rules, conventions, and codes of a society.

argumentum ad baculum a Latin phrase meaning "appeal to the stick." It attempts to justify the use of coercion to bring about a desired outcome.

ethical dilemma a situation where conflicting values, obligations, and expectations create a problem with no clear-cut answer as to what is right or wrong.

ethical restraint expressions such as etiquette, politeness, and self-restraint that guide us in our interpersonal relationships.

ethics the development, evaluation, and study of behavior boundaries and expectations within personal and professional interactions with others.

fire administrators people responsible for taxpayer money, signing contracts, disciplining employees, and writing policies affecting stakeholders both in the fire service and the community at large.

innate existing in, belonging to, or determined by factors present in an individual from birth.

morality a doctrine or system of moral conduct.

moral relativism a concept that right and wrong are predicated on context.

objectively in a way that is not influenced by personal feelings or opinions.

personal ethics the basic principles and values that govern interactions among individuals.

subjective characteristic of or belonging to reality as perceived rather than as independent of mind.

tolerance sympathy or indulgence for beliefs or practices differing from or conflicting with one's own.

viral post a blog or social media message that is continuously shared by everyone who receives it.

Challenging Questions

To check your understanding of this chapter's material, answer the following questions. It is highly recommended that you discuss your viewpoints with fellow students, peers, coworkers, and friends to discover their opinions as well.

- Is it fair that fire fighters are often held to a higher standard of behavior than their counterparts in other professions?
- It is not uncommon for fire fighters to be scrutinized or even sanctioned for off-duty behavior. Do the ethical obligations of being a fire fighter extend beyond the station house door? If so, why? If not, why not?
- Assuming that all particular fire fighters are essentially ethical and/or moral, why study professional ethics?
- If you have not done so already, read the quote by Bill Moyer on the first page of this chapter. Moyer's quote refers to the interdependency of each of us on the goodwill of others. What is the social necessity of mutual cooperation at a societal level?
- What individual had the greatest influence on your ethical philosophy? Briefly describe what you learned from that person.

Case Study Conclusion

Revisit the case study at the beginning of the chapter. Spend a few minutes considering the questions posed at the end of the case study. In light of the information shared in this chapter, have any of your original observations changed?

The case study touches on some basic questions regarding ethics. When are our actions ethical or unethical? When are we, the ethical, the responsible? Is there a difference between personal ethics and professional ethics?

Certainly, the described behavior of fire fighter Jones is troubling. But as an equal partner in the business, fire fighter Jones may be entitled to access the business's funds as he deems fit. The fact that Smith takes exception to Jones's behavior may be as much a dispute over authority as ethics. There is simply not enough information about this aspect of the situation.

A more complicated question is whether or not Jones's alleged theft has any ethical implications for you as a battalion chief. You have received a complaint of apparent misbehavior, although it is off duty. Conduct unbecoming of a fire fighter can have significant implications for a department and, certainly, the spat between Smith and Jones can spill over into the shift itself. As an officer, you likely have an ethical responsibility to either gather more data and act appropriately or report the issue to your supervising officer. This obligation exists, if for no other reason than the protection of the department.

Another facet of this scenario is whether or not Jones's alleged unethical behavior is a breach of fire department responsibility. If it is true, is it punishable? As described, there are no legal issues involved, but can the department's expected behavior standards be higher than what the law requires of private citizens? This is a matter in constant dispute, but, as we will learn throughout the rest of this text, there is a valid argument that, yes, that is the case.

Chapter Review Questions

1. Define the term "ethics."
2. List three incentives for people to be ethical.
3. List at least three qualities that make firefighting a unique profession as it pertains to ethics.
4. What did Socrates mean when he said, "An unexamined life is not worth living"?
5. What is meant by the term "argumentum ad baculum"?

References

Archie, Lee C. n.d. "Philosophy 302: Ethics: Why Be Moral?" Accessed December 14, 2015. http://philosophy.lander.edu/ethics/why_moral.html.

Cooper, T. L. 2012. *The Responsible Administrator.* New York: Jossey-Bass.

Merriam-Webster. 2003. *Merriam-Webster's Collegiate Dictionary.* Springfield, MA: Merriam-Webster.

Ruggiero, V. R. 2009. *Thinking Critically about Ethical Issues.* Philadelphia: Mayfield Publishing Company.

Sandin P. 2009. "Firefighting Ethics: Principalism for Burning Issues." *Ethical Perspectives* 16, no. 2, 225–251.

Judging Behavior

"Ethics is the activity of man directed to secure the inner perfection of his own personality."

—Albert Schweitzer

OBJECTIVES

After studying this chapter, you should be able to:

- Assess the "building blocks" of behavior judgment.
- Evaluate the relationships among social norms, morality, ethics, and the law.
- Discern the fluidity of social values.
- Explain the concepts of justice and fairness.
- Understand Western culture's theories about the origin of moral concepts.
- Ascertain the difference between professional and personal ethics.
- Interpret the three social intentions from which the law arises.
- Determine what makes the fire service culture unique.

Case Study

Lieutenant Smith runs a photography studio in his spare time. In addition to taking wedding and anniversary pictures, he also sells original photography. One day a citizen calls the fire chief's office to complain that Smith has a website selling pornographic pictures of local women. The caller insists that the website is inappropriate and is appalled that a local fire fighter would be engaged in such immoral behavior. Further, if the website is not taken down immediately the caller will make this a very "public" issue.

In investigating the site, the chief finds that indeed, Smith's website features "art" prints of numerous nude models. While the chief does not consider himself an art critic, he finds that while some of the photos are tasteful, some can only be described as overly explicit. The chief also finds that Lieutenant Smith clearly identifies himself as a local fire fighter in the biography section of the website. In speaking with Lieutenant Smith, he is adamant that the photos are art and not pornographic. He further asserts that he has every right to run his business off duty without interference from the fire department, and he doesn't care whether or not some citizen doesn't like it.

- Is the citizen's assertion that the website is not in keeping with the moral standards of the community relevant?
- Does Lieutenant Smith have a personal right to engage in legal behavior even if it is considered risqué by the community?
- Does Lieutenant Smith have an ethical obligation to refrain from behavior that reflects poorly on the department and his fellow fire fighters?
- Does the fire chief have an ethical obligation to protect the reputation of the department?
- Does the public have a right to expect fire fighters to behave consistently with common moral expectations?

Introduction

The issues of morality, ethics, community standards, and taboos are sometimes clear-cut, but they are often not. When does a behavior cross over from provocative to indecent? At what point does personal choice threaten community well-being? What can the public justifiably expect from its public employees? For the fire chief, interpreting and balancing community standards can be difficult and risky.

For the study of ethics to have any practical value, it must ultimately lead to guidance as to what is good, right, just, or proper behavior. But are these things the same? Can something be proper but not right? Can something be right but not good? When judging behavior, we rely on one or more concepts, which include **social norms**, **morals**, **ethics**, and **laws**. Each of these concepts is unique yet intertwined.

When discussing ethics, it is not uncommon to use the terms "morals," "ethics," and "values" interchangeably. In most daily situations you can use any or all of these words to get your point across. That distinctions exist, and understanding their differences, is elemental in the study of ethics, morality, and philosophy.

> **Listen Up!**
>
> Although people use the terms "morals," "values," and "ethics" interchangeably, it is important to note that they have different meanings.

These distinctions go beyond semantics, and their importance reach further than academics. For an individual to effectively evaluate the "rightness" of an action, he or she must consider the criteria being used to evaluate it. For instance, an action may be normal yet unethical; it may be legal yet immoral. Most of the time the concepts of morality, ethics, legality, and normalcy are more or less aligned. However, life has the unfortunate habit of presenting us with situations where we must choose between the lesser of two "evils" or the greater of two "goods."

In this chapter, we will introduce some concepts that you can use to evaluate the "correctness" of human behavior. In addition to defining them, we will examine how they have developed, how they overlap, and how they differ.

Social Norms

The term "social norms" actually refers to several concepts that combine to define "normal behavior" within the context of a particular social group or culture (Crossman, 2016).

Folkways

Folkways are unofficial practices and rituals that define routine and everyday behaviors. More often than not they mark the distinction between rude and polite behavior. Examples range from the polite—waiting your turn in line, opening the door for someone carrying bags, or covering your mouth when you sneeze—to the life shaping: what you eat, how you dress, and how you communicate (Sumner and Keller, 1906).

Folkways exert a form of social pressure on us to act and interact in certain ways, but they do not have moral significance and there are rarely formal consequences or sanctions for violating one. Regardless, folkways influence nearly every part of our social interactions, and we cannot overstate their importance (Crossman, 2016). People often perceive the violation of common customs as signs of aggression or even insanity. Consider the rules regarding a crowded room. When we find ourselves sharing a small space with several people, it is usually expected that we position ourselves shoulder to shoulder. Positioning yourself face-to-face with someone in a crowded area is nearly universally viewed as familiarity if you know each other or as aggressiveness if you do not. Clothing rituals provide numerous examples. Imagine the reaction you would receive from wearing a swimsuit to church or a baseball uniform to a business meeting. In some European countries, it is normal for men to wear small tight-fitting "Speedo-style" swimsuits. In America, where long, loose-fitting board shorts are the norm, the Speedo is most often viewed as inappropriate and comical.

Although folkways typically are not officially enforced within a community, violation of them can cause embarrassment, ridicule, and, in some cases, even disputes.

Mores

Some folkways become so ingrained in a culture that they take on a significance beyond the customary. Behavior expectations can become associated with the concepts of right and wrong; these are called **mores** (pronounced *more-ayes*). Mores are stricter than folkways, and people feel strongly about violating them. Sanctions for violating mores may likewise be more serious. Often, mores are enforced by law, and even then, community reaction can be vehement. Examples of social mores might include sanctions on public nudity, racism, adultery, and acts of violence. As you will learn later, there is a fine line between social mores and the concept of morality. Typically, mores arise from social convention whereas morality tends to arise from religious or universal concepts of good and evil. Walking around naked in public or cheating in business may be considered wrong, but those actions are typically not associated with the concept of depravity or wickedness. In these instances, the action is likely to be condemned and will reflect poorly on the individual, but may not necessarily define the individual's social standing. Some mores, however, are so strongly held that their violation identifies the individual as antisocial or an outcast. We call these taboos.

Taboos

A **taboo** is a group's strongest negative norm. It is a strict prohibition of behavior, and its violation may elicit outrage or disgust. Violation of taboos may elicit extreme reactions including shunning, legal proceedings, and even violence.

Typical examples of taboos may include polygamy, pedophilia, and incest. However, taboos are not restricted to sexual practices. In some cultures, the use of alcohol or eating pork may be considered taboo. Cannibalism is a nearly universal taboo. Like mores, taboos often have moral implications, yet they are distinct concepts.

Mores and taboos may be thought of as an academic description of what a culture may consider right or wrong. The judgment is made by social consensus, whereas morality springs from a concept of ideal behavior. Sociologists do not make a definitive distinction between the concepts of mores and taboos with that of morality. Theologians and many philosophers, on the other hand, do make this distinction. In the study of ethics, this distinction becomes important when evaluating an action's relative correctness. In short, sociologists look at social norms in terms of what people do, whereas in the study of morality we evaluate what people ought to do (Marinie, 2003).

After reading the previous descriptions, a certain amount of frustration may arise from a lack of clarity. A new student to ethics may fairly complain that there is little difference between folkways, mores, and taboos, or at least that it is difficult to discern where one begins and the other ends.

Clarification

What is the difference between social mores, taboos, and morality?

- Mores and taboos are typically used as academic terms to describe cultural behavior.
- Morality is typically used to describe ideal behavior.
- Mores and taboos describe what people do, whereas morality describes what people should do.

Unfortunately, this is the nature of human interaction. What is a taboo in one culture may be only risqué in another. In addition, people may view the same action differently at different times. Slavery is, by today's standards, a taboo and immoral, yet it was a social norm in the not-too-distant past. The same can be said for nonmarried couples living together. Just a few decades ago, sex outside marriage was considered immoral and taboo. Certainly, there are still some who reject the practice, but in Western culture current social convention treats it as relatively normal. This example also illustrates the inconsistency of behavior standards among cultures. As stated earlier, cohabitation is currently treated with relative indifference in Western society, whereas in many Middle Eastern and South American countries, it is still strongly censured.

Social norms evolve and are constantly tested. They are not universal, nor are they fixed. In the 1950s there were dozens of words and phrases considered unfit for television, now there are very few. What drives change in social norms? The consensus answer seems to be changes in moral principles.

Morality

There is a Latin phrase, "***malum in se***", meaning "wrong" or "evil in itself." The phrase is used to describe conduct believed to be sinful or inherently wrong by nature, independent of regulations or social conventions governing the conduct (Yogis, 2003). Moreover, morality can be attached to thought and emotion even in the absence of action. The Judeo-Christian tradition considers lust and hatred immoral whether acted upon or not. It is also important to note that people can attach moral value to actions independent from consequence. A behavior can be immoral even if the act harms no one else. Moral principles are often described as universal concepts. This is not to say that moral principles are universally agreed on, but rather that an action is deemed moral or immoral independent of intent, social acceptance, or effect. As you will learn later, this is a defining characteristic that separates it from ethics.

Where do moral imperatives come from? In Western culture, there seems to be no clear consensus on the origins of moral concepts. The debate focuses on four general theories.

The first theory is known as the *divine command theory*, which contends that morality originates from God. An action is moral and good because God wills it.

The second theory is known as the *theory of forms*. It was initially put forth by Plato and later advanced by the humanists of the 18th century. This theory states that there is a natural force that governs the systems of the universe. This force is responsible for both the laws of physics and those of metaphysics. This force seeks to create balance, and from that comes the universal concepts of good and bad (Brackman, 2014).

A third theory is called the *natural law theory*. It was advanced by Thomas Aquinas in the early part of the 13th century and further endorsed by Charles Darwin and later E. O. Wilson (Wilson, 1975). It proposes that the concepts of good and evil are hardwired into human beings. Aquinas believed that goodness was the natural state of man, whereas 19th- and 20th-century scientists believed morality was an innate trait, a product of evolution Figure 2-1.

Figure 2-1 Italian Dominican theologian St. Thomas Aquinas (1225–1274) was one of the most influential medieval thinkers of scholasticism and the father of the Thomistic School of Theology.

The fourth theory, *social conditioning*, comes from the field of social psychology. It states that moral concepts are a result of social conditioning. Dr. B. F. Skinner most famously argued this theory, suggesting that morality stems from the manifestation of social need (Prinz, 2016). An example of this is the argument that moral imperatives restricting mating outside marriage were the result of the practical need of our ancestors to identify the parentage of the offspring. This implies that the concept of morality is more or less arbitrary. If morality arises from practical need, then the moral imperative is irrelevant absent that original need. Further, you could argue that the moral imperative is only as relevant as the individual's interpretation of the validity of its antecedent need.

The arguments supporting and refuting these theories are numerous and have been fueling debate among philosophers, clergy, and social scientists for generations. What seems clear is that wherever moral concepts come from, they play an important part in human interaction.

Clarification

1. Divine command theory: A thing is moral (good) because God wills it.
2. The theory of forms: A natural force governs the physical and metaphysical universe. Similar to the Eastern philosophy of karma
3. The natural law theory: the concepts of good and evil are hardwired into human nature. Goodness is the natural state of man.
4. Social conditioning: a social psychology theory stating that morality is a manifestation of social need and created by culture.

Ethics

As you learned in the first chapter of this book, you can define **ethics** as: *The development, evaluation, and study of behavior boundaries and expectations within personal and professional interactions with others*. Similar to social norms, ethics seeks to define boundaries of acceptable human behavior. Similar to morality, ethics seeks to establish accepted standards of right and wrong. Where ethics differs is that it is rooted in the evaluation of how an action affects others.

As you will see in later chapters, there are numerous philosophies regarding ethics and subsequent ethical theories. What all of these theories have in common is that they evaluate the correctness of an action as it relates to its direct effect on others, or in its compliance to responsibilities assumed within human relationships.

Clarification

Moral standards are rooted in their compliance to a universal understanding of right and wrong springing from religion or natural law. By contrast, ethical standards are based on an action's effect on human relationships and endeavors.

Texture: Eky Studio/ShutterStock, Inc.; Steel: © Sharpshot/Dreamstime.com

It is important to note that within ethics there are two distinct applications (Figure 2-2). The first is **personal ethics**, which applies to an individual's personal code of behavior in dealing with others socially. The second application of ethics deals with professional standards of behavior. **Professional ethics** are standards or codes of conduct set by people in

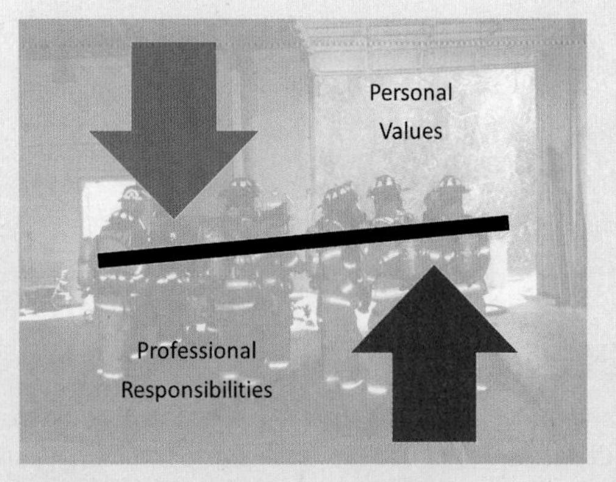

Figure 2-2 Personal and professional ethics are connected concepts that sometimes conflict.

Courtesy of Scott Walker.

a specific profession. Professional ethics are distinct from personal ethics in that they are imposed externally, are usually specific, and, most importantly, may conflict with personal ethics or even morality codes. Consider the example of a lawyer who is bound by professional ethics to vigorously defend a suspect who he is confident has committed an act that violates his sense of morality.

The following example illustrates distinctions among morality (good), personal ethics (right), and professional ethics (required). Bob is a fire chief in a department of 200 fire fighters. He is a 40-year-old, recently divorced Catholic who is engaged in an intimate relationship with his secretary. As a Catholic, a theologian would likely argue that Bob's intimate relationship is adulterous and therefore immoral on the grounds that the Catholic church does not recognize divorce. Ethically, Bob has entered the relationship in good faith, has not misled his new partner, nor made promises he does not intend to keep. It is an honest relationship between two consenting adults, and as such is in keeping with his personal values and ethics. Professionally, the relationship might be considered unethical as Bob is her supervisor. By its nature, the relationship is imbalanced. He has control over her professional outcomes, which can affect her private life. This may well limit her options and degree of influence within the relationship, and so the circumstances are unfair to her. It also implies several potential ethical problems regarding his responsibilities as a manager in dealing with not only her but her coworkers as well.

As we can see, morality, personal ethics, and professional ethics are very different. Yet they are intertwined, and one rarely considers one without considering all.

Law

The law is distinct from the concepts of morality and ethics, yet it is also intertwined with both. The law arises from three separate social intentions. The first is described by the Latin phrase ***malum prohibitum***, which means that an action is wrong because the law says so. In this case there is no moral or ethical value attached to compliance or noncompliance. As an example, wearing a seatbelt is required by law; however, compliance or noncompliance does not have moral or ethical significance. These sorts of laws are made by the government to ensure the smooth and safe operation of society. We comply with them partly because they make sense to us, but mostly out of a sense of social contract. We realize intuitively that there must be order if a society is to function.

The second class of law is described by the Latin phrase ***justitia socialis***. The term refers to a body of law that seeks to maintain justice and fairness within society. Examples include laws that protect individual rights, and tort law. In the United States, examples start with the Bill of Rights and continue through the Civil Rights Act, the Americans with Disabilities Act, and even consumer protection laws. These laws are often controversial, and enforcement can be difficult at times because they are often not universally accepted as necessary. Compliance with and enforcement of fire codes are a prime example of these sorts of laws. Fire codes are tied to this concept of social contracts, which infer an ethical obligation toward fairness, justice, and social responsibility as demonstrated by the maintenance of a safe property. Building owners may feel fire codes overreach or are an unfair burden.

Finally, the law often reflects widely accepted moral values within a society. In the previous section, you were introduced to the term ***malum in se***. It describes a principal wherein a choice is considered good at face value. Similarly, there are laws prohibiting actions that are considered wrong at face value, including assault, murder, and crimes against property. Moreover, legislators often seek to enforce moral standards that support religious doctrine or social tradition.

Clarification

Laws can be grouped according to their intention.

1. *Malum prohibitum:* A thing is illegal because the law prohibits it. Typically there is no ethical or moral attachment to compliance.
2. *Justitia socialis:* A body of law that attempts to regulate fairness and justice within society.
3. *Malum in se:* Laws prohibiting actions that are wrong at face value, such as murder and robbery. They may also reflect legislative attempts to compel compliance to moral standards.

Over the years these laws have included prohibitions against interracial marriage, same-sex marriage, liquor sales on Sunday, pornography, and, in some cases, the publication of books deemed subversive. In these cases, laws are reflective of consensus values, although they are rarely universally accepted and are often challenged and ignored. The role of morality in the law has always been subject to intense public debate and constant modification.

The distinctions among law and ethics, morality, and social norms are fairly straightforward **Figure 2-3**. Laws are generated by government

Folkways	Morals	Ethics	Law
• Define expected behavior • Unofficially enforced • Arise out of tradition and custom • No moral consequence attached	• Define good and bad, righteous versus wicked • Based in universally accepted truths • Arise from religion or philosophy • Unbounded	• Define right, fair and just • Based on human interaction • May be socially enforced • Arise from personal or professional values	• Define legal and illegal • Enforced by state • Arise from morals, social need, or social values • Clearly defined and limited

Figure 2-3 The distinctions among law and ethics, morality, and social norms.

agencies. Laws are enforced by officially sanctioned penalty. In contrast, personal ethics, morality, and social norms spring from numerous origins such as religion and social custom.

A less obvious but important distinction is that laws are finite. An action is either legal or it is not. That which is not forbidden by law is inherently legal. This is not true of ethics, morality, or social norms. A behavior may be deemed immoral or unethical even if not discussed in any text, teaching, or scripture. Most social norms are not written down anywhere; they just simply are accepted as *normal*. This creates a significant distinction—morals and norms are open to individual interpretation, whereas the law is clearly defined and not open to interpretation.

The law also has clear "bright lines" that define time and locale. An action may be legal on Monday, become illegal on Tuesday, and legal again the following month. Similarly, something may be legal in one jurisdiction and illegal in another, as demonstrated by discrepancies in restrictions on marijuana use in various states.

Because laws are finite and ethics, morality, and norms are not, the argument that an action is legal is rarely sole justification for the defense of a questionable behavior. Consider the national firestorm set off in 2017 by the owner of a pharmaceutical company that purchased the exclusive rights to a common form of insulin and then proceeded to drastically raise the price. His actions were completely legal and from a purely business point of view, even smart. Still, he was publicly villainized and there were calls, many for punishment, because his actions were unjust.

Justice and Fairness

Justice and fairness are unique concepts that are so interdependent that it is difficult to define one without referencing the other. The terms are often used synonymously, yet they have different meanings and applications within their relationship to ethical behavior.

You can define fairness as judgment or equal distribution without bias or regard for an individual's needs or feelings. If you and I were to share a sandwich, the fair distribution would be for each of us to get half. On the other hand, you can define justice as a standard of rightness based on circumstance. In the more common vernacular, justice is giving someone her or his due. In the example of our sandwich, if I had lunch just prior to being offered the sandwich and you had not eaten, it may be more just for you to receive more or even the whole sandwich.

You can also distinguish justice and fairness by their scope of application. Fairness is often applied to the treatment of individuals and individual cases, whereas the concept of justice can be applied to groups with more abstract consequences (Velasquez, Andre, Shanks, and Meyer, 1990). Fairness is typically the application of equal treatment of individuals within equal circumstance, whereas justice tends to be based on principles of fairness to groups of individuals of unequal circumstance.

Clarification

Justice is a conceptual standard of rightness and fairness based on an ability to judge on contextual good without reference to one's feelings or interests. Fairness is the ability to make judgments that are not overly general and are specific to a particular case.

The distinction between justice and fairness can be an important one. Let us apply it to a common issue of contention in the fire service, diversity Figure 2-4. Assume for a moment that two candidates are seeking promotion. One is a white male whom we shall call Bob. The other is a minority female whom we shall call Barb. Both candidates are qualified as demonstrated by objective testing. Unfortunately, in Bob and Barb's department, minorities are significantly underrepresented within the officer ranks. As part of the department diversity initiative, Barb is given preference points and promoted instead of Bob. From the perspective of pure fairness, Bob may argue that he was treated unfairly. He may point out that he is as qualified as Barb, and based on that should receive equal consideration. An individual defending the diversity initiative may well argue that promoting Barb is an exercise in justice. While

Figure 2-4 The distinction between fairness and justice is important and sometimes complex, especially when it comes to concepts such as diversity.

© kali9/E+/Getty

acknowledging that Barb and Bob are not being treated the same, justice is served not to Barb but to the minority community at large by her promotion. In this example, Bob is seeking fairness, while the authority determining promotions is seeking justice.

Clarification

To understand the rationale of diversity initiatives, it is necessary to distinguish between the concepts of fairness and justice.

Texture: Eky Studio/ShutterStock, Inc.; Steel: © Sharpshot/Dreamstime.com

The concepts of justice and fairness are typically not directly tied to ethical or moral standards, yet they are critical to our understanding of ethics and morality. Fairness represents the basis for most individual interaction, and justice is the foundation upon which societies function. In writing the U.S. Constitution, its framers made it clear that government cannot function without justice. As individuals, we also tend to place a high value on fairness and justice. The respect for these concepts can be a prime motivator in behavior that is deemed either ethical or moral. Although you may act unfairly in certain circumstances, your actions may not be immoral or unethical (giving preference points). Yet acts of unfairness and injustice offend our common sensibilities. As such, our respect for fairness and justice will actively shape our opinions of others.

The sense that we have been treated unfairly or unjustly is often a prime motivator for behavior deemed unethical, immoral, or illegal. Our sense of outrage at unfair treatment is often a catalyst for the rationalization of improper actions known as *moral disengagement*. Moral disengagement is a process whereby conscience or self-censure are suspended by convincing ourselves that unique or unfair circumstances somehow justify behavior that is otherwise considered unethical or immoral. We will explore how moral disengagement happens and how it ultimately affects behavior in detail later in this book.

Culture Fluidity and Ethics

Morals and ethics are subject to change, but those changes are relatively gradual and lack any clear demarcation Figure 2-5. Not many generations ago

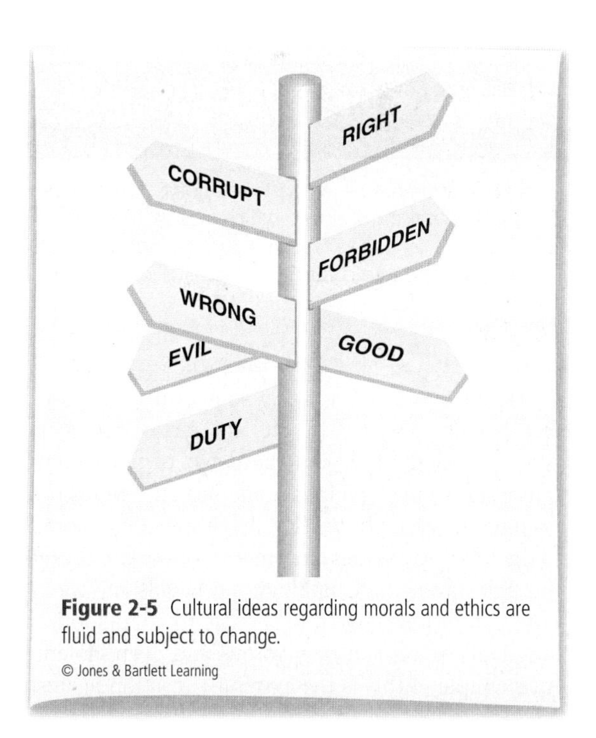

Figure 2-5 Cultural ideas regarding morals and ethics are fluid and subject to change.

© Jones & Bartlett Learning

young women who became pregnant out of wedlock were subject to scorn, so much so that it was not unusual for these women to leave their families and have their children in far-off communities. Today, the moral stigma attached to unwed pregnancy is greatly reduced and often nonexistent. By any measure, it is certainly viewed more liberally than it was by those of our great grandfathers' generation. When did this change occur? There is no specific date recorded where American culture collectively decided to change its mind on unwed pregnancy. These changes come gradually and in increments. To this day, there are still some who are uncomfortable with the concept of procreation out of wedlock. As with time variables, morality and ethics often vary by location even within the same country. Generally, states like California, Oregon, and New York have a reputation for being less bound by tradition than states such as Alabama or Mississippi. Even within the same state, urban areas are typically considered more liberal than rural areas. Whereas the law clearly defines where behaviors are expected, morality and ethics tend to be relative to the specific culture of a population.

Clarification

Law clearly defines behavior expectations, unlike moral and ethical standards, which may vary greatly from culture to culture. Variance in social norms can cause conflict when cultures interact.

Herein lies an underlying source of much conflict. Beginning with the great explorers of 500 years ago, and continuing with the dawn of mass communication and globalized economies, the world has become increasingly smaller. As people are exposed to the folkways, values, and morals of other cultures through travel and technology, our beliefs and traditions can be challenged. Resistance to adaptation of long-held beliefs can be vigorous and even violent. An example of this is the extremist reaction against Western culture encroachment into some of the conservative cultures of the Mideast and Near East.

Issues with cultural changes in morality and ethics are not always external. They often give rise to internal strife as individuals seek to make sense of an ever more complicated world. Postmodern theorists assert that ever-increasing cultural diversity tends to homogenize belief systems (Cooper, 2012). As a result, populations are gradually losing faith in what were once regarded as universal truths. This gives rise to an increasing adoption of **situational ethics** (Derrida, 1985). We will explore this topic later in detail; however, situational ethics can be briefly described as a belief that intent and outcome shape ethical expectation. This is a departure from the 19th century when most Westerners believed that ethics were rooted in duty, obligation, and compliance to established principals.

Similarly, contemporary philosophers and pundits are expressing increasing concern regarding current trends in business ethics that seem to be less focused on humanistic virtues and increasingly based on **game theory ethics**. This is a belief that ethical obligation is negated when all participating parties have no expectation of ethical compliance. In other words, it's okay to lie if everyone's doing it and assumes that you are too (Vanderschraaf, 1999). For many, this is inconsistent with personal views on ethics and morality. In order to accommodate this inconsistency, individuals must separate their personal values from their professional values. For some, this can give rise to conflicts of conscience.

In general, changes of folkways, morals, traditions, and their associated ethics are often accompanied by growing pains. Whether such changes cause social upheaval or individual stress, a single constant in our social values is that they are likely to change. And so too will our ideas about what is normal, ethical, moral, and fair.

Implications for the Fire Service

To this point, our discussion of folkways, norms, morals, and ethics has been relatively general. You

Figure 2-6 Teaching new officers the fundamentals of ethics and leadership is critical to the continued success of the fire service.

©Tyler Olson/Shutterstock

may fairly ask, How do these concepts affect the fire service? The fire service possesses a unique culture. While very much part of the community it serves, it also has its own culture, with its own traditions, norms, and folkways. For the most part, traditions and norms reinforce positive values that influence our fire departments for the better. Unfortunately, some can be obstacles to our better selves.

Consider some of the traditions and norms that help form the typical fire department culture. Fire service has a history steeped in a paramilitary tradition. It has a healthy respect for chain of command and unity of command. The fire service also has a tradition of maintaining discipline. Our culture places great emphasis on routine, teamwork, and group identity. All of these qualities conspire to reinforce adherence to rule and policy. There is also an emphasis on accountability, which tends to promote ethical behavior.

Unfortunately, scalar organizations relying on chain of command can also create bureaucratic issues, particularly the potential for disconnection between high-ranking officers and company officers. Organization structure places tremendous responsibility on company officers for the reinforcement of culture, values, and ethics Figure 2-6. Our junior and most inexperienced officers are the heart and soul of the department. In short, battalion chiefs rarely know what really goes on in stations.

Fire fighters' work is often dirty and dangerous. It can be physically demanding and intense. Fighting fires benefits from adrenaline and controlled aggression. Not surprisingly, the fire service has a culture that celebrates physicality, bravery, and team identity. Those qualities are quite useful for the task of fire suppression; however, suppression represents only a small part of the typical fire fighter's time. The same qualities can be an obstacle to "appropriate" behavior. A testosterone-fueled work culture lends itself to poor safety decisions, harassment, hazing, and exclusion.

Fire service is rightly proud of its traditions. Those traditions can reinforce positive attitudes and foster a sense of belonging to something bigger than oneself. Conversely, traditional environments can also "normalize" bad behavior. Resistance to change, prejudice, and a focus on department welfare over public welfare are all untoward results of a glorification of tradition Figure 2-7.

It is important for the fire service to recognize that it has a unique culture with its own norms, history, moral values, and unique ethical responsibilities. Understanding these elements and how they shape our view of good versus bad behavior is exceedingly important in maintaining an ethical culture.

Figure 2-7 Tradition must never interfere with doing the right thing.

© Glen E. Ellman

WRAP-UP

Chapter Summary

- The "building blocks" of social behavior include social norms, morality, ethics, and law. Each of these in one way or another establishes parameters by which behavior is judged. They are unique in their origin and scope, yet they are intertwined and affect each other.
- With few exceptions, most building blocks are not universal in that they are rarely agreed on by everyone everywhere and throughout time.
- The differences among professional ethics, personal ethics, and morality were detailed. Morality was loosely defined as a concept of good or evil, usually rooted in religion or some other higher authority. Morality also has the condition of being universal.
- Ethics differs from morality in that it is rooted in interpersonal relationships. A thing is ethical or unethical depending on its impact on others and its conformance to obligation and accountability. Our responsibilities are obligations relative to others.
- Universality does not mean that something is universally accepted, but rather universal in application. A thing is seen as universally, moral or immoral, regardless of circumstance or intention.
- Ethics can be further subdivided into professional ethics and personal ethics. Professional ethics are behavior codes expected of a particular profession; personal ethics are relative to our social interactions and obligations.
- The concepts of justice and fairness were compared and contrasted, and their relationship to an ethical theory was introduced.
- The fluidity of our perception of behavior was discussed. The concepts of morality, ethics, normalcy, and justice are all subject to variance depending on time, place, and culture.

Key Terms

ethics the development, evaluation, and study of behavior boundaries and expectations within personal and professional interactions with others.

fairness judgment or equal distribution without bias, or regard for an individual's needs or feelings.

folkways the unofficial practices and rituals that define routine and everyday behaviors.

game theory as applied to ethics, the belief that an otherwise unethical action is justifiable if there is no expectation of fairness.

justice the quality of being equitable, compliant with law, or acting with respect to individual rights.

justitia socialis a body of law that seeks to maintain justice and fairness within society

laws a collection of policies enforced by government to regulate the behavior of the governed.

malum in se a Latin term meaning "wrong" or "evil in itself." This is a fundamental concept in understanding morality.

malum prohibitum a thing is illegal because the law prohibits it.

morals universal principles of goodness usually attributed to religion or some other higher authority.

mores behavior expectations associated with the concepts of right and wrong. Terms like "decency" and "morality" are often connected to social mores.

personal ethics the basic principles and values that govern interactions among individuals.

professional ethics professionally accepted standards of personal and business behavior, values, and guiding principles.

situational ethics ethical theory in which right and wrong are judged by context and intent.

social norms pattern of behavior in a particular group, community, or culture that is accepted as normal and to which an individual is expected to conform.

taboos a strict prohibition of behavior. Violation of taboo will likely elicit social censure.

Challenging Questions

To check your understanding of this chapter's material, answer the following questions. It is highly recommended that you discuss your viewpoints with fellow students, peers, coworkers, and friends to discover their opinions as well.

- As a fire fighter, can you conceive of an instance where an action may be considered ethical yet conflict with department policy (law)?
- Have you given significant consideration to which value set you place the greater credence: the judgments of your particular religion (if any) or your own personal beliefs? As an example, the Catholic church holds that remarriage after divorce is a sin, yet many Catholics divorce and remarry with a relatively clear conscience. Do you believe morality comes from a higher source or from within?
- Is normal important? Is conformity to social standards an inherently important obligation in order for people to live together peacefully, or is individuality a right that must be honored? This chapter's opening quotation is from Albert Schweitzer. What is meant by inner perfection? Explain how the study of ethics can help an individual to attain inner perfection.
- For over 200 years there's been great debate in American society regarding the role of government in maintaining moral standards. One side argues that the government imposing moral and religious beliefs is an unconstitutional imposition on personal freedom. Another group has argued that we are a nation under God and it is the obligation of government to ensure the morality of its citizens. Critique both sides of the debate, giving at least two arguments supporting and opposing both positions.

Case Study Conclusion

Revisit the case study at the beginning of the chapter. Spend a few minutes considering the questions posed at the end of the case study. In light of the information shared in this chapter, have any of your original observations changed?

The ethical quandary faced by the fire chief is rooted in several issues. Is the behavior of Lieutenant Smith immoral, taboo, or simply provocative? How does a person determine immorality? Even if the behavior is immoral, does the fact that it is legal justify it? If the website is concerning because it could cause scandal, does the department interest in reputation management supersede the fire fighter's rights to privacy and self-determination?

WRAP-UP

Mores can be difficult because the community's collective values are usually much more conservative than the individual thoughts and behaviors of those who collectively espouse them. Is running a website featuring nude pictures outside mainstream values? In most communities, the answer would be yes. Is it immoral? Again, most community standards would say so even though individual morals tend to be somewhat more subjective.

A fire department is a taxpayer-funded organization, and as such, citizens have a right to expect the department to behave consistently within community standards. However, that expectation does not necessarily apply to the individual behavior of fire fighters off duty. While the public may not discern a significant difference between the fire fighters and the fire department, the law does. In short, public employees are private citizens as well. Generally, the courts protect the fire fighter's rights to engage in activities that are outlandish if otherwise legal. However, those protections are not absolute. If actions are measurably detrimental to the efficiency of the department, they can be subject to discipline.

Is Lieutenant Smith behaving unethically? This is a much more challenging question.

Assuming that everyone involved is a consenting adult, there is likely no ethics violation. However, Smith's behavior may reflect poorly on the fire department, which may have ethical consequences. Assuming that Smith is acting within his personal rights and makes no deliberate attempt to damage the department's reputation, it is likely unfair to hold him accountable for reputation damage as a result of ethically neutral and extensively legal activity. In other words, it's likely not fair to hold fire fighters ethically responsible for the opinions of others. However, the fact that Smith identifies himself as a local fire fighter on the website inserts the department into the issue and so changes the ethical scope. By identifying himself as a member of the department, he tacitly involves the department and becomes responsible for any effect on the department. If Smith's actions are causing direct harm to the department or to his fellow fire fighters, then he is responsible ethically for those results.

What is the chief's responsibilities? Assuming that there is no specific rule restricting off-duty behavior, the chief's options may be somewhat limited. Clearly, the chief has an obligation to maintain the reputation of his department. He also has an ethical obligation to treat Smith fairly and not infringe on his personal rights. These two ethical responsibilities are not at complete odds with each other. At the very least, the chief would be within his ethical rights to insist that Smith remove any mention of, or association with, the department from the website. While the chief may not have the authority to force Smith to remove the pictures, he is within his ethical rights to strongly encourage him to do so by appealing to his sense of responsibility to the department and its fire fighters.

Chapter Review Questions

1. What is a taboo?
2. How do ethics and morality differ?
3. What does the term *justitia socialis* mean?
4. Differentiate between the concepts of *malum in se* and *malum prohitum*.
5. What are situational ethics?

References

Brackman, L. 2014. "Where Do Ethics Come From?" Accessed May 10, 2016. http://www.chabad.org/library/article_cdo/aid/342501/jewish/Where-Do-Ethics-Come-From.htm.

Cooper, T. L. 2012. *The Responsible Administrator: An Approach to Ethics*. New York: Jossey-Bass.

Crossman, A. 2016. "What Are the Differences Between Folkways, Mores, Taboos, and Laws?" http://sociology.about.com/od/Deviance/a/Folkways-Mores-Taboos-And-Laws.htm.

Derrida, J. 1985. *Margins of Philosophy*. Chicago: University of Chicago Press.

Marinie, M. M. 2003. "Social Values and Norms." In *Encyclopedia of Sociology*, edited by R. J. V. Montgomery and E. F. Borgatta, 2828–40. New York: Macmillan Reference.

Prinz, J. 2016. "Morality Is a Culturally Conditioned Response." *Philosophy Now* I, 115.

Sumner, W. G., and A. G. Keller. 1906. *Folkways: A Study of the Sociological Importance of Usages, Manners, Customs, Mores, and Morals*. Boston: Ginn and Company.

Vanderschraaf, P. 1999. "Game Theory and Business Ethics." *Business Ethics Quarterly* 9, 1–9.

Velasquez, M., C. Andre, S. J. Shanks, and M. Meyer. 1990. "Justice and Fairness." *Journal of Issues in Ethics* 3(2).

Wilson, E. O. 1975. *Sociobiology: The New Synthesis*. Cambridge, MA: Harvard University Press.

Yogis, J. A. 2003. *Canadian Law Dictionary*. Hauppauge, NY: Barron's.

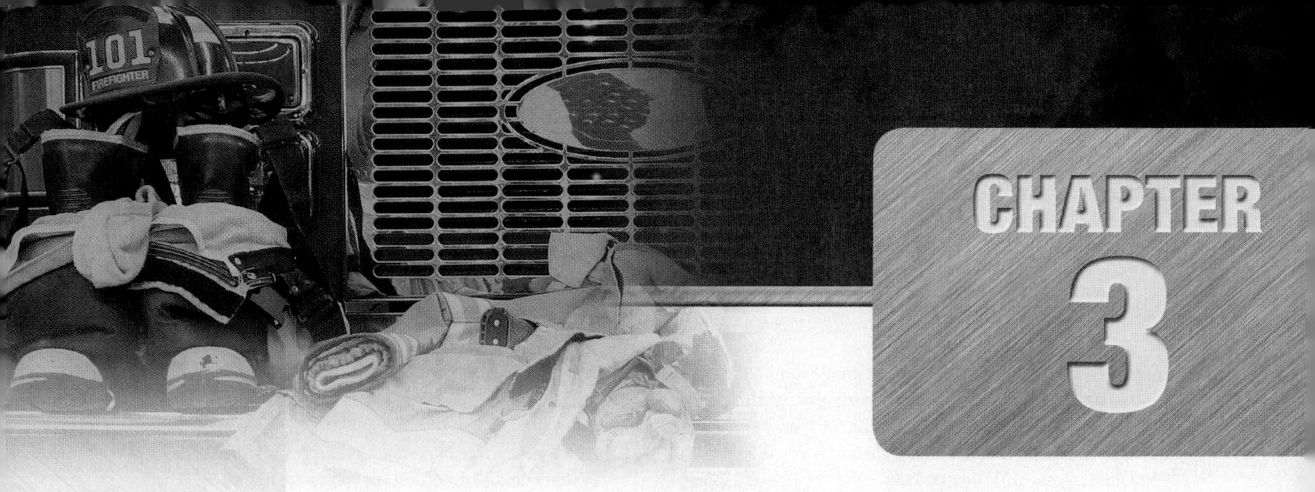

CHAPTER 3

Influencing Behavior

"Education without values, as useful as it is, seems rather to make man a more clever devil."

—C. S. Lewis

OBJECTIVES

After reading this chapter, you should be able to:

- Discern the role of human need in motivating behavior.
- Outline how responsibility shapes and motivates behavior.
- Define virtue, values, beliefs, and attitudes.
- Expound the components and function of our moral GPS.
- Assess our internal processes that maintain equilibrium between our values, beliefs, and responsibilities.
- Interpret the three most important learned beliefs.

Case Study

Captain Jones's daughter is about to play in the city championship tournament for her junior soccer league. He always enjoys his daughter's games, and he knows that it means a lot to his daughter to have him there.

The issue is the game falls on a shift day, and as luck would have it the shift is on minimum manning that day. As a result, he is unable to use any of his vacation or personal days. If he wishes to attend the game, he will have to "call in sick." Smith values his honesty and reputation above all else, and the idea of falsely using a sick day does not sit well with him. He also knows that his absence will cause his crew to run shorthanded, and a less experienced officer will have to "step up." He has repeatedly told his crew that falsely using sick days is a policy breach and shortchanges the rest of the crew. This frustrates him because due to staffing shortages this type of problem exists far too often. As Jones considers his options, the following goes through his mind.

1. Falsely using sick days is dishonest, but really it's just the white lie. After all, they are his to use.
2. He has repeatedly told his crew not to do it.
3. This is a special occasion; his daughter may never play in a championship game again.
4. It's not right that the department is understaffed, causing these kinds of problems.

- What values and beliefs are causing Captain Smith to suffer from internal conflict?
- Smith obviously feels a sense of obligation to both his family and to his coworkers. How should he prioritize them?

Introduction

In Chapter 2, *Judging Behavior*, we explored how behavior is judged relative to the concepts of right and wrong. As we learned, human interaction is complex and very often judged subjectively. We tend to evaluate the actions of others based first on their conformity to social expectations and then how we as individuals perceive an action, outcome, and intent.

In judging our own actions, however, we are guided by a more complex system of behavioral influences. These influences can be synergistic, antagonistic, and competitive. In this chapter, we will explore the influences that both shape our behavior and motivate it.

In the interest of clarity, we will divide the topic into two general discussions. First, we will discuss those elements in our lives that tend to initiate behavior. Think of these as actions we feel we *must* do. Second, we will explore those parts of our self-identity that tend to shape our behavior. Simply put, those factors that tell us what we *ought* to do.

Behavior Motivation

You likely already realize that values, beliefs, and attitudes shape our behavior. The form and extent of their influence are, to a large degree, determined by the motivation behind our behavior. Certainly, some of our behavior is reflexive, or random, in the sense that it has no specific purpose. Yet most of our conscious behavior is motivated by one or more influences. Generally, these influences can be described as either inwardly focused or outwardly focused. **Inward motivations** are those that seek to meet perceived needs or wants. **External motivations** are those that are rooted in a sense of responsibility to someone or something (Miller, 2012).

Clarification

Behavior influences can either be inwardly or outwardly focused.

- Inward: actions that seek to meet perceived needs and wants
- Outward: actions rooted in a sense of responsibility

As you will see, the ethical and moral significance of motivation can be substantial; as a result, understanding motivation is necessary for understanding ethics.

Needs and Wants

The concepts of needs and wants are easily distinguished from each other in definition; however, in practical application, lines can be blurred. In dealing with the subject of ethics, it is important to understand that distinguishing between needs and wants is fundamental in attaching ethical significance Figure 3-1. The common understanding of a **need** is that of an object or condition for which negative consequences will occur in its absence. As an example, you need food because, without it, you will die. Conversely, a **want** is a behavior motivator based on the perception that achievement will bring a highly desired outcome. You want pizza because it tastes good (Maslow, 1943).

Not all needs and wants are created equal. Some needs are objective and straightforward while others are more inscrutable. As needs become more abstract, they can be more difficult to separate from those things that we want badly. The relationship between needs and wants can be confused because the loss of potential gain or personal satisfaction is sometimes emotionally equated to the consequence of an unmet need. As you will see, the ethical significance between the two is substantial.

Needs, wants, and responsibilities rarely exist in a vacuum. As we live our daily lives we often prioritize competing interests. On a warm sunny day in spring you may *want* to go play golf; however, your *need* to earn money may compel you to go work. The relationship between competing priorities can be categorized into one of three states.

Bear in mind that needs, wants, and responsibilities are not mutually exclusive. Any of the three may interact with each other or with another similar to themselves; in other words, needs can interact with other needs or with wants or responsibilities Table 3-1.

We have described needs as things that are essential to the avoidance of negative consequences. However, human activity certainly involves more than the avoidance of negative consequence. Sociologists and philosophers will point out that, as humans, we have needs of attainment beyond those of avoidance. We not only need to avoid pain and suffering, we also need to add enriching elements to our lives in order to attain quality of life. To accommodate this element of humanity we must modify our understanding of need. For purposes of understanding ethics, let us define **human need** as *something that a human must have in order to live a recognizably human life.*

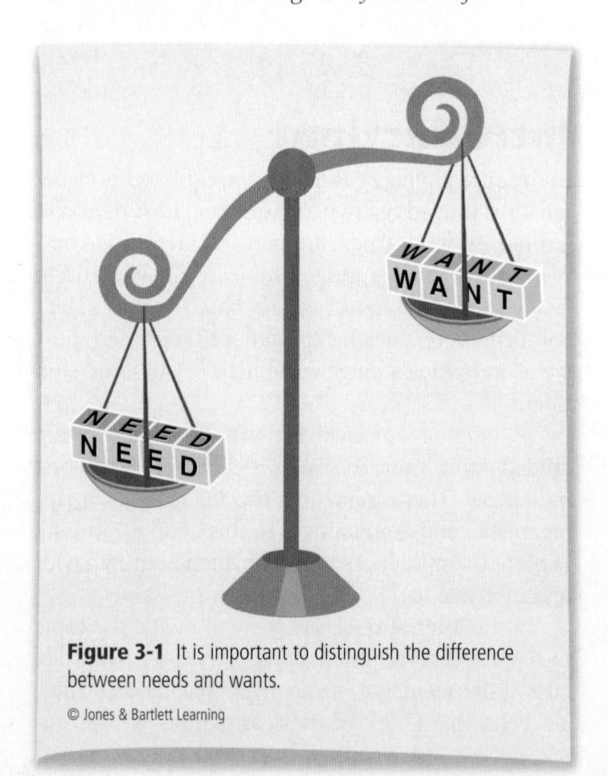

Figure 3-1 It is important to distinguish the difference between needs and wants.

Table 3-1 States of Competing Priorities

State of Competing Priorities	Relationship Between Competing Priorities	Examples of Priority Interactions
Independence	The actions spurred by needs, wants, and responsibilities have relatively little impact on each other.	I want to go to a shift party. I want to go to bed early.
Competitive	Needs and wants are mutually exclusive. The pursuit of one necessarily excludes the possibility of the other, or at the very least has a negative impact on its counterpart.	I want to go to a shift party. I am committed to go to a training class at the same time.
Aligned	An action springing from a need or want and supports some other need.	I want to go to a shift party. I am committed to cook at the party.

In 1943, Abraham Maslow published a paper titled, "A Theory of Human Motivation" (Maslow, 1943). In his work, he sought to identify, categorize, and to some extent prioritize human needs Figure 3-2. He put forth a now famous theory known as Maslow's *hierarchy of needs,* which states that as humans we have needs that can be categorized as physiological, safety, social, esteem, and self-actualization. His contention is that these needs are listed in their order of *primacy*. While there is general agreement regarding his identification and classification of human need, there has been considerable debate over his concept of primacy. In discussing ethics within the construct of motivation, however, primacy seems particularly applicable. There is a practical relationship between the degree of ethical attachment and the level of motivational need being served. For example, it is more or less universally understood that stealing is not ethical. Yet a starving individual caught stealing food in order to survive is viewed substantially differently than one who steals a fashionable pair of sneakers in order to maintain social status. The need for acceptance and social status is recognizable and even understandable; yet it does not have the dismissive weight of the survival instinct. Both individuals are committing the same act, both are meeting recognized needs, yet one action is typically viewed as more ethically justifiable than the other. It would appear then that the question of ethics becomes more complex as needs become more complex. Our needs for physiological substance and relative safety seem straightforward and relatively understandable. Most of us would feel comfortable excusing some behaviors in the face of protecting one's life and safety, yet some needs are more complex and so can be more subjective. Our need for friendship, achievement, respect and autonomy are real and they are powerful motivators. Unfortunately, their meanings are more complex and subjective, and so the issue of related ethics also becomes more complicated. The term **interest** in sociology refers to the relationship between need and attainment. A person has an interest in something that is directly related to a need or the acquisition thereof. You have a need for health, hence you have an interest in medical services. The **proximity of interest** to a need affects our understanding of ethical attachment. If a hungry person steals food, we see an immediate connection and may likely find the action relatively excusable. Yet if a hungry person steals money in order to obtain food, the action becomes somewhat less excusable. More to the point, a person who is not currently hungry who steals money in order to avoid future hunger is likely to elicit no sympathy whatsoever. is an example of *proximity* in determining the relative ethics of an action Box 3-1.

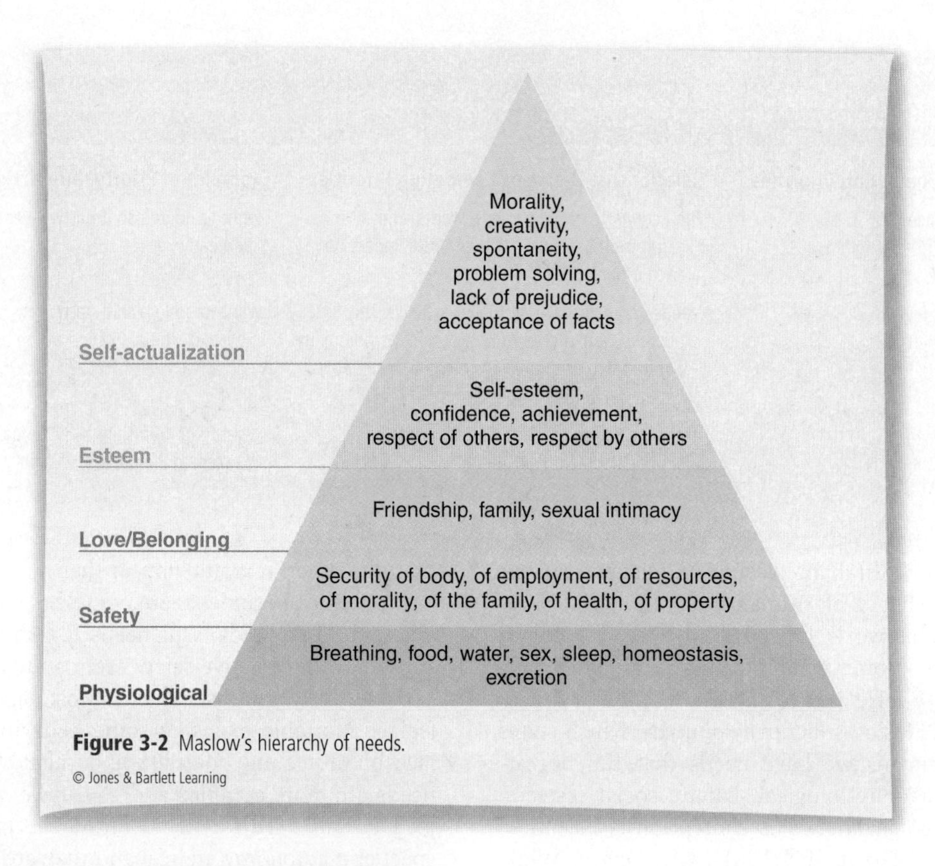

Figure 3-2 Maslow's hierarchy of needs.

BOX 3-1 Attachment of Ethical Significance to Need Attainment Is Relative to Several Factors

- Urgency of need: Deficiency needs tend to be more ethically relative. Stealing food for survival versus stealing money to attain desired goods.
- Wants are less ethically significant than need.
- Proximity of interest to action: Stealing food to survive versus stealing money to buy food to survive.
- Proximity of need: Stealing food to survive now versus stealing food to survive next winter.

Wants

In discussing ethics, it is useful to consider a *want* as a desire for an object or condition for which there is no interest. As such, wants are rarely justification for behavior. Yet our desire for various objects and conditions is undoubtedly motivation for behavior. Remembering Maslow's hierarchy of needs, it is useful to think of a similar hierarchy of wants. At a physiological level, we *need* food to survive, but we may *want* variety and excess. Here the distinctions between needs and wants are relatively easy to make. Compare this with the need for self-esteem. A positive self-image and confidence are certainly necessary to quality of life, yet the related (but subordinate) compelling want for recognition, fame, and credit have no direct relationship or interest. The need to achieve self-esteem can be viewed as "good." Recognition, on the other hand, may boost self-esteem but is not necessarily essential to maintain self-esteem. As such, it is ethically neutral, in that recognition in itself cannot justify any given action. The problem is that it can be difficult to discern where self-esteem ends and the need for recognition begins.

In general, wanting something is essentially ethically neutral. That is to say that the desire for one thing or another is, at face value, neither consistent with, nor inconsistent with, personal values or social expectations. It is, however, important to understand that the pursuit of needs is typically more consistent with acceptable behavior than is the pursuit of wants. Therefore, distinguishing between the two becomes important in understanding the ethical relationship between motivation and action.

Ethics and Responsibility

Needs and wants reflect internal motivations, but not all of our actions are related specifically to our needs and wants. We often enter into social relationships, and the consequence of those relationships is that we assume responsibilities to others. The concept of responsibility is essential in understanding ethics in that it not only motivates behavior but also is a condition of **ethical attachment**. To understand ethical attachment, assume for a moment that you wake up to find yourself alone in the world. There are no other human beings anywhere. Can you describe a circumstance whereby any action you take would be unethical? There is no one to cheat, there's no one to lie to, and no action you take can harm another person. If your actions have no impact on another individual, it would logically be considered ethically neutral or unattached.

You may argue that you have an ethical responsibility to your environment. For instance, your personal values may compel you to kill animals only in cases of self-defense or for food. In this case, your ethics is bounded not in a sense of responsibility to another person but to an idea of environmental harmony. This is by strictest definition an expression of moral values not ethics. In this case your self-determined relationship with your surroundings reflects ideals but lacks responsibility. A fundamental difference between ethics and morality is that morality is rooted in values that are considered "good" at face value, regardless of their effect on others. Morality is independent of intention, consequence, and obligation, whereas ethics is rooted in responsibility, obligation, and consequence. If a second person were with you, the issue changes considerably. You may have a responsibility (with ethical attachment) not to waste food sources that may be needed by your companion. Hence, killing for sport may become unethical in cases of limited food supply.

The concept of responsibility is complex, as is its relationship to ethics. Responsibility is comprised of several relationship variants. The first is accountability.

Accountability

Accountability is a form of responsibility focused on other people. There are many people in our lives to whom we are accountable, including our family, friends, and coworkers. The number of relationships for which you are accountable and the depth of that accountability will vary by your social network and commitments. Each bear some degree of ethical attachment. You are accountable to others for certain actions; those actions are deemed ethical to the degree that they satisfy those expectations (Cooper, 2012). For example, as a fire fighter, you may feel an ethical obligation to your fellow fire fighters to be competent in core skills. This sense of accountability arises from the fact that their lives may depend on your performance on the fire ground. The degree to which you meet that ethical obligation will be measured by how well you meet their expectations.

Clarification

One measure of ethical behavior is that of **accountability**. The degree to which you meet that ethical requirement will be measured by how well you meet the behavior expectations of others.

Obligation

In the previous example, we used the term "obligation" in relation to competency. **Obligation** is a responsibility to some task or ideal. This can be equated to a sense of duty. As you can see in our example, we are accountable to our fellow fire fighters to meet our obligation (duty) to be competent. As fire fighters, we may feel we have many duties, including competency upholding tradition and furthering the department mission. Nearly everyone in a society feels some sort of obligation to uphold social norms and meet behavior expectations. Our obligations and their attendant accountabilities not only motivate our behavior but significantly influence our judgment of the behavior of others.

Our perceived conformance to behavioral obligations is in some degree the basis of ethics, and so separating the concepts of obligation and accountability goes beyond semantics. Each has a certain ethical weight. Because an obligation is a responsibility to a thing, a task, or a concept, it may be considered primary in its ethical imperative when compared to accountability. This is because accountability measures responsibility to people's judgment. In other words, obligation measures the depth of a responsibility, whereas accountability is a method of assuring performance of an obligation. For example, society holds parents accountable for the care of their children. That expectation is rooted in an assumed obligation to be a caring and loving parent. Your accountability of action has less ethical significance than does the obligation of the act. A parent who feeds his or her child only out of fear of arrest is likely judged less ethical than one who feels an ethical obligation to feed his or her child but, due to uncontrollable circumstances, does so inadequately (Cooper, 2012). Accountability is more or less external and tends to be well-defined or objective, whereas obligation comes from within and is subjectively influenced by values and beliefs.

Objective and Subjective Responsibility

Each of our responsibilities, whether they be an obligation or some form of accountability, is either objective or subjective. An **objective responsibility** is imposed externally, usually with accountability attached. It is usually clearly defined and represents expectations imposed on us within our relationships. An example of an objective responsibility is our requirement to obey the law or tell the truth. Conversely, a **subjective responsibility** represents the imposition of expectation from within. Parents may feel a subjective obligation to save money for their child's college education. This sense of subjective responsibility arises from our values and beliefs, whereas objective responsibilities arise from the values and beliefs of those with whom we interact and to whom we are accountable (Cooper, 2012).

Shaping Behavior

Our Internal Guidance System

At one time, you may have heard the term "moral compass." It is an apt metaphor; like a compass we have an internal mechanism that guides us (Figure 3-3). It is developed first by our earliest teachers and later shaped by our own experiences. Assuming that there are no other influences, a compass will point north. As such it does not give us directions. Rather, it gives us a point of reference by which we may choose a path. Similar to an actual compass, our moral compass provides a reference to a best path. Circumstances may influence whether we choose to follow it or not.

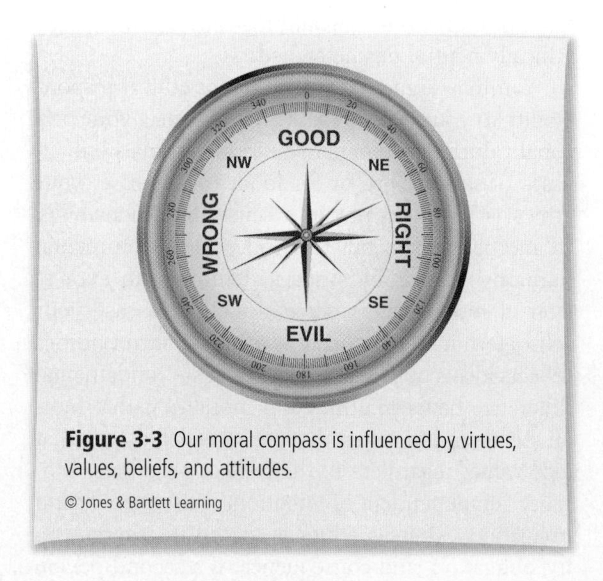

Figure 3-3 Our moral compass is influenced by virtues, values, beliefs, and attitudes.

A compass is a simple device influenced by the reliability of magnetic north, and normally that is the only influence on it. As such, it dependably points north. Our internal compass, however, is affected by many factors that vary in relevance and intensity. These include our sense of virtue, values, beliefs, and attitudes. These factors all influence where our compass may point.

Virtues

A **virtue** is a habit or an acquired character quality that is deemed universally good. Philosophers and most religious doctrines believe that living a life based on virtue is fundamental to human happiness and enlightenment (Crisp and Slote, 1997). A virtue can be moral or can be intellectual. A moral virtue is one that is virtuous in action. Examples may be generosity, honesty, and courage. An intellectual virtue is one that is virtuous in thought, such as modesty, humility, curiosity, and acceptance.

A key quality of a given virtue is that once acquired it becomes a character element. For instance, a person who develops wisdom is generally identified as being wise. Compare this with ethics or morality, which tend to apply to individual actions. Certainly, the totality of our actions tends to identify us as ethical or moral people, but this represents the sum of our actions and not any individual action. Virtues, however, can singularly identify our character.

The Greek philosopher Aristotle is generally credited as being the founder of the school of virtue ethics. He postulated that a person who possessed many virtues tended to act in ways that are consistent with moral principles and so was an ethical person. Aristotle also argued that the greatest virtues are those that are also most useful to other persons. This is a fundamental principle in understanding the role of virtue. Virtues are not developed in isolation; they reflect the values of the community, culture, and family in which we grow up. Because virtues reflect community standards, they shape how we act within a community (Velasquez, Andre, Shanks, and Meyer, 1988).

Virtues (or lack thereof), coupled with values, make up a significant portion of our self-identity. They shape our view of the world around us, and so are the bases for the more complex parts of our moral compass such as beliefs and attitudes. We will revisit virtues as an ethical system in their own right in Chapter 4, *The Philosophy of Ethics*.

Values

Values are a collection of virtues, principles, standards, or qualities that an individual or group of individuals hold in high regard. In the common vernacular, values are those qualities and principles that we consider to be of worth and importance.

The development of a value system is an important part of our maturation process. The process begins in earliest childhood and continues into old age. The earliest and most important shapers of our values are parents and teachers. The term "our formative years" reflects the importance that social psychologists place on value shaping during early childhood. It should be remembered, however, that our values are malleable. They evolve and are refined throughout our lives. As we mature, the influences of religion, art, and peers continue to shape our value system. Most importantly, personal experience, as interpreted by our understanding of events, does much to shape our value system (State of New South Wales, 2009).

The thoughts and ideas that hold primacy in our view of the world are as varied as the individuals who they affect. Commonly held values may include family relationships, patriotism, friendship, prosperity, safety, personal recognition, and heritage. An important common value set is composed of virtues related to interacting with others; the virtues of honesty, compassion, loyalty, and generosity form the core of this value system.

Clarification

Our values are shaped by many factors: parents, peers, workplace culture, teachers, religion, art, popular media, major events, personal experiences.

In discussing values, it is common for them to be presented in an exclusively positive light. In truth, values may be a positive or negative catalyst in ethical and moral behavior. Placing great value on recognition and respect can, in turn, motivate both high moral standards or a compulsion for negative behavior such as lying and cheating. Valuing one's heritage can motivate the keeping of tradition and history; it can also motivate misplaced divisiveness and racism. Values are seldom entirely "good" or "bad." It is how we act upon them that defines our behavior. Here is where the relationship between virtues and values comes into play. Our adherence to virtues defines the parameters of actions we feel are acceptable in support of our values. For example, my sense of values may place great importance on personal reputation, yet it is the relative strength of my adherence to the virtue of honesty that will define how I will handle difficult situations where my reputation may be at stake. Our values act as a filter in how we see the world. They tend to focus what is important to us, and to some extent, they shape what we see as truth.

Beliefs

Beliefs are those things that we believe to be true. The connection between truth and ethics is essential. Our actions are almost always based on an anticipated outcome shaped by our understanding of the world. Our opinions about the actions of others are based on an assumption of understanding. Our views of our own actions and those of others must be based on truth if they are to be ethically relevant. As a result, our understanding of ethics is intrinsically related to our understanding of truth. Our beliefs are formed and refined by several influences. Three of the most important are learned beliefs, cultural influences, and personal experience (State of New South Wales, 2009).

Learned Beliefs

To function as a society, there is a necessity for truth. As a result, even the most cynical of us tend to believe most of what we hear and see as being relatively true. Consider the chaos that would arise from a complete suspension of belief. The simple act of buying cereal in a supermarket is impossible if you do not believe that, in fact, there is cereal in the box. Belief in a common history and belief in common truths are the foundation that holds a society together. Because we are inclined to believe most of what we are taught at face value, those lessons form a great deal of our belief system. Our understanding of history, heritage, and social order are based on lessons we are taught in school and in our adult consumption of books and media. We accept that the world is round because sources we consider credible tell us so. Beliefs from learned behavior account for the majority of those things we believe to be true; however, they are also least likely to be deeply imprinted. Beliefs based on learned behavior are relatively easily modified by new or more credible knowledge. We typically lack emotional investment in learned beliefs, especially compared to cultural influences and experience.

Cultural Influences on Beliefs

Some beliefs are held as true because they are ingrained in our cultural understanding or rooted in tradition. Some things are believed true simply because they are. Our department only promotes from within. Why? Because we always have. In a sense, cultural influences represent beliefs based on emotion and faith.

The American fire service has a long and rich history steeped in tradition **Figure 3-4**. It is a fundamental part of our success; however, it has at times impeded our progress. In many cases, we have been slow to embrace technology and new philosophies in operational tactics. Unfortunately, we have sometimes also been slow to adapt to social changes as well. Often that resistance has had ethical implications. It is fair to say that many departments have been somewhat slower than society, in general, in embracing diversity issues. As an example, women remain significantly underrepresented within our workforce. Many women report instances of harassment and disparate treatment within fire departments. The fact that a large population is being treated differently demands ethical consideration. As a matter of justice, disparity necessitates ethical justification or

Figure 3-4 The fire service is steeped in traditions—and can have either a positive effect or negative effect on values.

condemnation. Most recently the emergence of the LGBT community into the social mainstream is presenting significant challenges to the traditional beliefs of the fire service. Western culture is progressively reassessing its views on gender and marriage. As a publicly funded institution, the fire service will likewise have to reassess the ethical implications of personal rights versus long-held cultural beliefs.

Personal Experience and Beliefs

Perhaps more than any other influence, personal experience both shapes new beliefs and reinforces existing ones. Assume for a moment that a male fire fighter believes that women have no place in the fire service. If his department hires a female fire fighter who proves to be competent and capable, he is much more likely to modify his position based on his personal experience as opposed to reading articles about successful female fire fighters. Conversely, if the same fire fighter works for a department that has hired a female fire fighter who turned out to be less than competent, her performance, or lack thereof, is likely to strongly reinforce and justify his previously held beliefs, even in the face of other evidence to the contrary. We place great value on what we see and hear, even if rationally we may be aware of evidence that challenges our beliefs.

To further understand beliefs, we can categorize them as descriptive, evaluative, or prescriptive. A **descriptive belief** represents a truth as we see it. An example of a descriptive belief might be "fire fighters save lives and protect property." An **evaluative belief** is expressed as a judgment rooted in a descriptive belief. For example, understanding that fire fighters save lives and protect property, I may make the evaluative statement that "firefighting is essential to the well-being of the community." A **prescriptive belief** is an expression of justifiable outcome and is based on either descriptive or, more likely, evaluative beliefs. An example of a prescriptive belief would be "local government should fully fund fire departments."

Our understanding of truth is, of course, an important part of our assessment of right and wrong. As humans, we tend to seek unity between our values and beliefs. What we hold true (beliefs) are closely tied to our values. A fire fighter with a good deal of seniority may likely place value on experience. As a result, that fire fighter's belief system may likely include the descriptive belief that experience bestows knowledge, and the evaluative belief that senior fire fighters are better qualified for promotion than younger "untested" fire fighters. This may likely lead to the prescriptive belief that seniority points should weigh heavily in the promotion process. Failure by the department to recognize seniority may lead to an attitude of frustration and disenfranchisement. This will likely have a significant impact on the fire fighter's beliefs or, in other words, his or her attitude.

Clarification

Belief system examples:

Descriptive belief: Fire fighters save lives and protect property.
Evaluative belief: Firefighting is essential to the well-being of the community.
Prescriptive belief: Local government should fully fund fire departments.

Attitudes

Attitudes are a collection of feelings, beliefs, and behavior tendencies directed toward specific people, groups, and ideas. They are a reflection of the many influences within our belief systems, including family, friends, and experiences. Our attitudes shape behavior in that they synthesize our general concepts of right and wrong, the justifiable and the inexcusable.

Attitudes have three components as they relate to behavior. In concert with held values, they define the parameters of our personal ethics. The **cognitive component** encompasses your thoughts and beliefs about a particular subject. As stated previously, the cognitive component represents what you believe to be true. The **affective component** refers to your emotional response to an event, person, or idea. The **behavioral component** of your attitudes defines "proper" responses to life experiences as they relate to your attitudes. The relative appropriateness of a response can be internally judged by anticipated outcome (past experience, learned behavior), emotional context, and, most importantly, values. Let us again refer to the fire fighter who believes that promotion points should be given for seniority; failure to receive what he or she believes to be fair treatment may cause frustration or anger. The fire fighter may assume that he or she has no legal recourse based on past experience, and this may influence that person's responses. If he or she places a high value on pride, the individual may vocally dismiss the promotion process as unfair in order to save face and may even use that as justification for future attempts at cheating. Conversely, if he or she places a high value on achievement, the individual may feel motivated to identify those tactics that will best lead to promotion and adapt appropriately (Chaiklin, 2011). Changes in behavior may likewise require a change in attitude. If the fire fighter is to successfully invest him- or herself in education in preparation for testing, the fire fighter may also have to adjust his or her belief that experience is the prime source of learning. Absent a change in beliefs and attitudes, the person will find him- or herself in the unhappy position of pursuing something that he or she finds a waste of time. This can cause great frustration, which the person can only resolve by either giving up on the idea of promotion or by changing his or her views on education.

Clarification

Attitudes have three components:

Cognitive component: your thoughts and beliefs on the subject
Affective component: how an object, event, or person makes you feel
Behavioral component: how an attitude influences your behavior

As stated previously, our self-identity requires consistency among beliefs, attitudes, and values. **Cognitive dissonance** is defined as the feeling of discomfort we feel when we perceive an inconsistency among values, beliefs, and attitudes (Festinger, 1957). The amount of dissonance felt depends on several factors; the two most important variables are the amount of inconsistency between beliefs and the relative value placed in the belief. Where a person feels a high degree of dissonance, there is a greater need for realignment of beliefs and attitudes. Individuals will seek to remove dissonance by a number of methodologies:

1. **They may develop new beliefs that refute the dissonant belief or behavior.**
   Fire fighters who learn that high-fat diets lead to increase chances of diabetes may discount the validity of the information or seek out and embrace dietary theories that argue the health benefits of high-fat diets.
2. **Reduce the importance of a conflicting belief.**
   For example, a chief officer who worries about his health might be upset in hearing high-stress jobs shorten life spans. Because he works in a high-stress job, it is difficult to change his behavior in order to reduce his feelings of dissonance. To negate his feelings

of discomfort, he may instead find some way to downgrade his situation by believing that his other healthy behaviors make up for his stressful professional lifestyle.

3. **Change the conflicting *behavior* so that it is consistent with other beliefs or behaviors.** Adapting a conflicting belief is the most effective way to deal with dissonance; however, it can be exceedingly difficult to adjust a person's beliefs and attitudes, especially when they are tied to core values. A further impediment is produced when a belief is contradictory to a need or significant want. The fire fighters mentioned previously will find it easier to change their attitudes than their lifestyles.

In our example of seniority points for promotion exams, our fire fighter's behavior choices are shaped by his or her beliefs and attitudes; however, they are also greatly influenced by his or her priority of needs. In this case those needs are either the need to save face or the need to achieve Figure 3-5.

The concept of cognitive dissonance is very relevant in the understanding of ethics. Our ability, or lack thereof, to align desired action with beliefs and values is a critical influence on our ethical decision making. It quite literally refers to our ability to "look ourselves in the mirror in the morning." The urgency of need and the relative strength of the involved values and beliefs determine the malleability of our ethical principles.

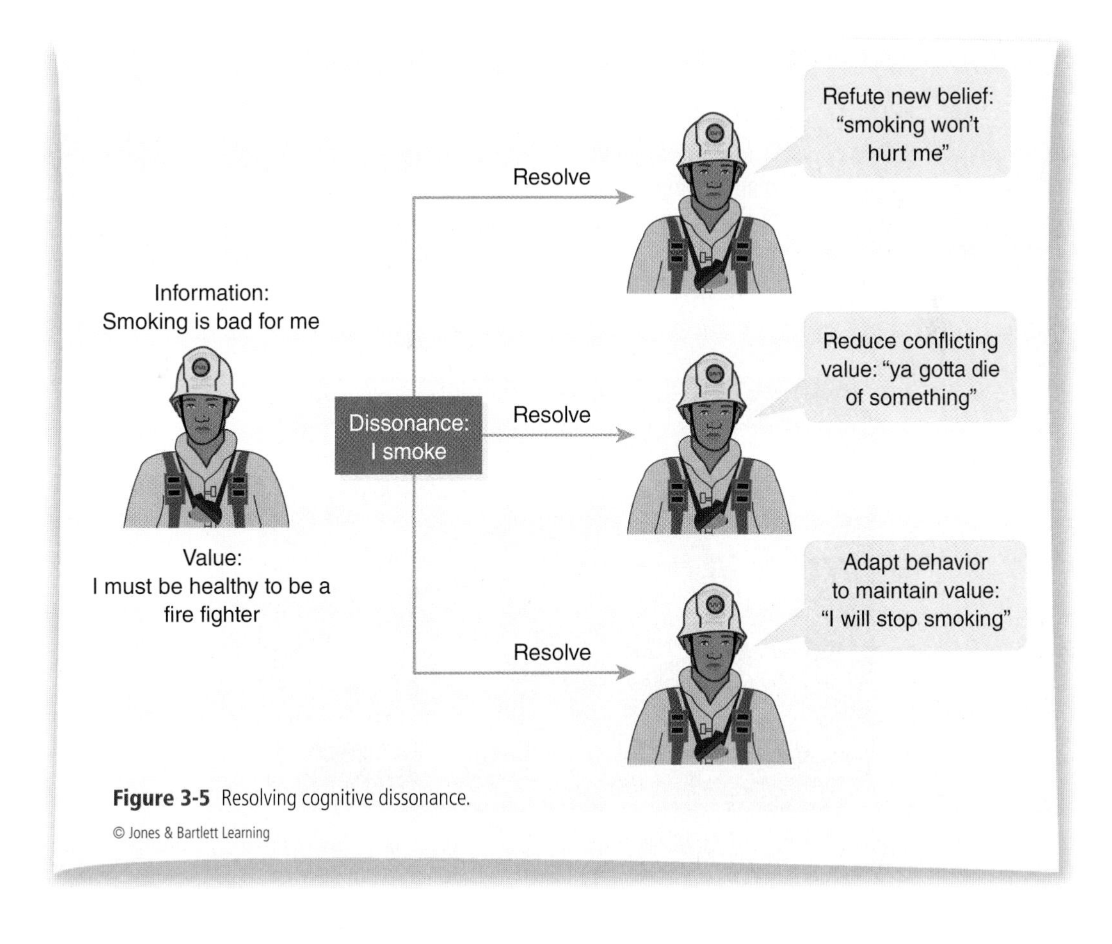

Figure 3-5 Resolving cognitive dissonance.

The Moral GPS Unit

In this chapter we have explored the elements that tend to motivate our behavior as well as guide it. For the sake of clarity, they have been discussed as individual elements; however, it is important to understand that each of these not only affects behavior but may affect each other as well. They work in both cooperation and competition to influence our direction, much as a compass is affected by the presence of numerous magnetic fields. The processes are so complex that the analogy of the moral compass may be more accurately replaced by that of *a moral GPS unit* Figure 3-6.

The process starts when we are encouraged to act either from a sense of responsibility or an attempt to meet some need. Typically, the pressures of responsibility and the pursuit of need satisfaction are both present. They may work in alignment or may be contrary to each other. Fire apparatus operators may feel an internal sense of obligation to maintain their driving skills because they are aware of the responsibility to protect the safety of those fire fighters who ride on their rig. Likewise, maintaining competency is consistent with ensuring their continued assignment as engineers, thus meeting a need for job security and perhaps the need for respect from their fellow fire fighters. In this case, needs and responsibility are aligned.

Conversely, a fire fighter who is aware of a fellow fire fighter reporting to work under the influence of alcohol will likely feel an obligation to report the behavior; however, he or she may also view the act as whistle-blowing and as a threat to satisfying his or her need for social acceptance from fellow fire fighters. In this example, there is a conflict between responsibility and need.

How we choose to react is in large part determined by our attitudes and beliefs. Which needs and obligations are served may depend on our attitude toward the elements of the situation. This selection process is shaped in turn by our values. Typically, we seek consistency between values and beliefs; however, in instances where the urgency of need fulfillment or the weight of responsibility dictates actions inconsistent with values, beliefs may have to be modified. Likewise, where experience forces a reevaluation of belief, values may be modified.

When you judge the ethics of others' behavior you are balancing your sense of values, beliefs, and attitudes against your perception of the urgent need and/or responsibility in the situation. As an example, most of us would agree that torture is inherently

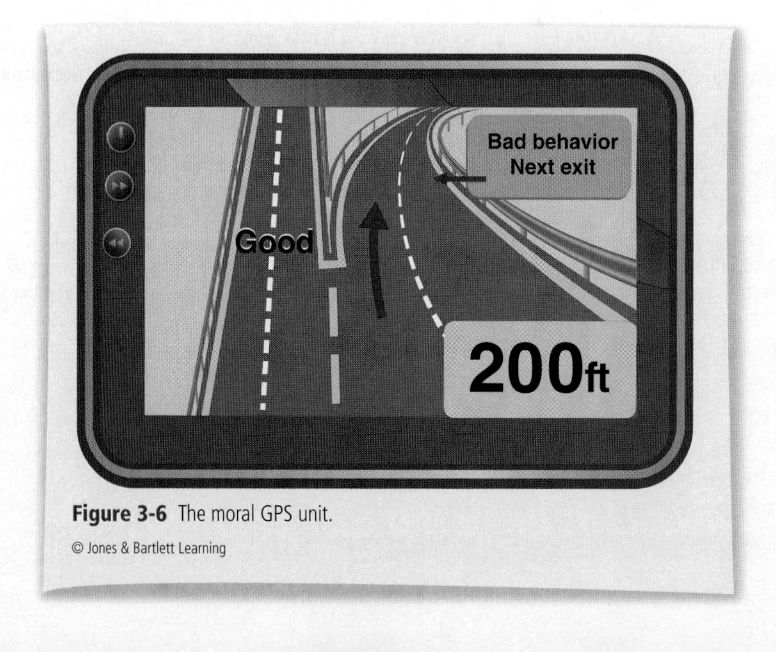

Figure 3-6 The moral GPS unit.

Clarification

Needs, wants, and responsibilities can act independently, competitively, or in alignment.

- Independent: I may want to go to a shift party tomorrow. I need to go to bed early tonight.
- Competitive: I want to go to a shift party. I am committed to go to a training class at the same time.
- Aligned: I want to go to a shift party. I am committed to cook at the party.

wrong, yet circumstances may modify our views on the ethics of torture. Some may argue that torture is a necessary evil (ethically justified) in cases where a terrorist attack threatening the lives of hundreds might be prevented. Our viewpoint on the scenario is determined essentially by the primacy of our values, in this case, the objection to torture versus our subjective obligation to protect an innocent life. In judging our own behavior, we undergo a similar exercise. Considerable internal influences compete and even conflict when perceived responsibilities and needs conflict with each other, or conflict with belief systems. Further complicating the matter is that, as our hierarchy of needs and responsibilities becomes more emotional and cerebral, it becomes increasingly difficult to distinguish between those things that we perceive as needs versus those things that are more accurately defined as wants. In short, what we may perceive as necessity may be judged entirely differently by others who were not influenced by our internal value and belief system.

Our moral navigation apparatus is a complex instrument Figure 3-7. It provides suggestions rather than guidance, and, even then, suggestions can be malleable to the point of unreliability. Imagine an apparatus in which a series of gears act to shape your choices of action. These gears represent your values, beliefs, and attitudes. Now imagine that in this machine the gears not only affect your choices but affect each other. As a result, the gears can change size and position. This makes the process of making ethical choices very complex.

In response, ethical philosophies and theories have been devised to clarify and simplify the process. These ethical systems become increasingly important as our lives become increasingly complex through the addition of responsibilities and evolving attitudes.

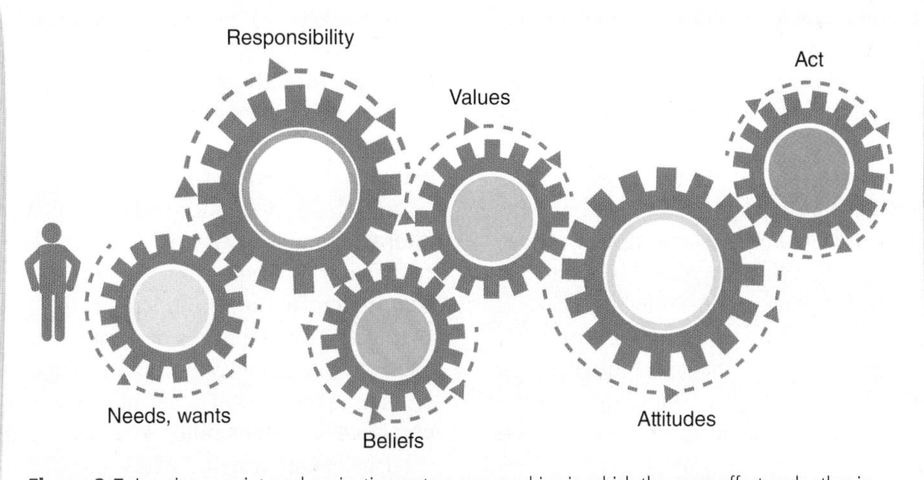

Figure 3-7 Imagine your internal navigation system as a machine in which the gears affect each other in addition to your choices. Imagine further the gears can change size and position.

WRAP-UP

Chapter Summary

- There are several factors that influence our behavior as well as our interpretation of others' behavior. These include motivation and our internal process of balancing the influences of our values, beliefs, and attitudes.
- We are motivated by a complex organization of human needs. These needs can range from the most basic physiological needs of food and water to the very complex need for self-actualization.
- Wants can be as compelling as needs and even confused for them. Working either alongside with, or in opposition to, our needs and wants are our responsibilities.
- Responsibilities can be expressed as internalized and subjective feelings of obligation or as externally imposed expectations from those people to whom we are accountable.
- Human motivation shapes our impression of behavior, yet it is important to understand that behavior is further modified by an individual's belief system.
- Beliefs arise out of our sense of values and are tempered with experience and lessons learned from trusted teachers and sources.
- Our values tend to organize and prioritize our beliefs especially when they may not be completely aligned with our needs and obligations.
- Human behavior is compelled and moderated by the continual adaptation and modification of the relationships among motivation, core values, and belief systems. Not only do they affect behavior, but they affect each other as well. It is by these relationships that we judge our own behavior—and through our perception of others' beliefs and motivations we come to judge the actions of each other.

Key Terms

accountability a form of responsibility focused on others' expectations.

affective component the emotions, such as anxiety, sorrow, or excitement, that a person has regarding the object.

attitudes a collection of feelings, beliefs, and behavior tendencies directed toward specific people, groups, and ideas.

behavioral component the way attitude influences how we act or behave.

beliefs those things that we believe to be true absent proof.

cognitive component a person's belief or knowledge about an attitude or object.

cognitive dissonance the feeling of discomfort we feel when we perceive an inconsistency among values, beliefs, and attitudes.

descriptive belief a truth as we see it.

ethical attachment a condition in which an action has ethical implications. A condition of attachment is that an action has an impact on the well-being of others.

evaluative belief a belief that a particular thing is good (has value) or is bad arising from descriptive beliefs.

external motivations influences that are rooted in a sense of responsibility to someone or something

human need something that a human must have in order to live a recognizably human life.

interest condition that is directly related to a need or the acquisition thereof.

inward motivations influences that seek to meet perceived needs or wants.

need an object or condition for which negative consequences will occur in its absence.

objective responsibility imposed externally, usually with accountability attached.

obligation a responsibility to a thing, task, or ideal.

prescriptive belief an expression of justifiable outcome and is based on either descriptive or, more likely, evaluative beliefs.

proximity of interest the relative immediacy of an action as related to a need.

subjective responsibility the imposition of expectation from within.

values a collection of virtues, principles, standards, or qualities that an individual or group holds in high regard.

virtue a habit or an acquired character quality that is deemed universally good.

want a desire for an object or condition for which there is no interest.

Challenging Questions

To check your understanding of this chapter's material, answer the following questions. It is highly recommended that you discuss your viewpoints with fellow students, peers, coworkers, and friends to discover their opinions as well.

- What are the fire fighter's primary obligations?
- Try to recall the last time you had a belief challenged by an event or discovery. Did you feel compelled to find arguments with the new information or discovery? Did you modify your beliefs in light of the new information?
- Have you ever taken the time to actually test your beliefs? In other words, have you ever given thought as to why you believe what you believe or why you feel like you feel about a topic?
- If you have not done so already, read the quote by C. S. Lewis at the beginning of this chapter. He seems to be saying that education without morals is dangerous. Others might say that education is a universal good. Critique both points of view and express at least two arguments supporting and opposing Lewis's view.
- Describe a current event in which there are opposing beliefs or attitudes that have ethical implications. Your task is not to convince the reader of one side or another but rather to explain the beliefs at work and their ethical implications.

Case Study Conclusion

Revisit the case study at the beginning of the chapter. Spend a few minutes considering the questions posed at the end of the case study. In light of the information shared in this chapter, have any of your original observations changed?

The case of Captain Jones is a fairly straightforward example of competing values. As is often the case, those values are also not aligned with his sense of obligation. Further, Jones is being made even more uncomfortable by the fact that his sense of obligation is in direct conflict with his own personal needs and wants.

Jones's personal values include honesty and integrity. These would, of course, urge him not to use the sick days. He is also uncomfortable with the fact that he is considering an action that is in direct violation of his crew's expectations. He correctly feels that using the sick days would be disingenuous.

WRAP-UP

However, his commitment to his daughter and family obligations are of equal importance to him. Like most of us, he has family responsibilities and is seeking to find a balance between two very important parts of his life.

In an attempt to "ease his conscience," Jones is rationalizing the relative ethics of using sick days by attempting to offset blame for his actions by pointing to the fact that the department is understaffed. By doing so, he is attempting to minimize the conflict between his values and his wants by making the department duplicitous in creating the problem. These rationalizations are quite normal but nonetheless irrelevant.

Further complicating Jones's thinking is that he is interjecting emotion into the equation. His rationalizations are being fueled by frustration, desire, and possibly even a sense of guilt.

The key ethical element boils down to two simple facts. Captain Jones wants to see his daughter play soccer because it's important to him and to his daughter, but to do this he must violate his own sense of right and wrong and betray the principles he extols to his subordinates.

In this harsh light Jones's use of sick days is not in keeping with his personal values and so is unethical.

It would be most helpful for Smith (and all of us) to remember that others viewing his behavior are not emotionally invested in Smith's need for social acceptance or his sense of empathy. By being outside observers, others will judge actions based on their perceived sense of obligation.

Chapter Review Questions

1. How are beliefs acquired?
2. Who developed the theory of hierarchy of needs?
3. Attitudes have three components as they relate to behavior. What are they?
4. Wants differ from needs in that they lack what?
5. Of the two, obligation and accountability, which is described as having ethical primacy?

References

Chaiklin, H. 2011. "Attitudes, Behavior, and Social Practice." *Journal of Sociology and Social Welfare* 38, no. 1.

Cooper, T. L. 2012. *The Responsible Administrator: An Approach to Ethics*. New York: Jossey-Bass.

Crisp, R., and M. Slote. 1997. *Virtue Ethics*. New York: Oxford University Press.

Festinger, L. A. 1957. *Theory of Cognitive Dissonance*. Stanford, CA: Stanford University Press.

Maslow, A. H. 1943. "A Theory of Human Motivation." *Psychological Review* 50(4): 370–96.

Miller, SC 2012. *The Ethics of Need: Agency, Dignity, and Obligation*. New York: Routledge.

State of New South Wales, Department of Education and Training. 2009. "Personal values, belief and attitudes." Accessed June 2, 2018. https://sielearning.tafensw.edu.au/MCS/CHCAOD402A/chcaod402a_csw/knowledge/values/values.htm.

Velasquez, M., C. Andre, S. J. Shanks, and M. Meyer. 1988. "Ethics and Virtue." *Journal of Issues in Ethics* 1, no. 3.

The Philosophy of Ethics

"Whatever is my right as a man is also the right of another; and it becomes my duty to guarantee as well as to possess."

—Thomas Paine, *Rights of Man*

OBJECTIVES

After studying this chapter, you should be able to:

- Interpret the methodology of testing an ethical theory.
- Recount the ethical branches of study, including metaethics, normative ethics, and applied ethics.
- Decipher the concepts of objectivism and subjectivism as applied to ethics.
- Determine the strengths and weaknesses of moral relativism.
- Discern the important philosophical orientations of normative ethics: virtue ethics, consequentialism, and deontology.
- Assess how to apply ethics to the fire service.

Case Study

Fire fighter Smith and fire fighter Jones are members of their department's public education program. They hold these volunteer positions in addition to their normal fire fighter duties. In May, they were sent to a fire education conference at a resort and convention center. The conference ran three days and consisted of several workshops on various topics. Smith attended the conference for the first time; Jones had attended several times before. To Smith's annoyance, Jones treated the conference as a paid vacation. He attended few, if any, sessions and spent two full days on the golf course. Further, through creative manipulation of receipts, he managed to fund his bar bill with the department's charge card.

When Smith confronted Jones, he defended the actions as "payback." Jones said he donates a lot of time and effort to the program, and this is his reward. He also claimed the chief is aware of what happens here but does not want to acknowledge it.

Smith is not surprised by Jones's behavior. He does not like or trust Jones and finds Jones to be a loud braggart and self-promoter.

As Smith considers Jones's behavior and what to do about it, he has the following thoughts:

1. Reporting Jones may result in Jones's removal from the team. This would be a positive in his opinion.
2. Reporting Jones's abuses may cause a general suspension of conference travel for the team. Smith found the conference programs very useful, and the team would suffer from their discontinuance.
3. It is also possible that Jones was right about the intention of sending people to the conference. If this was true, he might upset the chief by making a big deal of Jones's behavior.
4. Smith feels Jones's behavior is dishonest, and his ignoring it amounts to condoning it. He feels this is contrary to his personal values.
5. As a fire fighter, Smith feels he has a duty to report the behavior and let the chief do what he feels is best.
6. Other than getting Jones off the team, Smith sees no real benefit in reporting Jones. The only loss to the department was money, and it cannot be recouped. Regardless of how he feels about Jones, Smith believes that as a whole, the public education team will suffer by reporting Jones. His conscience aside, Smith thinks that ignoring Jones's transgressions will likely be the path of least damage.

- Are there any issues not identified by Smith?
- What should be Smith's primary concern—his personal views of right and wrong, his duty to the department, or seeking the best outcome to a bad situation?
- Consider what methods you may use to sort out your options if you were Smith.

Introduction

In Chapter 3, *Influencing Behavior*, we reviewed several factors that motivate and influence human behavior. We discovered that, in most cases, our behavior is influenced by more than one factor at any given time. We also found that both external and internal influences can reshape the elements that guide our behavior. These elements and conditions conspire to make the human behavior complex and sometimes baffling. Not surprisingly, a good deal of consideration has been dedicated to organizing and understanding the principles of human behavior. These studies form the foundation of psychology, sociology, philosophy, and, of course, ethics.

In this chapter, we will briefly review some of the more significant theories of ethical behavior. This chapter will divide the study into four general categories. We will first explore the methodology of testing an ethical philosophy or theory. Next, we will review the branches of ethical study. We will then move on to the general orientations of ethical theories including a survey of applicable theories. We will especially focus on those elements that pertain most directly to the fire service.

What Is Truth?

This question often causes students to cringe out of fear that they are about to be exposed to an in-depth and mind-numbing discussion on philosophy. Relax, you are not going to be asked to contemplate the meaning of life. However, the question, What is truth? is legitimate if we are going to study ethical theory. If you were to personally explore the various ethical theories explained in this chapter, and we would recommend you do so, you will find that each has its detractors and its supporters. Similar to human behavior, each attempt to analyze behavior is complex, and often theories overlap with each other. Still, each of us must consciously or subconsciously embrace some sort of ethical principles if we are to interact successfully within society. This begs the question, By what methods can we evaluate the truth of ethical philosophies?

Clarification

Explore some of the various related philosophies introduced in this chapter. A great resource is *The Internet Encyclopedia of Philosophy* located at http://www.iep.utm.edu/home/about/.

For an ethical theory to be of use, it should meet two general standards; first, it should be valid and, second, it should be applicable. Therefore, it is reasonable to expect an ethical theory to accomplish the following:

- It should set forth a series of ethical and moral principles.
- The theory should show how those principles are justified.
- The theory should guide people to a life of ethical and moral excellence.

To accomplish these expectations, the theory must be consistent and complete. Consistency is demonstrated by a lack of contradiction or incompatibility between the principles of the statement. In demonstrating completeness, the theory must be comprehensive, applying to all moral and ethical situations without dependency on another theory (Russow and Curd, 1989). For example, an ethic built on the values of both fairness and justice may be found to be inconsistent. Fairness would indicate that all fire fighters should receive the same pay for doing the same job regardless of seniority, whereas the concept of justice may well support pay grades based on seniority and experience. As such, the theory is inconsistent. In another example, an ethical theory based solely on adherence to duty may not address unjust outcomes arising from extenuating circumstances. In this case, you may find the theory incomplete.

In consideration of the requirements listed previously, an ethical theory may be challenged by one of three methods:

- It can be invalid or mistaken in truth.
- It can be incomplete.
- It can be inconsistent.

Keep these principles in mind as we review the various schools of ethical thought and philosophy. It is noteworthy to remember that you can apply the same tests to practically any statement of principle, or belief, and you can use them to judge any position, not only an ethical theory.

Branches of Ethics

Ethics is a broad concept in both theory and application. The study of ethics can take us down several paths of inquiry depending on the type of questions we ask. For instance, we may inquire about the history of various ethical philosophies and their origins; this area of study is **metaethics**. We may wish to compare and contrast the various beliefs and moral values of different groups of people; this is **descriptive ethics**. We may wish to investigate the moral and ethical implications of a particular action such as assisted suicide; this is an exercise in **applied ethics**. Last, we may seek to determine the proper way to live life through the development of an ethical code. This is an exercise in **normative ethics** Figure 4-1.

Each branch of ethical study contains different questions and unique implications regarding applicability. This chapter, for instance, is essentially an exercise in metaethics, whereas the overall focus of this book is a study in normative ethics. Understanding the types of ethical study is most useful in reviewing the individual ethical philosophies discussed later in this chapter.

Metaethics

Metaethics explores the history and meanings of concepts, words, and values associated with ethics. It is a study of the study of ethics. It can focus on abstract notions such as, What do people mean by good and right? or Where do morals come from? and Are there universal ethical principles, or are they relative depending on the situation? As fire fighters, we tend to be problem solvers. Emergency services tend to shape our thinking in terms of black and white, and right or wrong. We deal in facts and probabilities. This mind-set can make the study of metaethics seem unapproachable and even unnecessary. Certainly, we feel more comfortable with the question, What is the right thing to do? than Why is it the right thing to do?

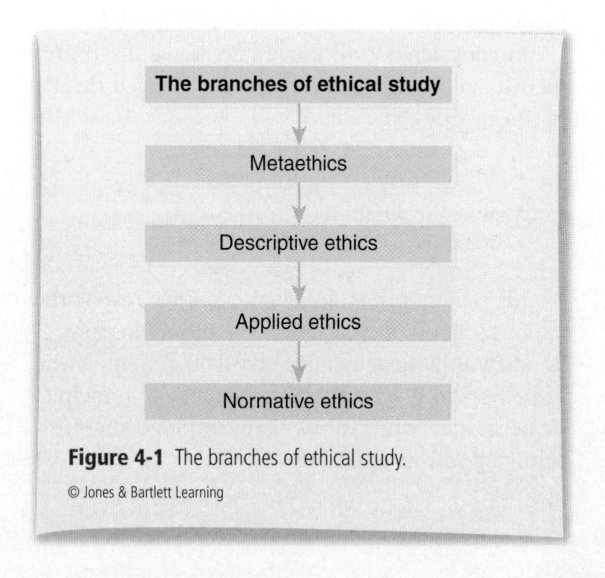

Figure 4-1 The branches of ethical study.

Objectivism and Subjectivism

One of the most fundamental questions addressed in metaethics deals with the source of morality and, hence, ethics. There are two points of view regarding the source of morality; the first is called **objectivism**, and the other is **subjectivism**.

Proponents of objectivism hold that moral values are universal, and they exist outside human conventions. They further insist moral principles are fixed and not subject to a change in interpretation. Objectivism assumes that moral principles apply to all rational creatures regardless of culture or the time in which they live (Gowans, 2015). Opinions as to the actual source of objective truth have varied. The Greek philosopher Plato argued that moral truth, like mathematics, was an abstract entity that existed in a spirit-like realm Figure 4-2. The 13th-century Italian Catholic philosopher, Thomas Aquinas, asserted that universal moral truth came from the will of God. This theory is known as the **divine command theory**. Aquinas also stated that moral virtue was a natural condition of nature, as a reflection of God's will.

Proponents of subjectivism assert that humanity develops moral values out of practical necessity or strictly by social convention. It is noteworthy that believers in subjectivism do not deny the existence of moral truth, only its universal nature.

There are two versions of subjectivism. The first is known as **individual relativism**, which holds that moral values are created by individuals, relative to

Figure 4-2 Plato was a philosopher in ancient Greece. He is widely considered the most pivotal figure in the development of philosophy.

their place, time, and situation. The second is known as **cultural relativism**, which maintains that morality is based on a culture's or social group's expectations (Gowans, 2015).

The importance of metaethical questions should not be overlooked. We must ask whether objective principles of morality exist or not, and, if they do, from where do they come. These questions are important because they shape the foundations of your individual values and beliefs. As a result, they shape your ethical orientation. The question of objective versus subjective interpretation of moral truth leads to several possible foundational beliefs.

Absolute Objectivism

Proponents argue that there are universal principles of good that exist as a condition of nature. From an academic perspective, the existence of God is not necessary to make the argument. If He does exist, God wishes us to follow these principles because they are good. Further, even if God does not exist, the universal nature of goodness still requires conformance to its principles. The concept of absolute objectivism has led to several questions that have been debated for centuries. Opponents to the theory ask, If God does exist and God is good, then are there rules or conditions by which God's actions are judged (as good)? By extension, if there are rules that apply even to God, is it not counterintuitive to the notion of an all-knowing and powerful God that there are rules that supersede the will of God? Opponents also ask if God does not exist, then from where do universal moral principles come and what suggests conformance to universal principles of morality (Firth, 1952)?

Divine Command Theory

A more common interpretation of absolute objectivism is that of the divine command theory, which suggests that God must exist. Further, a thing is good or bad because God wills it, and He commands us to obey His will (Austin, 2013). Many philosophers take issue with this position based on the following question, If anything is good or bad based on God's will, then can God change his will? If by the will of God something can be bad at one time and good at another, by definition how can it be a universal moral truth?

Divine Subjectivism

Divine subjectivism suggests that God exists; however, He has no expectations of behavior. God does not punish or reward behavior based on compliance to rule but rather to the purity of spirit. Humankind must exercise free will, and the "goodness" of an action is determined by intent. Perhaps one of the oldest metaphysical questions arises from the realm of divine subjectivism: If God, by definition, must be good and all-powerful, then how can a good and all-powerful God allow morally bad things to happen?

Clarification

Divine subjectivism is most often associated with deism as practiced by Voltaire, Thomas Paine, and Benjamin Franklin.

Cultural Subjectivism

God may or may not exist and His existence is irrelevant to morality. Moral truth arises from standards of behavior created by members of the society and imposed on those members (Gewirth, 1980). The most common critiques of this position are, first, if God exists how can He be morally irrelevant? Second, if moral standards arise from the consensus of society, how does a person account for the individual acts of morality contrary to societal standards that change societies' views? For example, the common belief of 18th-century citizens was that slavery was moral. If moral truth comes from consensus, then it would follow that those who opposed slavery were essentially "wrong." Further, how can moral evolution occur if there are no principles of truth beyond human consensus?

Absolute Subjectivism

There is no moral authority beyond humanity. Moreover, there is no moral truth beyond that held in the individual's conscience (Gewirth, 1980). The most common critique of this position is that it makes morality and ethics essentially irrelevant. If a thing is good or bad based on the intent of the individual, then arguably all things are good.

The issue of where moral truth comes from is obviously important in shaping our view of ethics. If I believe that there are universal moral principles that apply to all persons at all times, then I am obligated to understand what those principles are and to base my life decisions on them. On the other hand, if I believe that moral truth is subjective and that actions are good and right based on circumstance, then I am obligated to understand what defines right and good and how best to apply those principles. These are weighty questions on which great minds have debated for centuries, yet each of us must consider for ourselves the truth of morality.

For instance, if I were to make the statement that "murder is wrong," many of you would agree. As an objectivist, you may assert that taking a life is wrong in all cases because it violates natural law. Or you may assert that killing is wrong unless God wills it. A subjectivist, however, may argue that the ethics of taking a life is relative to cultural standards or an individual code of conduct. For instance, 21st-century American culture widely condones capital punishment and, as such, an executioner is theoretically acting morally in the performance of his or her job. Conversely, an absolute subjectivist would argue that the morality of a government executioner taking a life is entirely dependent on how the executioner views the act.

Descriptive Ethics

Descriptive ethics is a form of scientific research. It seeks to understand and catalog the attitudes of individuals or groups of individuals. Descriptive ethics can uncover the beliefs and values that are used to determine the relative "rightness" or "good" in a particular action (Icheku, 2011). As we judge either our actions or those of others, it is important to understand by what criterion we are making those judgments. In Chapter 1, *Introduction*, you read the quote by the Greek philosopher Socrates, "An unexamined life is not worth living." Understanding the beliefs, values, and attitudes influencing our view of right and wrong are critical in the pursuit of living a full and rich life.

Applied Ethics

The study of applied ethics seeks to identify a morally correct path as applied to a specific circumstance. For example, over the last 10 years, several fire departments have found themselves embroiled in controversy regarding "pay to spray" programs. These are instances where fire protection organizations provide service to neighboring areas on a contract basis. On several occasions, departments have responded to requests for assistance from "customers" only to find that the caller had not paid for the service. As a result, the responding department did not put out the fire at the individual's home. Pundits have questioned whether this response was an exercise in ethical business practices or whether it was a breach of the basic values intrinsic to the American fire service. This example demonstrates the nature of applied ethics because it seeks to apply ethics theory to a particular problem. The three commonly identified disciplines of applied ethics include decision ethics, professional ethics, and macro ethics (Chadwick, 1997).

Decision ethics is a genre of applied ethics in which various normative theories are applied to the decision-making process. Decision-making ethics focuses on personal decisions as they relate to personal values and attitudes.

Professional ethics tends to look at the application of preexisting codes of ethics specific to occupational activities. Professional ethics are subdivided into numerous occupational areas such as medical ethics, legal ethics, bioethics, business ethics, and research ethics.

Macro ethics is a term coined in the 20th century; it focuses on global issues such as the geopolitical ramifications of wealth distribution, technology emergence, international business practices, and environmental issues.

Normative Ethics

The study of normative ethics investigates questions arising from the consideration of how an individual should morally act. Normative ethics differs from applied ethics in that it seeks to identify general standards of behavior based on an ethical philosophy, whereas applied ethics tends to compare the application of ethical philosophy to a particular problem. Normative ethics is essentially an exercise aimed at determining the guidelines by which we should live our lives and, to some extent, by which we judge the moral context of others' life choices (Fieser, 2016).

Consider the following example: A fire unit leaves its district to go to a grocery store for that night's supper. Because they are leaving their district for unauthorized business, they do not notify dispatch of their movements. They go to the store and return without incident. Are their actions ethical? If not, is the action wrong because it breaks policy or because it is not in keeping with the virtue of honesty? Are their actions acceptable because there were no negative consequences? Would their actions be more acceptable if their battalion chief condoned the activities? The study of normative ethics has a long history of discourse as to what makes an action ethical or unethical. The arguments rest on adherence to one of three principal philosophies Figure 4-3. First, **virtue ethics** is the belief that ethics is rooted within compliance to particular virtues. Second, **consequentialism** is the theory that the ethical judgment of an action should be determined by its outcome. Finally, **deontology** holds that the ethics of an action should be judged relative to its compliance with a code of conduct.

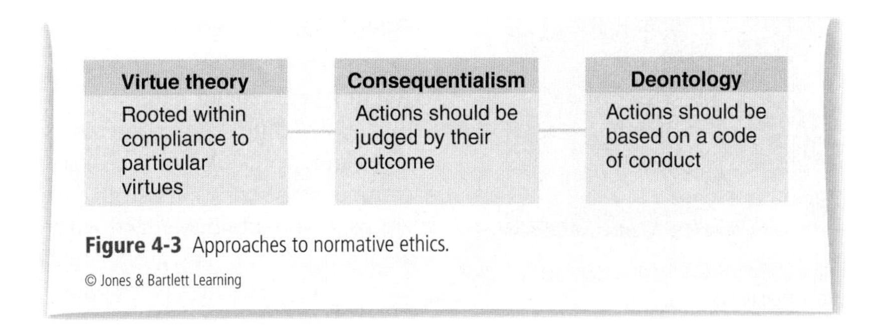

Figure 4-3 Approaches to normative ethics.

Exploring Normative Ethical Theories

Virtue Theory

You were introduced to the concept of virtues in Chapter 3, *Influencing Behavior*. Several Greek philosophers, including Aristotle, have asserted that understanding and complying with the demands of virtuous behavior forms the basis of ethical actions. Virtue theory or virtue ethics is essentially based on the development of good habits of character, such as personal integrity (Fieser, 2016). It follows that once you have made personal integrity a habit, you will then habitually behave with integrity. Because integrity initiates ethical action, the ethical "right" thing to do is that which a virtuous person (one with integrity) would do Figure 4-4.

Figure 4-4 Aristotle, 322 BC. A Greek philosopher and leading proponent of virtue ethics.

© Popperfoto/Getty

The Greeks divided virtue into two categories. **Moral virtues** are those based on character and acquired through learning. They include concepts such as justice, prudence, restraint, integrity, loyalty, and sincerity. **Intellectual virtues** are inherent in an individual's nature. In other words, you were born with them or not. Intellectual virtues include wisdom, empirical knowledge, reason, and skills in the arts. The Greeks believed that the truly virtuous individual would possess both moral and intellectual virtues.

Eudemonism

Aristotle asserted that virtues are good habits that are acquired through moral education and are formed primarily in youth. As such it is the responsibility of parents and teachers to instill virtues in the young. Aristotle further argued that one of the functions of virtue is the regulation of emotion and that self-control was essential to being in a state of virtue. This school of thought is known as **eudemonism**. Aristotle submitted that virtues are exercises in moderation and that extremes in behavior (including virtuous activities) are, at their heart, vices. For example, honesty is a virtue, and a person who is dishonest as a matter of habit is treacherous. On the other hand, honesty in the extreme can be hurtful and destructive to others. Honesty can be a weapon and, as such, its use could be a vice if unmoderated. A principle of eudemonism ethics is that an act in and of itself does not reflect virtue if done for selfish reasons. Virtues require more than cognition; they must be part of a person's character and pursued for their own value (Hursthouse, 2013).

Agent-Based Theories

Agent-Based Virtue Ethics

Agent-based virtue ethics is rooted in the idea that each of us has an internal sense of what is right and wrong. It holds that a virtuous person's commonsense intuition tells that person which traits are good and admirable, and what action would be a correct action in any given situation. Another important element of this ethical philosophy is that the motives of an individual spring from within his or her character, and so values are essential in understanding an action's rightness.

Figure 4-5 The core mission of the fire service is most compatible with care ethics.

Care Ethics

Care ethics is the third commonly described form of virtue ethics. It is based on the importance of being a positive force within a social group Figure 4-5. The guiding principle behind care ethics is that persons are understood to have varying degrees of dependence on and interdependence with one another. According to the theory, it follows that if we are all interdependent, then it is in our best interest to act in ways that support each other, particularly the weakest among us (Gilligan, 1982). As you will see later, this premise borrows heavily from the concept of social contracts expressed by consequentialists such as Thomas Hobbes and John Rawls, as it asserts a duty for each of us to contribute to the welfare of society. This is also consistent with the deontologist theories of John Locke, which you will read about later. What separates care ethics from other ethical theories is that its primary principle is that of interdependence. Our actions are a consequence of being an agent of a greater self, whereas the consequentialists and deontologists tend to believe that each of us is an independent agent acting on his or her behalf. Two of the prominent advocates of care ethics are Carol Gilligan, who founded the principle as part of an overall philosophy of feminism, and Joan Tronto, who refined Gilligan's ideas into a slightly wider and more mainstream philosophy Figure 4-6. The theory of care ethics generally recognizes four elements that constitute care morality (Tronto, 2015). These are:

- Attentiveness: Care requires recognition of the needs of others and acknowledgment that those needs are relevant.
- Responsibility: Because we derive benefit from membership in a social group, we have an obligation to contribute to the welfare of that group.
- Competence: It is not enough to acknowledge a need or to accept responsibility for it without adequately addressing the problem causing the need.
- Responsiveness: The principle of responsiveness is based on the recognition of vulnerability and that most culture groups lack social equity. Wealth, power, and resources are typically not evenly distributed. Care ethics assumes that care is not guided by an expectation of reciprocity.

Virtue ethics represents one of the earliest forms of ethical thought in Western civilization. However, interest in virtue ethics declined in the 19th century as industrialized society began to embrace the concepts of social duty and consequentialism. In the postmodern age, virtue ethics is being reassessed as a

Figure 4-6 Carol Gilligan. A feminist author known for her work in relationship ethics.

viable answer to increasingly complex ethical issues in a culture where social institutions such as religion and tradition are becoming increasingly questioned.

Consequentialism

Throughout our lives we routinely find ourselves judging moral responsibility through the consequences of our actions. What action yields the most good? This is a reflection of the ethical principle of consequentialism. In looking at the theory of consequentialism, some fundamental questions arise: Whose good and happiness is considered? and By what measure do we determine good? These questions have given rise to three distinct forms of consequentialism:

- **Utilitarianism**: An action is deemed right if it is most favorable to everyone concerned.
- **Ethical egoism**: An action is morally right if the consequences are favorable to the person performing an act.
- **Ethical altruism**: An action is right if the consequences are favorable to everyone except the person performing an act.

All three of these theories support the notion that consequences dictate the rightness of an action. However, they yield distinctively different results. Consider the example of a fire fighter who discovered that a fellow fire fighter was likely drinking on duty. He discovered this by searching the locker of the fire fighter without his permission. Under the theory of ethical egoism, the fire fighter would be ethically justified not to risk his job and the economic welfare of his family by reporting the drinking incident if it meant admitting that he had unlawfully searched someone's locker.

Under the theory of ethical altruism, the fire fighter would be obligated to report the incident to protect the welfare of his fellow fire fighters who may be injured as a result of a fellow fire fighter being inebriated on duty. Finally, under the theory of strict utilitarianism, our fire fighter would have to weigh the greater good for everyone involved and may decide it best to report the drinking incident anonymously. He may also be able to justify not reporting the incident at all if that would, in his opinion, serve the greater good.

Utilitarianism

The ethical theory of **utilitarianism** came to prominence in the 19th century as the result of the works of Jeremy Bentham and John Stuart Mill. It sprang from the then-radical theories of an egalitarian society free from class distinction. Utilitarianism rejected the notions of social duty based on social standing. Instead, utilitarianism proposes that the ethics of an action can most effectively be judged by its consequence. That which does the greatest good for the greatest number of people should be deemed as the greatest ethical good.

Jeremy Bentham (1748–1834) was an English philosopher who is often credited with originating the theory of utilitarianism Figure 4-7. He suggested that pain and pleasure were the only two relevant values in the world. Based on this proposition, he created the **rule of utility**, which asserted that good is that which brings the greatest happiness to the greatest number of people. His version of ethics makes all ethical actions relative to both the situation and the actor's interpretation of goodness in

Figure 4-7 Jeremy Bentham was an English philosopher and is regarded as the founder of modern utilitarianism.

addition to the amount of good resulting from the action. Critics argue that his theory lacks the consistency of application. As an example, consider the pedestrian who walks by a homeless person without helping him or her. In most ethical theories this act is neutral, it neither causes harm nor produces good. Bentham would hold that the action could be either good or bad depending on your intention. If you fail to act out of a lack of compassion, your action is unethical as it fails to produce a possible good. If, however, your lack of support for the homeless person is somehow rooted in an intention to instill self-reliance, it may, in fact, be ethical. As you learned earlier in the chapter, a theory is considered inconsistent if an action can be both ethical and unethical in the same circumstances. Bentham's approach to utilitarianism is often described as overly burdensome. In its strict interpretation, his theory argues that no act is ethically neutral. All acts either contribute to another's good or suffering, and taking no action is, in fact, promoting suffering or denying happiness (Mastin and Bentham, 2008).

John Stuart Mill later refined the theory of utilitarianism to assert that an individual must follow the rules whose intent are to provide for the common welfare, but there is no general obligation to act altruistically apart from following those rules Figure 4-8. Also noteworthy is that Mill distinguished the concept of utilitarianism from hedonism. He noted that not all pleasures are equal and that "good" is best defined by the quality of pleasure as opposed to the quantity. He held that moral and intellectual enrichment were of great value and that physical pleasure was of little consequence. Mill also contributed the notion that doing good was as much an act of self-interest as altruism. That is, by providing for the general welfare, the individual as well as the general population prosper. In today's political jargon, Mill's concept is present in the saying that a "rising tide lifts all boats."

An elemental principle of utilitarianism is that happiness is intrinsically valuable and that all other values should be judged according to their ability to promote happiness or conversely to eliminate unhappiness.

Figure 4-8 John Stewart Mill, 1806–1873. An English social philosopher and proponent of utilitarianism.

Egoism

The opposite of utilitarianism is **egoism**, which asserts that individuals ought to promote self-interest above any other value. Protagonists argue that acts required for self-preservation and life enrichment are the first requisites to universal welfare. Unless each of us first takes care of ourselves, we are unable to take care of each other. That we should act in our own self-interest is an exercise in **normative egoism**. Its supporters point to the theory of **psychological egoism** for validation. It suggests as a universal truth "that individuals will always act out of self-interest" (Fieser, 2016). Further, psychological egoism suggests that any act of altruism is an emotional response contrary to human nature. Supporters of normative egoism assert that failure to take care of oneself places the burden of care on others, which is also inherently unethical.

An ethics theory came to prominence in the early part of the industrial revolution called **rational egoism**. Rational egoism is most often associated with Ayn Rand's theory of enlightened self-interest, which rejects what she called the "Judeo-Christian heritage of sacrificial ethics." She asserted that placing someone's interests above your own was contrary to rational thought and only sprang from emotions of guilt. According to her ethic, the first principle of morality is that each man and woman must be the beneficiary of his or her actions and that each of us is obligated to act in our own rational self-interest as a condition of humanity Figure 4-9 (Baier, 1990).

Rational egoism has had some measure of populist support since the industrial revolution. It has become increasingly popular in the United States and Europe in association with a resurgence of conservativism. Ayn Rand's novel *Atlas Shrugged* is often quoted as a manifesto for the theory of enlightened self-interest.

Typically, egoism is rejected by most ethics philosophers on the critique that it is essentially unviable. Assume for a moment that you live in a small community in which water is obtained from a number of wells. Eventually, all the wells within the community dry up except for the one on your property. Out of concern for your welfare and that of your family, you decide not to share your well water with your neighbors because it is not in your best interest to do so. Eventually, logic would dictate that your neighbors will decide that it is in their best interest to forcibly take your water. You may then find yourself looking for some law enforcement agency to protect your interests. However, in doing so, the law enforcement agency would be acting against its self-interest because it members are also members of the community and they also need water. Further, as civil employees, it is in their interest to serve the needs of the many as opposed to protecting your individual interests. Some egoists may attempt to defend their position by stating that it is obviously in your self-interest to share the water to protect your life. However, you are still acting against your self-interest in sharing the water, whatever your motivations are for being altruistic. The mainstream ethical theorists agree that egoism leads to chaos if practiced by the majority of people within a society, and it is only practical if acted on by a select minority. This makes egoism problematic as an ethical theory, as it cannot be universally applied.

Figure 4-9 Ayn Rand was a Russian-American novelist and philosopher, known for her best-selling novel *Atlas Shrugged*, and for developing a philosophical system she called rational egoism.

Another type of egoism partially addresses the shortcomings of rational egoism. It is known as **conditional egoism** This theory asserts that acting in self-interest is ethically acceptable only if it leads to moral outcomes.

For example, an act is acceptable if it produces a betterment to society, and it is unacceptable if it causes harm. Further, it is not in a person's best interest to harm the society in which the person lives. An example of this thinking was the seminal work of Adam Smith titled *The Wealth of Nations* Figure 4-10. In it Smith writes, "It is not from the benevolence of the butcher, the brewer, or the baker, that we expect our dinner, but from the regard to their own interest" (Smith, 1776). The problem facing conditional egoists is this: If conditional egoism is dependent on a superior moral goal or the attainment of the general good through self-interest, then what defines the public good? Further, how does one balance the public good against the self-interest when they are unaligned?

Figure 4-10 Adam Smith was a Scottish moral philosopher and pioneer of political economy.

Altruism

Ethical altruism is a consequential doctrine that each of us has a moral obligation to help serve and benefit others. In complete antithesis to egoism, it asserts that self-interest is less important than the welfare of others. The theory of altruism exerts that an action is morally right only if its results are favorable to all parties concerned, except the individual. It is the exclusion of the individual actor's needs that differentiates altruism from utilitarianism. Box 4-1 shows an excellent example highlighting the difference between altruism and utilitarianism.

It should also be noted that altruism differs from care ethics in that it intimates a necessity for self-sacrifice not necessarily addressed in care ethics. Further, the act of nurturing associated with care ethics is considered a virtue, whereas in altruism it is considered a social obligation (Fieser, 2016). The presence of a social obligation creates a bridge of sorts to deontology (duty ethics), in that it asserts that each of us has an obligation to social order, even though the general origin of that duty arises from a moral imperative to create a "good" consequence.

Social theorists criticize altruism as an ethical principle because it inhibits initiative. It also places an unequal ethical burden on one group (those with surplus resources) more so than it does other groups (those without surplus). The first few paragraphs of this chapter stated that a viable ethical theory should apply to all persons equally.

In studying consequentialism, it seems logical that the definition of the term "good outcome" becomes important. Consider the case of an incident commander who is considering whether or not to deploy a crew into a burning building to accomplish a rescue. He or she has an ethical obligation to rescue a person inside if possible and an ethical obligation to protect the welfare of the fire fighters working for them (see the *ethics axioms* in the preface). After sizing up the fire, the incident commander proceeds with an interior attack and rescue. Unfortunately, soon after entry a partial collapse occurs. Crew members are injured, and the rescue fails. If we judge the ethics of the incident commander's actions by a consequence alone, we must conclude that the actions were not ethically right, as there was no good outcome. Yet the intent of the actions taken would appear to have

BOX 4-1 Altruism versus Utilitarianism versus Egoism: A Comparison Through an (Admittedly Unrealistic) Example

Three men, A, B, and C, are arrested and put in jail. None of the men have committed a crime. They each know nothing about each other except for the fact that they are all innocent.

A is pulled aside and given three choices:

1. He will be given one year in prison, B and C will each get ten years,
2. All three will each be given five years in prison, or
3. He will be given 20 years in prison, B and C will each get 1 year.

Assuming that external factors are ignored (the happiness of the men's families, later consequences of each option, etc.), and that each man would suffer equally from each year spent in prison, Altruism would say to take #3, as he should live for others.

Utilitarianism would say to take #2, as that would minimize the amount of overall suffering.

(Normative) Egoism would say to take #1, as that would minimize one's own suffering.

https://reddit.com/r/philosophy/comments/eil73/altruism_vs_egoism_a_comparison/

Figure 4-11 Intent and outcome are critical in judging ethical actions.

been justifiable. Philosophers supporting consequentialism use the concept of **dual consequentialism** to argue that the rightness of consequences can be measured both morally and objectively. In the case of our fire ground, the objective consequence was not right, yet the morality of the *intended* consequence was justifiable Figure 4-11 (Portmore, 2011).

Another element in studying consequentialism is that of the proximity of consequence. **Proximity** is the relative value of a consequence either in time or scale. Let us assume you live in a neighborhood where everyone throws their trash in the ditch. As a result, the ditch is always full of trash. By throwing your trash in the ditch you are not creating a new problem or making an existing problem substantially worse. And so, your throwing trash in the ditch has no unethical consequence. Yet if everyone disposed of their trash properly, then the community would be cleaner and more sanitary. In this case, the actions of all contribute to a negative consequence. As a result, consequentialism has been modified to accommodate consequences on large groups acting through time. This is known as **rule consequentialism**. This theory states that an action is ethically correct only if it complies with rules that would necessarily have the best-intended consequences as opposed to following another rule or no rule at all (Haines, 2008). The groundwork for rule consequentialism was first laid by John Stuart Mill and later modified by Richard Brandt.

Deontology

Deontology is based on the belief that as human beings we have clear obligations to society. Further, we have obligations to the relationships in which we enter. Deontology is sometimes called Kantianism, in deference to Immanuel Kant, its most prominent advocate. His ethics are organized around the idea that we are bound by a universal imperative to respect the humanity of other people. As such, we are obligated to act in accordance with rules that affect all people equally. Finally, we must meet those obligations out of free will (Fieser, 2016). Kant's version of deontology focused on what he called **categorical imperatives**. A categorical imperative is an action that is required out of duty, whether it benefits the actor or not. Kant expressed his categorical imperatives in several ways, but perhaps the clearest is as follows: "*First, we should always treat people with dignity, and never use them as mere instruments*. Second, we *should act in a way consistent with how we would want all people to act in the same circumstance*" Figure 4-12 (Mastin, 2008).

Figure 4-12 Immanuel Kant, 1724–1804. A German philosopher and author. He argued that reason was the basis of morality.

Pluralistic Deontology

While Kant is certainly one of the most influential proponents of the deontology movement, he is not alone. W. D. Ross (1877–1971) described a theory of morality known as **pluralistic deontology** Figure 4-13. Similar to Kant, he believed that there were categorical imperatives. However, he asserted that there were more than two. Included in those imperatives were that individuals had the duty to keep promises, a duty to pursue justice, a duty to improve the conditions of others, a duty to self-improvement, and most importantly a duty not to injure others (Ross, 1930).

Figure 4-13 W. D. Ross. Best known for developing a pluralist, deontological form of intuitionist ethics.

Figure 4-14 John Locke, 1632–1704. Born in England, he was a proponent of personal responsibility and individual freedoms.

Natural Rights Theory

An important variation on deontology was authored by the British philosopher John Locke Figure 4-14. He proposed what has come to be known as the natural rights theory. Locke held that the very laws of nature gave people the rights of humanity. Further, it was unethical to harm anyone's life, health, liberty, or possessions. He felt these rights were endowed by God as a condition of birth. If this sentiment seems familiar, it is because the work of John Locke heavily influenced Thomas Jefferson and Benjamin Franklin, who would include those ideas in the U.S. Declaration of Independence nearly a century later (Fieser, 2016). Locke believed that ethical obligation is present regardless of the desirability of the expected outcome of our actions. An obvious example of this theory at work would be that of a defense attorney. An attorney has an ethical obligation to provide a robust criminal defense even if, in doing so, the attorney may set a guilty person free. The greater obligation is to the justice process and not to the consequences of any given trial.

Divine Command Theory

Earlier in this chapter, we introduced the subject of the divine command theory in relation to the metaphysical question, From where does morality arise? Divine command theory is also prominent within normative ethics, particularly deontology. Moral philosophers such as William of Oakham (1287–1347) and René Descartes believed that because morality springs from God's will, humans had a moral duty to obey that will. The rightness of any action depends on that action being performed in conformance with God's will and *only* out of a sense of duty to God's will Figure 4-15. It follows then that doing a thing that is consistent with God's commandments, but that is motivated by self-interest, is not necessarily "good." For example, let us assume that a fire fighter argues that he cannot work on Sundays because God wills him to keep the Sabbath holy. He may be doing the right thing by obeying God's will, but only if he is doing so because God commands it. If he is motivated by self-interest, his actions may be unethical even though they are consistent with God's will as he understands it (Mastin, 2008).

Figure 4-15 René Descartes, 1596–1650. Dubbed the father of modern Western philosophy.

Contractarian Ethics

Contractarian ethics is based on the concept that morality arises from social norms and expectations. As rational beings, we realize that rules of behavior are necessary to maintain social order. Thus, compliance benefits all members of the group as well as ourselves. It follows that as members of society we implicitly agree to a "social contract" to abide by those rules. Proponents of contract theory ethics assert that principles of right and wrong are valid only if reasonable members of society would agree with them. An important distinction needs to be remembered when considering contract ethics—the theory focuses almost exclusively on human relationships. Rules exist to protect individuals' rights and to promote the common good. This is reminiscent of utilitarian theory described earlier in this chapter. The distinction between contract theory and utilitarianism is in how right and wrong are judged. While both focus on actions that contribute to the common good, contract theory asserts that rightness is based on compliance with agreed-on rules that are intended to promote a common good. On the other hand, utilitarian theory asserts that rightness is based on the good produced by an act, not the culture's expectation of compliance.

Moral Relativism as a Normative Ethic

Early in this chapter we discussed metaethics. Part of that discussion focused on the philosophical question of whether moral truth is universal. You may recall that there were two schools of thought in response to this question. The objectivist point of view argues that morality is either a condition of nature (natural law) or determined by God's will (divine command). It further holds that moral truth is universal; in other words, it transcends culture and time, regardless of human interpretations. In contrast, the subjective view argues that moral principles are human inventions.They suggested that morality springs from human understanding, practical need, and shared values. As morality is rooted in a human construct, it is subject to evolution and contextual interpretation. They argue that relativism is the only rational approach to understanding the morality of human behavior. While these are metaphysical questions, they have practical implications for normative ethics.

Virtue ethics and deontology are both rooted in the assumption that right and wrong are tangible and unmalleable concepts. Consequentialism comes close to embracing a subjective view of moral goodness. However, it still relies on a shared understanding of morality as a concept of goodness and happiness. Although an action may be judged right or wrong relative to its outcome, that outcome is ultimately decided by a mutually accepted concept of "goodness." And so, ultimately, it too is objective in orientation.

The last 50 years have seen an increase in a normative ethical approach called **moral relativism**. It is rooted in the subjectivist belief that there are no universal moral truths and that we can only understand the actions of an individual or a culture in light of the circumstances that existed at the time. Relativists recognize that right and wrong exist; however, they do not accept the principle of universal concepts of right and wrong (Westacott, 2013).

George Washington is an example of the divide between objectivist and subjectivist principles and their effect on ethical orientation. Washington was (and is) considered an example of an extremely virtuous individual. Although the legend of the cherry tree is likely a myth, Washington was respected by his peers as an honest, intelligent, and socially aware individual. His personal history and writing indicate a strong sense of duty and commitment to the principles of universal human rights. Accordingly, Washington is immortalized as a virtuous man. Unfortunately, George Washington was also a slaveholder. By almost any manner of judgment this is incongruent with the moral standards he was purported to hold.

The contemporary understanding of morality nearly universally condemns slavery as immoral. It is both considered an affront to natural law and in opposition of God's will as we understand it. According to the objectivist theory that moral truth is universal and unchanging, a thing which is immoral today would in fact, always be immoral. As a result, George Washington's holding slaves was an immoral act.

A subjectivist would argue that his participation in slavery must be judged according to his experience and to the time in which he lived. They would argue that Washington, like the vast majority of his contemporaries, ascribed to the mistaken belief that African Americans were an inferior race, with the intellect and judgment of children. As difficult as it is for us to understand now, Washington, like his contemporaries, believed that African Americans were actually better off as slaves than as free individuals. As such, his stated belief was that slavery was moral. That belief was formed by his personal experience and was tempered by widely held beliefs at the time. Of course, today we understand his beliefs to be wrong, and his experiences were evaluated based on prejudicial misinformation, but ultimately his actions would be considered moral because they were consistent with moral values of the time and acted on without malice.

This defense of Washington is an exercise in cultural relativism and is rooted in the belief that actions are moral or immoral relative to the beliefs and traditions of a culture. The assumption is that each of us is the product of our upbringing and the

What Do You Think?

Assume that by some rip in the time space fabric, Thomas Jefferson and George Washington are brought before you to be judged for holding slaves in the eighteenth century. Do you believe that acting in accordance with commonly held beliefs of their time excuses their behavior? Or do you believe that they violated a fundamental principle of natural law and their behavior was as reprehensible then as it is now?

culture that surrounds us. Therefore, our actions are judged as moral or immoral based on their consistency with beliefs of right and wrong held at the time.

Those who reject the theory of cultural relativism argue that this excuses cultures from culpability for egregious actions. It also has a problem of consistency. Logically a thing cannot be moral and immoral at the same time and same place. If Washington's actions are moral because of the time and culture in which he lived, then it would logically follow that those who rejected the idea of slavery and fought to stop it (John Adams) must be morally incorrect. It would also hold that activists who manned the underground railroad "stole slaves" and so are unethical. This is because the rule of consistency discussed at the beginning of this chapter would insist that a thing cannot be moral and immoral at the same time.

Some relativists assert that it is not a culture that decides morality but rather the individual. This is moral relativism, which suggests that each of us has a responsibility to decide what is right or wrong according to our own personal set of values and beliefs. Further, we apply these beliefs on a case-by-case basis, meaning that correctness of action lies in the individual's conformance to whatever standard he or she holds most applicable.

Under this theory, two individuals may commit the same act but for different reasons and with different intentions, and one's actions may be moral while another's is immoral. As an example, President Truman deployed the atomic bomb on Hiroshima knowing that there would be catastrophic civilian injuries. The act may be considered moral if he did so for moral reasons, such as ending the war sooner, which might save millions of civilian lives. Conversely, the act may be immoral if he did so out of malice toward the Japanese.

At the root of moral relativism is the argument that ethical culpability must be contingent on motive, the actor's understanding of circumstances, and anticipated outcome. Those who reject moral relativism do so because they argue that it makes the meaning of morality useless. Under the theory of moral relativism, the architects of the Nazi atrocities could be excused if they truly believed (many did) that they were accomplishing a greater good (as they understood it). A more contemporary example can be found in modern-day terrorism. Religious extremists would theoretically be acting morally if they truly believed that God willed them to slaughter nonbelievers, even if their religion specifically condemned murder. Critics would argue that if any act is excusable, then any act is moral, and therefore morality is irrelevant (Westacott, 2013).

While moral relativism has its critics, it is gaining greater acceptance, particularly over the last half-century. One theory suggests that moral relativism is a result of the influences of mass communication. As cultures interact with each other more frequently, differences in cultural experience and norms must be assimilated into an increasingly complex society. Cultural relativism seems the most prudent adaptation to the accommodation of colliding cultural norms. Failure to recognize the validity of another culture's moral standards almost necessarily invites condemnation and eventual aggression.

The expressions "Live and let live," and "When in Rome, do as the Romans do" are reflections of cultural relativism. Others suggest that individual moral relativism is the result of a postmodern breakdown in cultural institutions. They suggest that increased cynicism toward government, religious dogma, judicial institutions, and cultural traditions has essentially forced individuals to develop their own sense of right and wrong (Baghramian and Carter, 2016). Nearly every issue, celebrity, and government official are robustly supported or attacked in an avalanche of opinion pieces. In the United States, the trust of government agencies, political parties, businesses, and media has diminished. As trusted sources of leadership and traditional values become less relevant in a more complex society, there seems to be a natural tendency toward situational ethical solutions.

Normative Ethics and the Fire Service

In Chapter 1, *Introduction*, we explored the question of whether the fire service was unique in how ethics applied compared with other public service professions. It will be instructive to revisit that issue specifically focusing on normative ethics.

If we review some of the more recognized ethical systems of other professions and organizations, we find that they are relatively straightforward. The military, law enforcement, and the legal profession are rooted in deontology. They each share the common ethical focus of adhering to rules and procedures. The medical profession certainly has rules, but the intent of those rules is to seek the best outcome for the patient. This is best exemplified by the first rule of medicine, which is to "do no harm." The utilitarian approach to medical ethics can also be exemplified in the application of triage. Medical rules require physicians to treat all patients and to never abandon a patient once treatment is initiated. However, medical ethics does allow physicians to withhold or delay treatment to some patients to provide the best outcome for the most patients.

General business ethics and the ethical code of the Boy Scouts are both examples of virtue ethics. In each, ethics are centered on personal traits that, if ingrained into the personality, will cause the individual to act ethically in any given situation.

The fire service is unique as a profession because it is diverse in service and organizational structure. We wear many hats and as a result are bound by complex ethical requirements.

At our core, we are an emergency response organization that exists to provide service to those in need Figure 4-16. This is an example of benefice ethics (care ethics). As such, the ethical system applied to the fire service must at least in part be rooted in consequential ethics. However, the fundamental responsibilities of a fire fighter (see *axioms* in the preface) are firmly rooted in Kantian ethics. Because we operate in a dangerous environment, virtues such as bravery, trustworthiness, and loyalty are greatly valued in our profession. Again, these are hallmarks typically associated with virtue ethics.

Beyond our role as fire fighters, many fire departments also provide emergency medical services. As such the same utilitarian principles associated with medical ethics must, of course, apply to us as well.

Figure 4-16 The fundamental mission of the fire service is to help others in need. Any consideration of fire service ethics must therefore be rooted in care ethics and in Kant's principle of duty.

The fire service is a paramilitary organization; most departments have a scalar organizational structure. We have clearly defined chains of command, standard operating procedures, and divisions of labor. This requires us to incorporate elements of deontology or duty ethics. Similar to law enforcement, fire inspectors enforce codes and regulations. In many cases inspectors have the power to censure. Arson investigators actively participate in the justice process and, in some cases, have the power of arrest. So it is apparent that deontology must be part of the fire service ethic.

It has been said that the fire service is all things to all people. Because we provide diverse services, the ethical principles required of fire fighters are complex and equally diverse. Virtue ethics, utilitarianism, and deontology are all relevant to us, and the challenge is in their application. The challenge is identifying and prioritizing the sometimes competing ethical requirements. For the fire service, the question is not which philosophic approach to ethics is the best but rather how much of which should be applied and when.

WRAP-UP

Chapter Summary

- The processes by which we assess ethical theory was reviewed. It was asserted that for an ethical theory to be of use it should meet two general standards: it should be valid and it should be applicable.
- It is reasonable for an ethical theory to accomplish the following: it should set forth a series of ethical and moral principles; it should show how those principles are justified; it should guide people to a life of ethical and moral excellence.
- The four branches of ethical study include: descriptive ethics, applied ethics, metaethics, and normative ethics.
- Descriptive ethics is the study of the ethical beliefs and practices of a given group.
- Applied ethics is the analysis of a particular ethical problem in light of individual ethical theories.
- Metaethics is the study of ethics. It asks questions regarding the nature of ethics, such as, From where do morals come?
- Normative ethics is the study of how people should act. Within the study of normative ethics, we explored virtue ethics. Virtue ethics is an ethical theory whereby actions are judged based on their compliance with ideal personal traits.
- Eudemonism is a virtue ethic theory that centers on those traits that cause the individual to "flourish." Those traits are categorized as either moral or intellectual. We flourish in character and productivity within society when we adhere to those virtues.
- Agent-based ethics assumes an action is "good" if it is consistent with virtues that reasonable people would agree are important.
- Care ethics is centered on the idea that the primary virtue any person can possess is that of benefice. An action is "good" if it improves or supports the lives of others.
- Deontology is a normative ethics theory that judges the rightness of action based on its adherence to duty.
- Kantian ethics suggests that duty arises from a categorical imperative to serve a universal good.
- John Locke argued that all people were born with fundamental human rights. As a result, we are all also burdened with adjoining responsibility. The quality of an action is, therefore, judged in its compliance with those duties.
- Social contract theory argues that each member of a society is bound by an implied contract to act in a way that promotes the general welfare and order of the group.
- Consequentialism is a branch of normative ethics in which the correctness of an action is defined by the impact of its outcome.
- Altruism asserts that the greatest good is that which is best for others. Ethics then is rooted in the service of other's needs.
- Egoism asserts that society is best served by each individual acting in that person's own self-interest. The best-known and most widely accepted version of consequentialism is that of utilitarianism. This theory asserts that the rightness of an action is judged by the amount of good to the most people.
- The concept of moral and cultural relativism was explored. This is a normative ethical theory supporting the concept that ethics is relative to the situation and to the individual's thoughts, beliefs, and intentions.

Key Terms

agent-based virtue ethics a system of ethics that is in the idea that each of us has an internal sense of what is right and wrong.

applied ethics a system of ethics that seeks to identify a morally correct path or action relative to issues in everyday life.

care ethics a system of ethics based on the importance of being a positive force within a social group.

categorical imperative according to Immanuel Kant, an action that is required out of duty, whether it benefits the actor or not.

conditional egoism assertion that acting in self-interest is ethically acceptable only if it leads to moral outcomes.

consequentialism the theory that the ethics of an action should be judged by its outcome.

cultural relativism the belief that actions are moral or immoral relative to the beliefs and traditions of a culture.

decision ethics a genre of applied ethics in which various normative theories are applied to the decision-making process.

deontology the ethics of an action should be judged relative to its compliance with a code of conduct based on certain categorical imperatives.

descriptive ethics a study to understand and catalog the attitudes of individuals or groups of individuals.

divine command theory assertion that universal moral truth came from God's will.

dual consequentialism consequences can be measured both morally and objectively.

egoism theory asserting individuals ought to promote self-interest above any other value.

ethical altruism consequential doctrine that holds that each of us has a moral obligation to help serve and benefit others.

eudemonism a virtue ethic theory that centers on either moral or intellectual traits that cause the individual to "flourish."

individual relativism moral values are created by individuals, relative to their place, time, and situation.

intellectual virtues virtues that are inherent in an individual's nature.

macro ethics a system of ethics focusing on global issues such as the geopolitical ramifications of wealth distribution, technology emergence, international business practices, and environmental issues.

metaethics a system of ethics that explores the meanings of concepts, words, and values associated with ethics.

moral relativism ethical culpability contingent on motive, the actors' understanding of circumstances, and anticipated outcome.

moral virtues virtues based on character and acquired through learning.

normative egoism acting in our own self-interest.

normative ethics a system of ethics that investigates questions regarding how an individual should morally act.

objectivism moral values are universal, existing outside human conventions.

pluralistic deontology theory of morality with imperatives stating that individuals have the duty to keep promises, a duty to pursue justice, a duty to improve the conditions of others, a duty to self-improvement, and most importantly a duty not to injure others.

professional ethics professionally accepted standards of personal and business behavior, values, and guiding principles.

proximity the relative value of a consequence either in time or scale.

psychological egoism suggestion of a universal truth that individuals will always act out of self-interest.

rational egoism assertion that placing someone's interests above your own was contrary to rational thought and arose only from emotions of guilt.

rule consequentialism rules that must have best-intended consequences.

rule of utility an assertion that good is that which brings the greatest happiness to the greatest number of people

subjectivism moral values are developed by people out of necessity or social convention.

utilitarianism the ethics of an action can most effectively be judged by its positive consequence to a majority of stakeholders.

virtue ethics the belief that ethics is rooted within compliance to particular virtues.

Challenging Questions

To check your understanding of this chapter's material, answer the following questions. It is highly recommended that you discuss your viewpoints with fellow students, peers, coworkers, and friends to discover their opinions as well.

- Of the ethical schools of thought described in this chapter, which do you believe apply most to the fire service?
- Do you think that there are universal moral principles, or do you believe that our view of right and wrong is unique to each group?
- What three virtues are most important for a fire fighter to possess?
- If you have not done so already, read the quote by Thomas Paine at the beginning of this chapter. He seems to be arguing that each individual has an obligation to protect the rights of others. Can you form two rational arguments supporting the social obligation referred to by Paine? Conversely, can you articulate to arguments in opposition to Paine's assertion that we each have a responsibility to others?
- Select a well-known movie in which the lead character faces an ethical dilemma. How does the character solve it, and what ethical philosophy is the character's solution most consistent with?

Case Study Conclusion

Revisit the case study at the beginning of the chapter. Spend a few minutes considering the questions posed at the end of the case study. In light of the information shared in this chapter, have any of your original observations changed?

This case study demonstrates some of the complexities associated with normative ethics. Fire fighter Smith wishes to do the right thing, but what exactly is meant by right may be subject to interpretation. Clearly, the actions of fire fighter Jones are offensive to Smith's personal values and demonstrate a general lack of character on Jones's part. Smith's internal sense of virtue does not condone Jones's behavior nor does he wish to enable it. As a result, he is inclined to report Jones's behavior. In this case, Smith's thinking is consistent with the theory of virtue ethics. Like most of us, Smith would like to see the best possible outcome for all involved. Reporting Jones's behavior will likely have negative consequences for the rest of the team, and perhaps the department at large. Other than satisfying his sense of duty and his own beliefs of right and wrong, Smith sees no particular upside to reporting Jones. Smith's concerns are consistent with that of consequentialism, and, as a result, he feels some compulsion to remain silent. Finally, Smith feels an obligation to uphold departmental rules, and the trust placed in him as a member of the department. This is an expression of duty ethics or deontology.

In sorting this out, consider the following: Smith's sense of right and wrong (virtue) and his sense of responsibility to the department (deontology) both suggest reporting Jones's abuses. The only compelling reason not to report is a fear of the negative consequences. While Smith's thinking is utilitarian in that his concern is primarily for the welfare of the team, it is one-dimensional. That single dimensionality is rooted in no actual good coming from Smith's silence, only avoidance of negative consequences. The negative consequences are assumed possible but not necessarily guaranteed. Smith's silence may postpone negative consequences if they exist but will likely not guarantee their future occurrence.

Chapter Review Questions

1. Differentiate among the terms "virtue ethics," "deontology," and "consequentialism."
2. What is normative ethics?
3. What ethical theory is most often associated with Immanuel Kant?
4. What school of ethics is most often associated with John Stuart Mill?
5. The question, What is truth? is most consistent with what branch of ethics?

References

Austin, M. W. 2013. "Divine Command Theory." *Internet Encyclopedia of Philosophy*. Accessed June 2, 2018. http://www.iep.utm.edu/divine-c/.

Baghramian, M., and J. A. Carter 2016. "Relativism." *The Stanford Encyclopedia of Philosophy*. Accessed June 2, 2018. https://plato.stanford.edu/archives/win2016/entries/relativism.

Baier, K. 1990. "Egoism." In *A Companion to Ethics*, edited by P. Singer, 197–204. Oxford, England: Blackwell.

Chadwick, R. F. 1997. *Encyclopedia of Applied Ethics*. London: Academic Press.

Fieser, J. 2016. "Normative Ethics." *Internet Encyclopedia of Philosophy*. Accessed June 2, 2018. http://www.iep.utm.edu/ethics/#H2.

Firth, R. 1952. "Ethical Absolutism and the Ideal Observer." *Philosophy and Phenomenological Research* 12, no. 3, 317–345.

Gewirth, A. 1980. *Reason and Morality*. Chicago: University of Chicago Press.

Gilligan, C. 1982. *In a Different Voice*. Cambridge, MA: Harvard University Press.

Gowans, C. 2015. "Moral Relativism." *The Stanford Encyclopedia of Philosophy*. Accessed June 2, 2018. http://plato.stanford.edu/entries/moral-relativism/#ForArg.

Haines, W. 2008. "Consequentialism." *Internet Encyclopedia of Philosophy*. Accessed June 2, 2018. http://www.iep.utm.edu/conseque/#SH1f.

Hursthouse, R. 2013. "Virtue Ethics." *The Stanford Encyclopedia of Philosophy*. Accessed June 2, 2018. http://plato.stanford.edu/archives/fall2013/entries/ethics-virtue.

Icheku V, Phil M. 2011. *Ethics and Ethical Decision-Making*. Bloomington, IN: Xlibris Corporation.

Mastin, L. 2008. *Deontology, The Basics of Philosophy*. Accessed June 2, 2018. http://www.philosophybasics.com/branch_deontology.html.

Mastin, L., and J. Bentham. 2008. *The Basics of Philosophy*. Accessed June 2, 2018. http://www.philosophybasics.com/philosophers_bentham.html.

Portmore, D. W. 2011. "Dual-Ranking Act-Consequentialism." *Commonsense Consequentialism: Wherein Morality Meets Rationality*. Oxford, England: Oxford Scholarship Online.

Ross, W. D. 1930. *The Right and the Good*. Oxford, England: University Press.

Russow, L. M., and M. Curd. 1989. *Principles of Reasoning*. New York: St. Martin's Press.

Smith, A. 1776. *The Wealth of Nations*. London: W. Straham and T. Cadell.

Tronto, J. C. 2005. "An Ethic of Care." In *Feminist Theory: A Philosophical Anthology*, edited by A. E. Cudd and R. O. Andreasen, 251–63. Oxford, England: Blackwell Publishing.

Westacott, E. 2013. "Moral Relativism." *Internet Encyclopedia of Philosophy*. Accessed June 2, 2018. http://www.iep.utm.edu/moral-re.

ADKINS
FAIRFAX COUNTY FIRE & RESCUE
PROTECTIVE
FIRE & RESCUE DEPARTMENT
0701
0702
EXTREME
MASK

SECTION 2

Ethics for Frontline Personnel

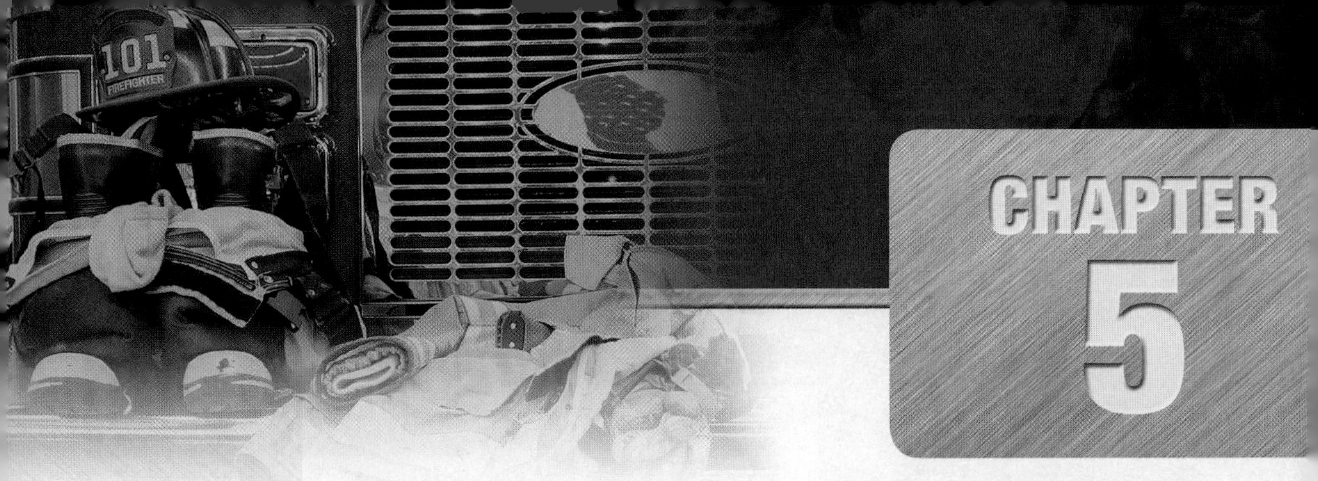

CHAPTER 5

Professionalism on the Line

"He that troubleth his own house shall inherit the wind: and the fool shall be servant to the wise in heart."

—Proverbs 11:29

OBJECTIVES

After studying this chapter, you should be able to:

- Exemplify the concepts of professionalism.
- Differentiate between licensure and certification.
- Explain the definition and purpose of value statements.
- Describe the role of professional standards of ethics in the fire service.
- Interpret the values of personal codes of ethics.
- Outline the 10 essential virtues of our service professionalism.
- Assess the impact of social media on the public's perception of the fire service as well as other professions.
- Differentiate between a profession and a trade.

Case Study

One evening during the Christmas season a group of fire fighters are sitting in their dayroom when Captain Jones posts an announcement asking for volunteers to deliver Meals on Wheels the following week. This has been a long-standing tradition at their fire department. The fire fighter's union contributes several hundred dollars to purchase prepared meals for the elderly and shut-ins. The fire fighters then donate an off-duty morning to deliver the meals.

Upon seeing the announcement, fire fighter Smith remarks, "This is a bunch of crap. Just once I'd like to come to work without somebody asking me to be some sort of Good Samaritan."

Captain Jones asks Smith what seems to be the problem. Jones replies, "This is a job and I get paid to come to work, do what I'm told to do, and go home. My free time is my free time to do what I want. I'm not paid to be a hero or a saint or even a good guy. I am tired of hearing about public service and giving back. I don't owe the public anything."

Captain Jones replies, "You're setting some kind of example for the kids that look up to you." Smith gets up and walks away saying, "I'm okay with that."

- Does fire fighter Smith have an ethical obligation to uphold the department's public image?
- Does Smith's attitude have any reflection on his professionalism?
- Does Smith's attitude represent a possible long-term problem for the company?
- Should Captain Jones address Smith's attitude?

Introduction

This chapter's title has a dual meaning. It is not uncommon to describe fire department operations in two categories, staff and line. Line personnel are typically assigned to emergency response, while staff are generally assigned to department administration. So one possible meaning of this chapter's title is the implied study of professionalism for line personnel, which is, in fact, the chapter's intention. This chapter will explore the various concepts of what professionalism means, especially as it applies to suppression personnel Figure 5-1.

You can also infer a second meaning—that professionalism in the fire service is at risk. If you're competing for your softball league's championship, the trophy is on the line. Every time you go in a burning building, you're placing your safety on the line. Likewise, every day that you and your brother and sister fire fighters report for duty, your department's reputation is on the line.

Section two of this text focuses on ethics and behavior standards relating specifically to response personnel. This chapter is an introduction or touchstone for the remaining chapters in the section. It will place the concept of professional ethics and standards of behavior in a context that properly frames the discussion in the ensuing chapters. Most importantly, this chapter will define professionalism as it applies to rank-and-file fire fighters. We will explore the attitudes, behaviors, and values consistent with the best traditions of the fire service. As you have learned in previous chapters, values and attitudes directly affect behavior and, as such, are critical in any ethics discussion.

Remember that it is difficult to discuss values and attitudes in relation to behavior standards without exploring the correctness of those attitudes. An important element of ethics is understanding our attitudes and beliefs through self-reflection. Likewise, an important element of leadership is the shaping

Figure 5-1 Every time fire fighters provide service they put their professional reputations on the line.

of attitudes. This chapter will explore values as they relate to professionalism in the fire service. The expression of those values reflect mainstream consensus as demonstrated by contemporary literature within the numerous fire service publications, websites, and blogs.

What Is Professionalism?

Common wisdom holds that the best place to start any endeavor is at the beginning. As such, it appears evident that any discussion of professional ethics in the fire service should start with a shared understanding of just what we mean by the term "professionalism."

Merriam-Webster's dictionary defines **professionalism** as: "The conduct, aims, or qualities that characterize or mark a profession or a professional person" ("Professionalism," 2015). Professionalism is identified as acting in a way that is consistent with the expectations of a given profession. An alternate and more usable definition is: "the skill, good judgment, and polite behavior that is expected from a person who is trained to do a job well" (Bullock and Trombley, 1999; "Professionalism," 2015). The distinction between the two is subtle but important. The first definition pins behavior to consistency within the defined profession. The second describes a standard of behavior independent of a given profession.

Is the Fire Service a Profession?

Is the fire service a trade or is it a profession? Surprisingly, although the question is fundamental, there is no clear consensus regarding the issue. Many traditionalists say that the fire service is essentially a trade. Although highly trained and burdened with great responsibility, fire fighters are essentially blue-collar workers who are taught specific tasks to be applied in specific situations. Many still hold that a college education is not actually necessary for fire fighters and that the basic skills needed for the career can be taught in training schools and fire academies.

Others assert that the fire service has evolved and continues to evolve into a skilled profession. They argue that the role of the fire fighter has changed dramatically in the last 40 years. They further state that the demands of specialties such as hazardous materials, code enforcement, inspection, data management, and labor law have far outpaced the standard curricula taught in fire academies. As evidence, they offer the fact that most fire fighters continue their education long after the academy by attending both academic programs and highly specialized advanced training programs produced by state and national institutions.

For the most part, the debate seems to revolve around education—the need for it and its role in the future of the fire service. While the role of education certainly is at the center of the argument over professionalism in the fire service, it is not the only consideration. Standards of behavior and professional ethics are important elements of professionalism and, unfortunately, are too often left out of the discussion of fire service professionalism.

A functional and often quoted description of a profession was put forth by Alan Bullock and Stephen Trombley in 1999 (Bullock and Trombley, 1999). They assert that an occupation becomes a profession with the development of formal qualifications based on education, the existence of an apprenticeship, and some sort of standard examination. They further state that professions have regulatory bodies who can control admittance to the profession and impose discipline for malfeasance and violations of ethical standards (Bullock and Trombley, 1999). Some scholars have added to the list that professions tend to garner high degrees of respect and prestige associated with high standards of professional and intellectual excellence Box 5-1.

In assessing the fire service's professional credentials against these standards, you find a mixed result. The fire service certainly enjoys a high level of prestige and respect based on excellence of performance. Perhaps no other profession places so much value on education specific to such a wide array of tasks as does the fire service. The knowledge base for fire fighters is comparable or even greater than many of the traditionally regarded professions. The fire service places a tremendous value on training and continuous education. The typical fire fighter medic will participate in over 100 hours of continuing education each year, covering numerous topics. At first glance, this makes a strong case for the fire service as a profession. But there are issues that complicate the argument.

BOX 5-1 Qualifying as a Profession

- The adoption of formal qualifications
- A recognized curriculum supporting qualifications
- Existence of an apprenticeship
- A formal qualifying exam
- A regulatory body
  - Require licensure
  - Set standards of competency
  - Create a code of ethics
  - Impose discipline

The fire service has numerous standardized qualifications based on National Fire Protection Association (NFPA) standards, and those standards are generally accepted and utilized by fire departments across the country. Although those standards exist, they are not necessarily required by any recognized central authority. NFPA 1001: Standard for Fire Fighter Professional Qualifications and numerous similar standards clearly define the minimum performance capabilities of a fire fighter (National Fire Protection Association, 2013), yet in many jurisdictions, an individual can be hired and enjoy long-term employment without meeting those standards. In some cases, an individual can become a fully vested fire fighter by meeting whatever unique standards are imposed by their particular department or jurisdiction. This is not to say that many jurisdictions do not require certification of personnel either by a state agency or national body. It is important to note that those requirements are provincial. There remains no requirement imposed by a professional oversight body.

Surprisingly, the fire service actually imposes greater consistency in fire fighter performance standards than they do that of chief officers. As with fire fighters, there are standards that define minimum qualifications for fire chiefs, yet they are rarely the criteria used for appointing a chief. Further, there are no requirements imposed on chief officers for continuing education.

The fire service has accepted professional qualifications and has numerous agencies that offer **certification** based on some standardized tests. However, certification alone does not meet the normally accepted standard for professional status.

The connection between educational requirements and professional ethics may not seem immediately obvious. However, it is significant. Both issues are rooted in a central professional authority that can set professional qualifications, confer license to practice, and impose standards of ethical behavior. The same authoritative professional organization that would impose standardized qualifications for competency should also be responsible for oversight of professional behavior as a condition of license.

A glaring disparity between the fire service and other professions is the lack of a cohesive, universal standard of ethical behaviors. Unlike doctors, financial officers, and lawyers who are guided by specific canons of ethics, the fire service has no universally recognized body of rules. Each fire department certainly has rules and regulations that are enforced by fire chiefs and commissioners. However, there is no national accrediting board that admits members and imposes behavior standards as a condition of membership. Further, no professional organization imposes discipline for malfeasance or ethical breaches.

An example of such would be the American Bar Association for lawyers or the American Medical Association for doctors. These professional organizations author and enforce specific standards of professional behavior. The standards are universal, and adherence to them is a condition of professional **licensure**. No such requirement exists for fire fighters of any rank, including fire chiefs. Partially filling this void are state and federal ethics regulations, which apply to governmental employees. Career fire fighters are typically municipal employees, and similar to any other government employees are prohibited from certain activities such as accepting bribes or using their positions for personal gain. However, governmental ethical standards represent an ethical minimum. They tend to focus exclusively on fiduciary issues such as conflict of interest and graft. They do not address many of the issues faced by fire fighters on a daily basis.

In comparing the fire service to other professions, we find that we have professional standards competency that may include certification, but we lack licensure. We have varied and important responsibilities yet no standardized measure of competency. We have professional organizations but no universal standards for admittance. We are controlled by rules and regulations yet have no codified body of ethics.

Are we a profession? It's a tough question to answer because the fire service is such a dynamic organization. It can be argued that a fire fighter or medic is a highly skilled and dedicated member of an occupation. While their skills encompass life-and-death issues and are extremely varied in scope and scale, their sphere of responsibility is limited. They are essentially under the direct supervision of their officers, who make most of the decisions. Conversely, high-ranking officers are less blue-collar workers and more like managers. Their actual responsibilities are not as varied as the people they supervise, but those responsibilities are much more complicated and often require subjective judgment. As such, there are greater ethical implications.

Clarification

Fire fighters often possess subsets of skills that require special certifications and have ethical responsibilities similar to that of other professions.

As a result, you may jump to the conclusion that suppression personnel are essentially blue-collar tradespeople, whereas chief officers are better defined as white-collar professionals. However, the aforementioned conclusions are not an exact fit. Suppression personnel often perform prevention inspections, and company officers are responsible for a myriad of management responsibilities in addition to suppression duties. Fire fighters are often paramedics, hazmat technicians, rescue specialists, and even juvenile fire counselors. These subsets of skills require specialized certifications and attendant ethics responsibilities similar to that of other professionals.

Is the fire service a profession? Unfortunately, the answer seems to be yes and no. However, regardless of the exact status of the fire service as a profession, the requirement for behavior standards is paramount and on par with any commonly recognized profession. You may fairly assert that the absence of a nationally recognized code of ethics and associated professional association does not so much disqualify the fire service as a profession; rather, it exposes a glaring need for professional development.

The Need for Professionalism

A continuing theme throughout this text is the dynamic relationship between the fire service and the public it serves. The adjective "dynamic" is apt in that the relationship has a significant impact on both parties, is subject to rapid change, and is influenced by a number of factors. Few occupations or professions enjoy the level of public support that is enjoyed by the American fire service, and rightfully so. The fire service is ingrained in the national fabric. Some fire departments can trace their origins to the pre–Revolutionary War years. To paraphrase a colloquialism, "The fire service is as American as apple pie."

Fire fighters routinely risk injury and even death providing help to those in need regardless of social, race, religion, or citizenship status. Communities of all sizes take great pride in their local fire departments and generally feel a sense of ownership in them.

More than 1 million Americans from all walks of life proudly identify themselves as fire fighters. Beyond providing needed public service, the American fire service also fills an important cultural niche; the fire fighter is seen as a champion of what we hold dear in the community. The qualities of self-sacrifice, commitment to others, and public service are embraced as important social threads that hold communities together. In a cynical world that has become increasingly devoid of heroes, many consider fire fighters as icons of trust and respect **Figure 5-2**.

> **Clarification**
>
> Few professions enjoy the level of trust and affection that is bestowed upon the American fire service, and few professions depend on the public's goodwill as does the American fire service.

Texture: Eky Studio/ShutterStock, Inc.; Steel: © Sharpshot/Dreamstime.com

Figure 5-2 Many consider fire fighters as icons of trust and respect.

Courtesy of Library of Congress, Prints and Photographs Division. LC-USZ62-108164.

The affection the public holds for the fire service certainly helps to build self-esteem among fire fighters but, beyond validation, the public's view of the fire service is critical to its ongoing success. The last 40 years have seen an increasing public disillusionment with once highly regarded professions. Lawyers, politicians, and doctors are all professions whose public standing has diminished in recent popular culture. Most recently, law enforcement has suffered from the hypervigilance imposed by social media. Still, people will always need lawyers and doctors, and so long as there is a government, there will be elected officials. As long as people are afraid of crime, there will be police forces, and, when crime rises, money will be spent to protect the citizenry. While each of these professions certainly values reputation, they all have a certain resilience to bad publicity because people uniformly understand the need for their services, and professional status stands somewhat apart from reputation.

The fire service, however, provides services most believe they will never need. Many consider the fire department to be an insurance policy of sorts. The public is glad we are there but they never really expect their house will catch fire. As a publicly funded service, the fire service exists because of the support and money received from elected officials who are responsive to the will of the people. Yet those officials and the public are not always wholly convinced of the need to fully fund the fire service. It is the public's affection that has prompted the continued funding and growth of the American fire service. Whereas the public feels they may *need* a doctor or a lawyer someday, they *want* a fire department even if they don't necessarily believe they will ever call for its services.

Evidence of the value of the fire service's reputation is plentiful. There is a long history of public outcry and demonstrations of support for fire departments in beating back staffing reductions and station closures. Congressional fire caucuses are highly regarded by elected officials in both of the major political parties. The International Association of Fire Fighters is one of the strongest lobbying voices in Washington, DC. They are well-funded, but political endorsements by the "fire fighters" are considered valuable as much for the fire fighter's reputation as for any money donated to any political action committees.

Fire fighters owe much to public support, and they have earned it. The fire service has been a positive force in American culture for over 200 years. Still, it should not be taken for granted. Recent history in law enforcement has demonstrated that the reckless actions of a few irresponsible people can shake public confidence in an entire profession. New realities in news reporting and social media have demonstrated that once confidence is lost, every indiscretion becomes national news, thus damaging the collective reputations of agencies across the country.

If a young fire fighter were to spend some time with a 70-year-old retiree, he or she would likely find that the career has changed greatly, but, surprisingly, it is not the technology or the methodology that has changed. A fire fighter who started in the early 1970s is familiar with GPS, multiple band radios, mobile computers, and thermal imaging devices. That generation saw the fire service integrate EMS, hazardous materials, and positive pressure ventilation. What has changed is culture.

The last 40 years have seen a great change in the behavior of fire fighters and in the expectations of the public. Practical jokes, language, and station house entertainment that were common in the 1970s are considered completely unacceptable today. The world has changed, as has the definition of professionalism. Major changes in societal attitudes have initiated changes in firehouse culture. Those changes are a reflection of changing public expectations, as well as public perception.

Clarification

Changes in societal attitudes have initiated substantial changes in firehouse culture.

Perception is driven in large part by how information is gathered and delivered. The conduit that connects the public to the fire service has traditionally

LOCAL NEWS

Firefighter Makes Death Threats: Investigation Underway

thebridge

Figure 5-3 Changes in societal attitudes have initiated substantial changes in firehouse culture.

Courtesy of Scott Walker.

been local media. However, the media as well as social media have greatly changed the dynamic of how people receive their information. Certainly, the means by which bad fire fighter indiscretions are reported and editorialized has changed greatly in the 21st century compared to the middle 20th century.

For example, in 2016, shortly before this text was written, a fire fighter responded in social media to comments made by President Barack Obama (Jackson, 2016). In his post, the fire fighter suggested (likely insincerely) that someone should shoot him (the president). Not so many years ago, the fire fighter would not have had access to as wide an audience as we do today with social media. If the fire fighter in question had publicly stated or written an insincere and unrealistic threat to the president, the issue may have been ignored. At worst, the fire fighter would have been quietly suspended. Had the local press chosen to cover it, it would've amounted to a one-time local story destined to fade into oblivion Figure 5-3. Today, the fire fighter's comments were seen by thousands; the story was placed online and potentially seen by millions as was the department's response. The story became well known in fire departments around the country and, to some extent, became an item of comment on social media. The story will also remain available for review for years to come, making it a constant embarrassment to the department.

Due to the scope of attention and the public's attitude toward poor behavior by public officials, what once may have been dismissed as foolish behavior has become essentially a scandal. There are several factors that drive how conduct breaches are reported and how the public responds to them.

Traditional Media

A major change has occurred in news delivery. Not so many years ago, local news tended to overlook stories about public officials' personal behavior. The introduction of profit motive into newsrooms has turned news delivery into, at least partially, an entertainment commodity. As a result, investigative reporting, human interest stories, and scandal have become essential elements of the news cycle. Public employees as well as public officials are under more scrutiny by news organizations because scandals sell. Of interest to the public are local stories of "bad" behavior.

Social Media

Social media has also played a major role in changing news reporting. Today every individual with a cell phone is a would-be news reporter Figure 5-4.

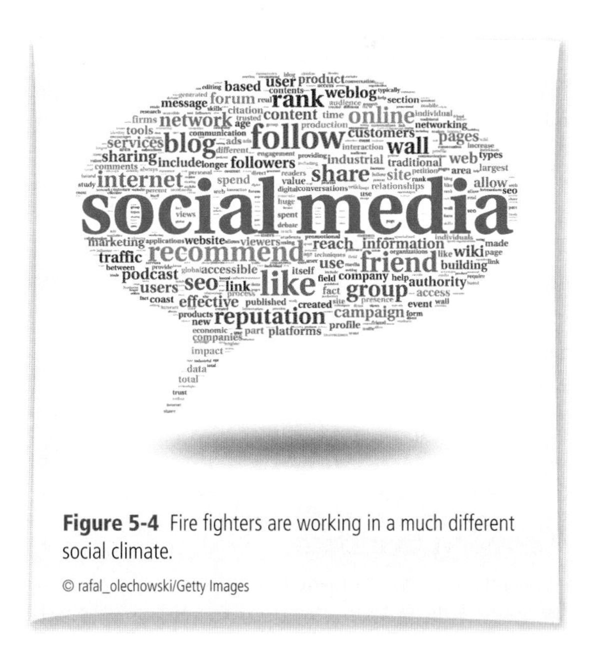

Figure 5-4 Fire fighters are working in a much different social climate.

© rafal_olechowski/Getty Images

Almost every action a fire department makes is under the scrutiny of the camera and with a few quick clicks, unedited video is available around the world. Often, "grassroots reporting" can be delivered out of context and be somewhat misleading. Still, it is widely distributed and often given the same credibility as are news stories produced by journalists. With the advent of social media, there is no such thing as a local news story; instantaneous distribution makes all news world news.

Not too many years ago, a fire department scandal had a relatively short shelf life. Local newspapers and television would likely cover it for a week or two before the public's attention was drawn to other stories. Unfortunately, stories posted on the Internet are permanent. If a researcher were to conduct an Internet search on the phrase "fire fighter scandal," he or she would likely see stories spanning many years. "Legacy" stories have the effect of constantly reopening wounds. Readers are constantly reminded of past transgressions, making it much more difficult for departments to move on.

Opinion as News

Modern media has changed the nature of scandal. There was a time when reporting consisted of who, what, where, and when. Today's media focuses much more on opinion; pundits seem obligated to share personal opinions about every story. Just as importantly, the advent of social media has begun a trend of news outlets inviting the public to weigh in on their opinions as well. Blogging, tweeting, and Facebook have created a culture of opinions, any of which may or may not be rooted in fact. The sad truth is social media has created a "Wild West" atmosphere when it comes to the sharing of information. Information is pervasive, as is misinformation. In a world where clicks and shares are currency, context is often ignored, facts are sometimes omitted, and sensationalism rules the day.

Growing Cynicism

The effect of the aforementioned changes in 21st-century culture has conspired to change the very nature of the public's relationship with institutions. Our citizens are bombarded daily by examples of bad behavior. As a result, they are becoming cynical about the behavior of their leaders, institutions, and fellow citizens. Unfortunately, they are predisposed to believe the worst and, as a result, the judgment of behavior lapses tends to lack restraint. Too often, a report of an accusation is tantamount to the report of guilt.

Unfortunately, in today's environment, even the behavior of one or two fire fighters can severely damage the reputation of their department and, more importantly, the profession itself. Sadly, we need only look to our brothers and sisters in blue as an unfortunate example. The constant stream of bad publicity for law enforcement agencies has created an inertia of sorts. As the public becomes more concerned about law enforcement behavior, it becomes more likely that stories of bad behavior are shared or even fabricated. Stories become trends and take on a life of their own. As a result, the many honest and professional law enforcement agents who are routinely risking their lives to protect us are tainted by the actions of a few.

So far, the fire service has managed to escape both the justifiable and unjustifiable scrutiny of the public. We remain a trusted and respected institution within our communities. However, it is apparent that trust is fragile.

A quick Internet search will reveal case after case of fire fighters behaving foolishly, unethically, or even illegally. Many pundits and authors are openly worrying that the current generation of fire fighters seems uninvested in reputation management. There are more than a few articles questioning whether or not the fire service is properly articulating its

Clarification

The fire sevice must be proactive in maintaining its reputation by actively promoting the highest standards of ethics and professional behavior.

tradition of values and public service to contemporary fire fighters.

Never before has the potential for the mistakes by a few individuals meant so much to so many. The fire service must be proactive in maintaining its reputation by actively promoting the highest standards of ethics and professional behavior. We can no longer afford to assume that our people are doing the right things out of an organic understanding of professional behavior. There are enough examples already out there to indicate that fire service leadership must be proactive in articulating values and expected behaviors. Later chapters will explore the specific and unique responsibilities of ethics management by officers of various ranks. It is important, however, to remember that reputation management is the responsibility of every member of the department. From the fire chief to the newest recruit, the public expects much from us, and we can ill afford to let them down.

Is the Fire Service Just a Job?

I witnessed an incident very early in my career that inspired the case study at the beginning of the chapter. A senior fire fighter was commenting on an inspirational poster that had been hung in our dayroom. It was typical of many such posters; depicting a fire fighter battling flames with a caption similar to "200 years of bravery, integrity, and service." His comments essentially criticized the poster for trying to make being a fire fighter into something that it was not. He essentially said that being a fire fighter was a job and nothing more. You come to work, you do what you're told, and you get paid. He did not want to be anybody's hero, and public service went as far as his paycheck. Our lieutenant was in the room and said nothing, so I assumed that he agreed. As a young fire fighter, I found it a bit unsettling to hear the comments along with my officer's tacit approval.

I took great pride in being a fire fighter and liked being part of something bigger than myself. I truly felt being a fire fighter was something more than just a job. After some soul-searching, I discounted his opinion. But I remembered it, and as I grew older and some facets of the job became routine, I tried to remind myself never to take being a fire fighter for granted.

It is not just a job. Our mistakes can cost people their most treasured possessions and, more importantly, their lives. If we are off our game, other people suffer. We bear tremendous responsibility not only to be competent but to act in a way that demonstrates our understanding of those responsibilities. The people we protect trust us with those things most dear to them. How can we be blasé about that?

If it were just a job, we would not wear fire department shirts and hats off duty, we would not have fire decals on our personal vehicles, and we would not collect fire fighter memorabilia. We do these things because we are justifiably proud to be fire fighters. We are proud not only for what we do but also because of the respect that is given to us by our fellow citizens. With that comes a responsibility to uphold the trust and reputation that so many other fire fighters have worked hard and honorably to build. Every fire fighter represents an ongoing legacy of those who have gone before us.

Clarification

The roots of professionalism and its associated ethical standards are entwined in an understanding of the responsibilities and values that are unique to firefighting.

Professional Values

In Chapter 3, *Influencing Behavior*, you learned that attitudes are shaped by values. Both attitudes and values are prime influences on our actions. What are the professional values of the fire service? Certainly, one way to determine the values of an organization is to review its **value statement**. A value statement is a formal declaration of what an organization considers

BOX 5-2 Value Statements

The Alameda County Fire Department

Members: We promote an atmosphere of trust and respect that encourages individual growth, participation, and creativity and acknowledges the achievements of our members.

Organization: We support an organization built on a foundation of initiative, collaboration, and commitment to efficiency, consistency and results, while attaining the goals of the organization.

Customer Service: We are dedicated to providing superior customer service.

Strategic Management: We plan for change and develop management strategies to meet the challenges of our future.

Regional Cooperation: We promote, encourage and participate in partnerships that provide all communities with the highest level of service.

Statesville North Carolina Fire Department Value Statement

The Statesville Fire Department achieves its mission and vision by building upon a system of values. We uphold professionalism as our core value. Our defining values also include: Integrity, Compassion, Service, Honesty, Stewardship, and Courteousness.

Professionalism: Our core value of professionalism defines who we are. We believe our chosen career is an upstanding and sound service to the community we serve. We take our role seriously and do all that we can to be a positive role model to the future generations.

Integrity: We believe in living by moral and ethical principles. We demonstrate our values by the way we [live] our lives in the public's eye and in the confines of the fire station walls.

Compassion: We believe in caring for our community members who are suffering from tragic events in their lives and do all that is in our power to assist in stabilizing the situation with a merciful attitude.

Service Excellence: We believe in providing the best possible service to the community where we live and work and do all we can to meet the needs of our neighbors through a humble, competent, well trained, and efficient team.

Cumberland Valley Volunteer Fireman's Association. Reputation Management White Paper. Oklahoma City, OK: IFSTA; 2010.

are its priorities and core beliefs. It is designed to assist management in making decisions consistent with core values and remind employees of behavior expectations (Rossi, 2015).

Box 5-2 provides some excellent examples of fire department value statements (Alameda County Fire Department, n.d.; City of Statesville North Carolina Fire Department, n.d.).

Both sets of value statements are typical examples of similar documents by other fire departments across the country. They speak to important concepts and express generally accepted ideals regarding professionalism. However, fire department value statements raise some questions: First, do they actually reflect the values of their organizations? Second, do they actually have any positive impact on their fire fighters' values and attitudes?

Values in the Fire Service

In reviewing the example value statements provided as well as those numerous examples found on the Internet, you are likely to find references to integrity, service excellence, teamwork, and compassion. Certainly, few fire fighters would disagree that these are all values worthy of aspiration, but are they actually the pervasive values found in most fire departments? If you were to speak candidly with fire fighters you'd likely hear a somewhat different list.

This discrepancy points to the existence of two sets of values at work within any given fire department. There are the **official values** of the department as expressed by value statements and mission statements. And there are the **unofficial values** held by the fire fighters themselves, which are a reflection of the department's culture and traditions Box 5-3.

Fire fighters have traditionally valued bravery, loyalty, tradition, respect, physicality, and competency. At face value, these, too, are worthy values. However, they differ in their orientation. The values expressed in most official statements are **externalized values**. They are aimed at support of the department's mission and its general service

BOX 5-3 Fire Service Values

Official
- Service
- Integrity
- Compassion
- Service
- Diversity
- Innovation

Unofficial
- Bravado
- Brotherhood
- Loyalty
- Competency
- Physicality
- Respect of Tradition

orientation. Externalized values tend to focus on responsibilities and obligations. Conversely, personal values typically expressed by fire fighters tend to be more internalized. **Internalized values** typically express personal needs and wants. To be sure, loyalty, tradition, and competency are values worthy of respect. However, they can also be self-serving and, as such, be a catalyst to unacceptable behavior.

Bravery is essential to the very nature of a fire fighter's work, yet if valued only for itself, it can support a macho culture of unnecessary risk taking. Loyalty to the department's mission is certainly a positive. However, fire fighter loyalty is usually focused on the fire fighters they most often work with. This can lead to "covering up" improper behavior out of a sense of personal loyalty. Valuing physicality can lead to initiatives in maintaining strength and physical stamina; this can have a positive effect on a fire fighter's health and safety. However, valuing physicality can also cause some fire fighters to dismiss education, or it can be a catalyst to exclude women from the fire service. Tradition is a treasured aspect of the fire service. It offers continuity and a sense of belonging. Unfortunately, tradition can also impede progress and cause fire fighters to cling to a status quo that serves them, but not necessarily the public welfare, well.

It is apparent that values, especially internalized values, can be a double-edged sword. To act in your self-interest is a common trait of all people, and in some cases a necessary one. However, if values are to promote professionalism, they must support concepts beyond self-interest. This requires a deliberate and concerted effort by the department to identify and integrate desired values into department operations.

Unofficial values are reinforced every day on apparatus floors, in training rooms, and around kitchen tables. Unfortunately, the department's official values are too often just flowery words on a website or in a policy manual. Proper values and their resulting attitudes must permeate an entire organization from the recruit academy to the strategic planning committee. The responsibility of conveying official values must be recognized and accepted at all levels of the organization, starting with chief officers and culminating with the fire fighters. Creating a culture whereby individuals contribute to something larger than themselves is not easy, but it is exactly what leadership is about. Later chapters will describe strategies for fire department leaders to pursue this goal. However, it should not be forgotten that the rank-and-file fire fighters must also accept responsibility for the level of professionalism within their department.

Clarification

For values to have meaning, they must permeate every level of the organization.

There is a scientific theory known as the **Hawthorne effect** (Draper, 2016). In its simplest interpretation, it suggests that the act of observing an action modifies it. The implication is that you cannot exist apart from your surroundings. This theory is applicable to professionalism in the firehouse. Every individual in the station environment affects his or her surroundings. Your statements and actions reflect your values, and they either contribute or detract from the professional environment. As a result, it is reasonable to assert that standards of professional behavior are the responsibility of every single member of the department.

Professionalism and Codes of Ethics

A **professional code of ethics** is a formalized statement of expected behavior created by a recognized authority and adopted by organizations within the profession. They should be based on the shared values of the profession and are intended to act as a guide for daily action. Value statements, as discussed earlier in the chapter, shape and reflect attitudes. Codes of ethics differ in that they are action based. Professional codes of ethics often serve as a template for individual organizations in their own development of organizational codes of ethics. Similar to a value statement, for a code of ethics to be relevant to an organization, it must be integrated into the fire fighter's daily routine. A profession's or organization's code of ethics must serve as a touchstone in policy enforcement, strategic planning, and management practices.

The Fire Fighter Code of Ethics

In 2010 the Cumberland Valley Volunteer Fireman's Association (CVVFA) released a white paper on reputation management (Cumberland Valley Volunteer Fireman's Association, 2010). Its intention was to raise awareness of ethical issues in the fire service and to draw attention to the need for the development of professional codes of ethics. In it the authors wrote:

> *The antidote to a false sense of entitlement is to rebuild the fire service's foundation of ethical behavior and ethical decision-making. One step in that direction is to establish a national Fire Service Code of Ethics as a guide for improved ethical decision-making.*

In 2011 the CVVFA in cooperation with the National Society of Executive Fire Officers released the **Fire Fighter Code of Ethics** (Figure 5-5).

Firefighter Code of Ethics

Background

The Fire Service is a noble calling, one which is founded on mutual respect and trust between firefighters and the citizens they serve. To ensure the continuing integrity of the Fire Service, the highest standards of ethical conduct must be maintained at all times.

The purpose of this National Firefighter Code of Ethics is to establish criteria that encourages fire service personnel to promote a culture of ethical integrity and high standards of professionalism in our field. The broad scope of this recommended Code of Ethics is intended to mitigate and negate situations that may result in embarrassment and waning of public support for what has historically been a highly respected profession.

Ethics comes from the Greek word ethos, meaning character. Character is not necessarily defined by how a person behaves when conditions are optimal and life is good. It is easy to take the high road when the path is paved and obstacles are few or non-existent. Character is also defined by decisions made under pressure, when no one is looking, when the road contains land mines, and the way is obscured. As members of the Fire Service, we share a responsibility to project an ethical character of professionalism, integrity, compassion, loyalty and honesty in all that we do, all of the time.

We need to accept this ethics challenge and be truly willing to maintain a culture that is consistent with the expectations outlined in this document. By doing so, we can create a legacy that validates and sustains the

Figure 5-5 Fire Fighter Code of Ethics.

International Association of Fire Chiefs. IAFC Position: Code of Ethics For Chiefs. IAFC website. http://events.iafc.org/Admin/ResourceDetail.cfm?ItemNumber=7067. Published 2003.

distinguished Fire Service institution, and at the same time ensure that we leave the Fire Service in better condition than when we arrived.

I understand that I have the responsibility to conduct myself in a manner that reflects proper ethical behavior and integrity. In so doing, I will help foster a continuing positive public perception of the fire service. Therefore, I pledge the following ....

- Always conduct myself, on and off duty, in a manner that reflects positively on myself, my department and the fire service in general.
- Accept responsibility for my actions and for the consequences of my actions.
- Support the concept of fairness and the value of diverse thoughts and opinions.
- Avoid situations that would adversely affect the credibility or public perception of the fire service profession.
- Be truthful and honest at all times and report instances of cheating or other dishonest acts that compromise the integrity of the fire service.
- Conduct my personal affairs in a manner that does not improperly influence the performance of my duties, or bring discredit to my organization.
- Be respectful and conscious of each member's safety and welfare.
- Recognize that I serve in a position of public trust that requires stewardship in the honest and efficient use of publicly owned resources, including uniforms, facilities, vehicles and equipment and that these are protected from misuse and theft.
- Exercise professionalism, competence, respect and loyalty in the performance of my duties and use information, confidential or otherwise, gained by virtue of my position, only to benefit those I am entrusted to serve.
- Avoid financial investments, outside employment, outside business interests or activities that conflict with or are enhanced by my official position or have the potential to create the perception of impropriety.
- Never propose or accept personal rewards, special privileges, benefits, advancement, honors or gifts that may create a conflict of interest, or the appearance thereof.
- Never engage in activities involving alcohol or other substance use or abuse that can impair my mental state or the performance of my duties and compromise safety.
- Never discriminate on the basis of race, religion, color, creed, age, marital status, national origin, ancestry, gender, sexual preference, medical condition or handicap.
- Never harass, intimidate or threaten fellow members of the service or the public and stop or report the actions of other firefighters who engage in such behaviors.
- Responsibly use social networking, electronic communications, or other media technology opportunities in a manner that does not discredit, dishonor or embarrass my organization, the fire service and the public. I also understand that failure to resolve or report inappropriate use of this media equates to condoning this behavior.

__________________________ Signature ______________ Date

Figure 5-5 *(Continued)*

Firefighter Code of Ethics

The International Association of Fire Fighters, in its Manual of Common Procedure and Related Subjects, contains this code which helps union firefighters uniformly remember their career mission and goals.

As a firefighter and member of the International Association of Fire Fighters, my fundamental duty is to serve humanity; to safeguard and preserve life and property against the elements of fire and disaster; and maintain a proficiency in the art and science of fire engineering.

I will uphold the standards of my profession, continually search for new and improved methods and share my knowledge and skills with my contemporaries and descendants.

I will never allow personal feelings, nor danger to self, deter me from my responsibilities as a firefighter.

I will at all times, respect the property and rights of all men and women, the laws of my community and my country, and the chosen way of life of my fellow citizens.

I recognize the badge of my office as a symbol of public faith, and I accept it as a public trust to be held so long as I am true to the ethics of the fire service. I will constantly strive to achieve the objectives and ideals, dedicating myself to my chosen profession--saving of life, fire prevention and fire suppression.

As a member of the International Association of Fire Fighters, I accept this self-ivmposed and self-enforced obligation as my responsibility.

Figure 5-6 The IAFF Code of Ethics.

The International Association of Fire Fighters, Manual of Common Procedure, Firefighter Code of Ethics, http://www.affi-iaff.org/images/shared/Documents /Firefighter%20Code%20of%20Ethics.pdf

In an attempt to reinforce the ideals expressed by the code and make the document as influential as possible, the authors encourage individual fire fighters to print the code, sign it, and display it in plain view.

Often organizations related to a profession will develop a specific code of ethics for their members. Examples of organizational codes include that of the International Association of Fire Chiefs (International Association of Fire Chiefs, 2003) and the International Association of Arson Investigators (International Association of Arson Investigators, 2017). The International Association of Fire Fighters included a code of ethics in its *Manual of Common Procedures* **Figure 5-6** (International Association of Fire Fighters, 2016).

As guidance documents, these are all useful. They are rooted in the traditional values of fire service and can offer significant guidance. More importantly, they represent first steps toward creating standardized ethics for the fire service. However, through no fault of their own, they have some shortcomings when compared to other professions' codes of ethics. The codes are based on shared values and reflect general best practices. However, they lack specific mandates for action or restrictions on behavior considered unethical. Compliance with these ethical codes is not a condition of licensure or membership in a professional association.

Personal Ethics Codes

Similar to an organizational ethics code, a **personal ethics code** is a formal declaration of values and priorities. It is based on an individual's perception of what constitutes ethical behavior. Why would an individual write a personal ethics statement? The following are some good reasons for creating one.

- The process helps an individual examine his or her values and attitudes. You may remember the quote from Socrates in Chapter 1, "An unexamined life is not worth living." The process of writing a personal ethics statement is a self-examination.

- The submission of a personal ethics statement is becoming a common component of promotional processes, scholarship applications, and grant submission requirements.
- If done properly, your personal ethics statement can be a powerful guide when facing ethical dilemmas.

How to Develop a Personal Ethics Statement

Writing a personal ethics statement may seem a bit daunting. Certainly, formalized self-reflection is not something we do on a regular basis. It is also true that writing about oneself is uncomfortable for most people. But the reality is that the exercise is not nearly as difficult as you might expect. At its core, the process is simply a listing of individual priorities.

Your individual ethics statement should contain three sections: an introduction, the body, and a conclusion.

In the introduction you introduce yourself to the reader. You may want to include an explanation of important ethical influences in your life, or begin with a quote by your favorite author that is consistent with the body of your text.

The body of your statement should include your core beliefs and your opinions about correct behavior. Again, this document is about you and it is highly recommended that you only include those things in which you strongly believe. You should list only your personal behavior priorities; however, the statement can include both professional and personal elements.

Typically, the conclusion of your ethics statement will include some explanation of the importance of ethics.

As you create an effective personal ethics statement, you may want to keep the following in mind.

- Your statement should be honest and unbiased. It is very much about what you value, not what you think others should hear.
- You should clearly state your moral beliefs and thoughts about right and wrong.

BOX 5-4 Example of a Personal Ethics Code

As a fire fighter, I understand that I shoulder great responsibility. In my hands lies the well-being of my fellow citizens and my brother and sister fire fighters. I understand that generations before me have built a legacy of professional integrity, bravery, and service. I accept that I am responsible for tutoring that tradition of excellence.

Excellence: I will always strive for excellence in everything that I do. I will give my best efforts to the public I serve, and to the men and women with whom I serve. Good enough is never good enough.

Integrity: I am a member of the most trusted profession. My actions must always be consistent with the fire service's highest traditions. Just as I strive to maintain my personal reputation, I shall dedicate myself to maintaining the reputation of my fellow fire fighters and my department.

Courage: I will endeavor to maintain courage of body and spirit. My profession demands bravery and I will not shrink from my duty. Further, I will have the courage to speak the truth and to do what is right.

Humility: I will remember that my success is shared by many, and no matter how much I learn, there is so much more that I need to learn. My best is ahead of me, and I owe my profession my best.

Being a fire fighter is a privilege. With privilege comes tremendous responsibilities. I owe my fellow fire fighters, my department, and my profession a standard of behavior worthy of our tradition.

- Even though the statement is about you, remember that you are writing for an audience. They want to know about you so keep it simple, honest, and on point.
- Resist the temptation to include statements that are not really important to you. Consider each included statement carefully. Does it really affect your daily life or reflect those values that are most important to you?

Box 5-4 is an example of a personal ethics code.

In reviewing the example in Box 5-4, you will notice that the content is very similar to that of a listing of virtues. As you may recall from the previous chapter, ethics is rooted in either anticipated outcome, compliance to duty, or personal virtues. It is not surprising that a personal ethics statement would be reflective of personal virtue. Aristotle asserted that virtue represented the habits of virtuous people and that we must strive to incorporate virtues into our lives to the extent that they become a habit. This is consistent with the goal of a personal ethics code—it must influence daily life. As you consider personal ethics, it may be useful to consider those virtues that constitute professionalism.

Professional Virtues

If professional behavior is rooted in the priority of our values, then our actions are a reflection of those things that guide our behavior. Box 5-5 is a list of 10 virtues that a fire fighter should possess. It represents an amalgamation of numerous articles, blogs, and web postings regarding character and leadership in the fire service. While individual authors may include or exclude a particular item, there is general consensus within the profession as to what constitutes important professional qualities.

BOX 5-5 Ten Virtues of Fire Fighters

1. **Integrity:** Truthfulness, honesty, and trustworthiness are critically important to fire fighters. We are entrusted with great responsibility by the public, and the nature of our work requires us to place a great deal of trust in each other. We are a team-based culture. Integrity is essential to a successful team, and its absence, by even one member, can be incredibly disruptive and counterproductive.
2. **Competence:** Very often our work has life-and-death consequences. When we accept the job of fire fighter, paramedic, fire officer, or fire chief, we are tacitly stating that we will endeavor to meet the challenges placed before us. In a profession where the stakes are so high, we owe the public and each other our dedication to excellence through education and training.
3. **Loyalty:** Certainly loyalty to our fellow fire fighters, to our housemates, our shift, and our departments is important. But never forget that our first loyalty is to our mission and to the public we serve. We exist to provide a public service. The needs of the department are secondary to the needs of the citizenry. This is essentially why we exist.
4. **Intellectual curiosity:** To be a fire fighter is to be a lifelong learner. It is true that we embrace tradition as it connects us to our past. But we must never let tradition interfere with our ability to adapt to new methods and technologies to better serve the public. The pursuit of excellence demands growth.
5. **Physical fitness:** It is a myth that firefighting requires brute strength. However, it is undoubtedly physically demanding. We have an obligation to ourselves, our families, our coworkers, and the public to keep ourselves in the very best physical shape that we can.
6. **Teamwork:** Being a team member means acknowledging that you are part of something bigger than yourself. The team's cooperative effort creates a whole greater than its parts. By placing the good of the department above self-interest, you contribute not only to its success but to your own.
7. **Generosity of spirit:** We are a public service provider. We exist to help those who are having the worst day of their lives. Competency alone cannot effectively serve the public. A fire fighter must possess compassion, tolerance, and humor. The absence of these can lead to bitterness, frustration, and poor professional behavior.

8. **Dedication:** Fire fighters must dedicate themselves to excellence, service, and integrity. Every training session is important, as is every inspection, every call, and every public education duty. This requires bringing your "A-Game" every day, not sweating the small stuff, and seeking consistent personal growth through professional self-development.
9. **Obedience:** By joining the fire department you tacitly agree to follow the rules and regulations of that organization. It is not the privilege of any fire fighter of any rank, including fire chief, to randomly decide which policies are important and which are not.
10. **Respect:** To be a successful fire fighter you must respect the department's traditions, the legacy of honor that is attached to the profession, your superior officers' authority, and most importantly the humanity of your fellow fire fighters and the public you serve. Without respect you act only in your self-interest, and this is inconsistent with the principles of the fire service and its reason for existence.

WRAP-UP

Chapter Summary

- The fundamental concepts of professionalism were introduced. Professionalism is defined as the skills, judgment, and polite behavior that is expected from a person who is trained to do a job well.
- The chapter differentiated between the elements that constitute a profession versus those of a trade. These include the adoption of formal qualifications, a recognized curriculum supporting those qualifications, a formal qualifying exam, and a regulatory body that can admit participants and impose professional standards of behavior as a condition of membership.
- Is the fire service a profession? While the fire service has a robust set of formal qualifications, those qualifications are not necessarily a requirement for employment. This is because the fire service does not have licensure. Professional qualifications such as the NFPA standards are essentially recommendations. Most importantly, the absence of licensure also means that there is no ruling body for the fire service.
- We are in essence a profession comprised of thousands of autonomous organizations. There's no standardized code of ethics, no behavior review board or licensure, nor any consequence for breaches of ethics.
- This chapter reviewed the differences between certification and licensure as they apply to the fire service and, ultimately, to professional standards of behavior within the fire service.
- While the fire service may not qualify, in some respects, as a profession, it is clearly apparent that standards of behavior for the fire service are significantly higher than that of the typical trade.
- The fire service is comprised of skilled practitioners who routinely engage in tasks with life-and-death consequences. They have significant ministerial duties in the areas of inspection and code enforcement that require professional judgment and discretion, along with their related ethical implications.
- The fire service is a public service provider representing a positive force in the community and, as such, represents an important thread in the American social fabric.
- Few organizations enjoy the respect and admiration of the public as does the fire service. The chapter reviewed the relationship between the fire service and the public and how its reputation has been elemental in its continued success.
- The chapter discussed the relationship between reputation management and 21st-century culture, particularly social media and the new business models of traditional news outlets. We explored the importance of maintaining the fire service's reputation in an ever-changing social climate that is becoming increasingly cynical and hypervigilant.
- The chapter explored professional and personal codes of ethics and their importance in maintaining an ethical culture and in maintaining the reputation of the fire service within the community.
- The chapter also focused on individual values and virtues critical to the fire service's success, both as an organization and for individual fire fighters.

Key Terms

certification the assertion by a certifying body that an individual has been exposed to training and tested competency.

externalized values values focused on responsibilities and obligations.

Fire Fighter Code of Ethics code of ethics developed by the Cumberland Valley Volunteer

Fireman's Association and adopted by the U.S. Fire Administration.

Hawthorne effect scientific theory that suggests that the observation of an action tends to modify it.

internalized values values focused on needs and wants.

licensure permission by a recognized authority to practice a profession or trade.

official values those values expressed within the department's mission and value statements.

personal ethics code a formal declaration of values and priorities.

professional code of ethics a formalized statement of behavior expectations created by a recognized authority.

professionalism the skill, good judgment, and polite behavior that is expected from a person who is trained to do a job well.

unofficial values values held by fire fighters that are a reflection of the department's culture and traditions.

value statement a formal declaration of an organization's priorities and core beliefs.

Challenging Questions and Exercises

To check your understanding of this chapter's material, answer the following questions. It is highly recommended that you discuss your viewpoints with fellow students, peers, coworkers, and friends to discover their opinions as well.

- Do you think the fire service will benefit from mandatory standards of qualification and licensure?
- What are the unofficial values of your fire organization?
- What steps can the fire department take to proactively manage its reputation?
- If you have not done so already, read the quote at the beginning of the chapter. What does the quote mean? How does it relate to professionalism?
- Using the instructions presented in the chapter, write a value statement for both your professional and personal life.

Case Study Conclusion

Revisit the case study at the beginning of the chapter. Spend a few minutes considering the questions posed at the end of the case study. In light of the information shared in this chapter, have any of your original observations changed?

The case study about fire fighter Smith and his attitude toward public service speaks to some fundamental issues regarding professionalism and individual rights. Most fire fighters would agree that Smith has every right to his opinion regarding public service, charity, and the donation of his free time. However, most fire fighters also recognize that one of the fundamental strengths of the fire service is its relationship with the public. Over the decades countless fire fighters have gone the extra mile to build that reputation. All of us, including Smith, benefit from those efforts. In this sense, an argument can be made that Smith is reaping the benefit of others' labor without contributing to it himself.

This chapter demonstrated that being a fire fighter has professional obligations beyond competency. We are, by our very existence, a public service organization. Whether Smith wishes to acknowledge it or not, there is an unwritten, but nonetheless important, contract between the fire service and the public. This reluctance to acknowledge that condition of employment reflects poorly on his professionalism, even though it may not have anything to do with his competency as a fire fighter. To be clear, there is a distinction to be made between professionalism and attitude. Smith would certainly be keeping his ethical responsibilities toward professionalism by participating in the Meals on Wheels program even if he does not like it. Remember, ethics is about actions and consequences.

Whether Captain Jones should address Smith's outburst or not depends on a couple of factors. First, does Smith's attitude represent a problem? Second, does Jones have a responsibility to modify Smith's attitude even if it's not an issue? Is there a leadership responsibility?

As to the first question: Can Smith's attitude pose a problem? The answer is yes. One disgruntled individual can influence others' attitudes. Jones should ask himself whether or not he would be happy if his entire company had the same attitude. If not, he needs to address the issue.

The second question is whether or not Jones has a leadership responsibility to address Smith's attitude, even if he does not think that Smith poses a risk to esprit de corps. Again, the answer is yes. As an officer, Jones has an ethical responsibility to represent the department's desire and culture. He also has an obligation to Smith to assist him in feeling part of the team and helping him to enjoy his career.

Chapter Review Questions

1. What is meant by the term "professionalism"?
2. What are some of the essential elements of a profession?
3. What is the difference between licensure and certification?
4. List three benefits the fire service derives from a positive public image.
5. List at least two ways that social media has influenced the public's perception of public institutions.
6. What is a value statement?
7. What two organizations developed the Fire Fighter Code of Ethics?
8. What is the difference between internalized and externalized values?
9. Why would someone want to create a personal ethics statement?
10. List at least three virtues associated with fire service professionalism.

References

Alameda County Fire Department. n.d. "Mission, Core Values & Philosophy." Accessed June 4, 2018. https://www.acgov.org/fire/about/mission.htm.

Bullock, A., and S. Trombley. 1999. "Professionalism." *The New Fontana Dictionary of Modern Thought*. London: Harpercollins.

City of Statesville North Carolina Fire Department. n.d. "Mission/Vision Statement." Accessed June 4, 2018. http://www.statesvillenc.net/Departments/FireDepartment/MissionVisionStatement/tabid/213/Default.aspx.

Cumberland Valley Volunteer Fireman's Association. 2010. "Reputation Management" White Paper. Oklahoma City, OK: IFSTA.

Draper, S. W. 2016. "The Hawthorne, Pygmalion, Placebo and Other Expectancy Effects: Some Notes." Accessed June 4, 2018. http://www.psy.gla.ac.uk/~steve/hawth.html.

International Association of Arson Investigators. 2017. "Code of Ethics." Accessed June 4, 2018. https://www.firearson.com/Member-Network/Code-Of-Ethics.aspx.

International Association of Fire Chiefs. 2003. "IAFC Position: Code of Ethics For Chiefs." Accessed June 4, 2018. http://events.iafc.org/Admin/ResourceDetail.cfm?ItemNumber=7067.

International Association of Fire Fighters. 2016. *Manual of Common Procedures*.

Jackson, Sarah. 2016. "Fire Fighter Under Investigation Over Social Media Post." *Wave 3 News*. Accessed June 4, 2018. http://www.wave3.com/story/32449282/firefighter-under-investigation-over-social-media-post.

National Fire Protection Association. 2013. *NFPA 1001 Standard for Fire Fighter Professional Qualifications*. Accessed June 4, 2018. http://catalog.nfpa.org/NFPA-1001-Standard-for-Fire-Fighter-Professional-Qualifications-P1388.aspx.

"Professionalism." 2015. *Merriam-Webster's Collegiate Dictionary*. Martinsburg, WV: Merriam-Webster.

Rossi, H. L. 2015. "7 Core Values Statements That Inspire." *Fortune*. Accessed June 4, 2018. http://fortune.com/2015/03/13/company-slogans.

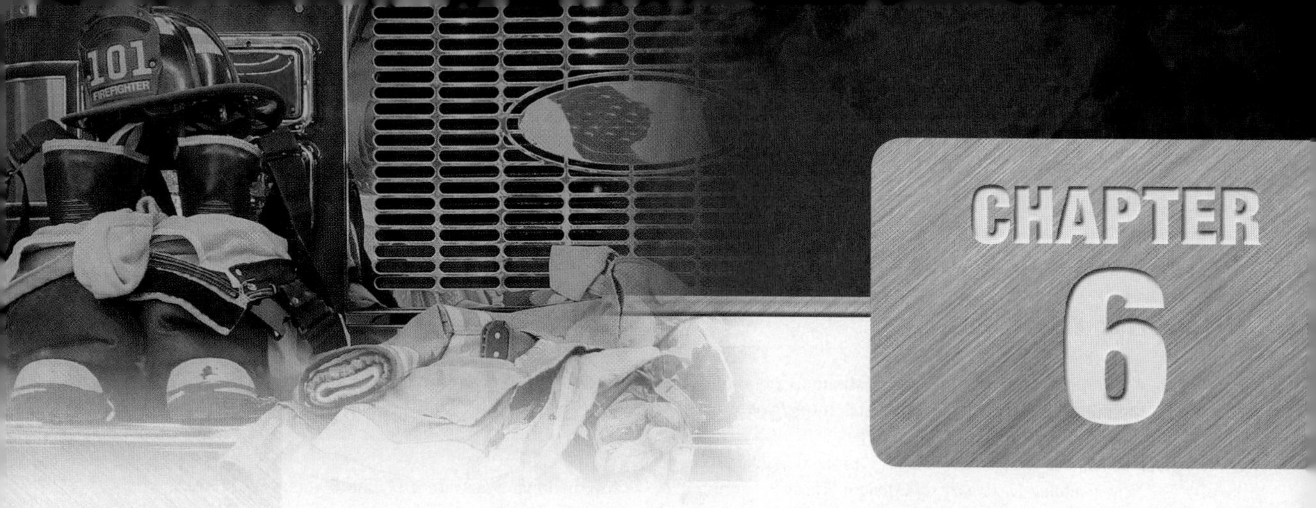

Fire Fighter Responsibilities

"I count him braver who overcomes his desires than him who conquers his enemies, for the hardest victory is over self."

—Aristotle

OBJECTIVES

After studying this chapter, you should be able to:

- Define the concept of professional responsibilities.
- Differentiate between responsibility and ethics.
- Assess the relationship between a department's mission statement and operational responsibilities.
- Differentiate between roles and responsibilities.
- Comprehend the objective responsibilities of fire agencies.
- Describe the objective responsibilities of fire fighters.
- Evaluate the ethics of customer service initiatives.
- Recount fire fighters' subjective responsibilities.
- Characterize the responsibilities of union leadership.
- Decipher the importance of respecting and protecting privacy, including HIPPA legislation.

Case Study

Fire fighter recruit Jones is in the tenth month of his one-year probation. Jones is assigned to Lieutenant Smith's company. In completing his monthly evaluation, Smith notes that fire fighter Jones needs improvement in hose operations. Specifically, he has shown a lack of aggressiveness in advancing hose lines to the seats of fires. However, Smith notes that Jones is making progress and feels comfortable that with additional training Jones can overcome these deficiencies.

Training Chief Campbell feels that the department should never have hired Jones because Jones is the mayor's nephew. Although he has no evidence to prove it, he is fairly certain that Jones was given special consideration for employment. As a result, he has little patience for Jones's deficiencies. He explains to Smith that Jones's reluctance to advance hose lines so late in his probation is indicative of his unsuitability for employment. Campbell instructs Lieutenant Smith to remove his remarks indicating that Jones can overcome his deficiencies with further training. Campbell further instructs Smith that his evaluation will be used as a basis for recommending Jones's termination from probation. Campbell intimates to Smith that he believes that the chief will be looking for an excuse to keep Jones on the department and that he would appreciate Smith's remaining quiet about their conversation.

Smith is not entirely comfortable with altering his evaluation. However, he respects Campbell's experience and opinion. Further, he is not entirely sure that Campbell may not be correct in his assessment regarding Jones's potential.

Two days later Smith is summoned to the chief's office. The chief explains that he has had a conversation with Training Chief Campbell and that Campbell believes that Jones's probation should be terminated. The Chief asks Smith point blank if he believes that Jones is beyond salvaging as a fire fighter. The chief assures Smith that he will not continue Jones's probation if Smith feels that Jones cannot be a good fire fighter. However, for political reasons, he does want to give Jones every benefit of the doubt. In that spirit, he asks that Smith be as optimistic as accuracy will allow regarding Jones's evaluation.

- What are Smith's responsibilities in this case?
- Which of Smith's competing loyalties should be given greatest credence?
- What path would you follow?

Introduction

Bob Dylan once shared a simple yet compelling message with us: "You gotta serve somebody" (Dylan, 1979). Our personal and professional lives are a complicated matrix of responsibility and obligation. The more our personal life flourishes and the more professional success we achieve, the more possessions we attain and the more responsibilities we shoulder. There is a fundamental relationship among the concepts of obligation and accountability and ethics. The consequentialism theory of ethics focuses on the outcome of an action in judging its relative ethical worth. Inherent in that assertion is the assumption that we have a responsibility to seek a just outcome for others. Deontology likewise recognizes that an individual has responsibilities and obligations, although it differs in that those obligations are rooted in a sense of duty.

In Chapter 4, *The Philosophy of Ethics*, we reviewed various notions regarding the origins of responsibility. Some argue that the responsibility for ethical behavior comes from God. Others suggest responsibility arises from a social contract that is based on mutual benefit, and others argue that social norms impress upon us the responsibility to adhere to ethics. Regardless of where responsibility comes from, it is obvious that there is a connection between responsibility and ethics.

Each of us has many responsibilities in our lives. Sometimes those responsibilities compete for our time, and sometimes responsibilities appear to demand conflicting actions. As individuals, we try to balance and juggle our many responsibilities, and we routinely make decisions regarding our personal priorities. These decisions help define us as individuals, and they are critical to personal and professional ethics.

The fire service is in the business of saving lives and protecting property. At our core, we are a service agency rooted in helping others. We deal with life-and-death situations, and, in most cases, we are the last defense of those in dire need. In this context, the depth of responsibility shouldered by fire fighters should be obvious. Yet the fire service is much more than fire and EMS response. It provides many services in addition to fire suppression, such as public education, fire inspections, and code enforcement. Beyond the services we provide, we are also a complex organization with varied internal and external stakeholders. This chapter will explore the relationship between ethics and the many responsibilities of the fire department.

Fire departments are, of course, made up of individuals. Just as a fire department has certain responsibilities, so do individual fire fighters. A significant portion of this chapter will explore the individual responsibilities of fire fighters from an ethics viewpoint. We will ask the questions: What does a fire fighter owe his or her department? What do fire fighters owe the public? And finally, what do fire fighters owe each other?

Throughout this chapter, there will be assertions made regarding responsibility and obligation. Those assertions apply to both fire departments and individual fire fighters. As you read the chapter, it is important to understand that those obligations and responsibilities are rooted in a **prime theory of obligation** that applies directly to the fire service (refer to the ethics axioms in the preface). The prime theory of obligation asserts the following:

- Fire departments have an ethical obligation to act in the public interest. That obligation is fulfilled by efficiently meeting its core responsibilities of life safety and property protection.
- The obligation for a department to efficiently fulfill its duties exists because of the nature of the work a fire department does, and the fact that it is supported by tax dollars.
- By joining a fire department, a fire fighter explicitly agrees to support the mission of that fire department and to refrain from any activity that hinders the department in meeting its obligations. That agreement imparts an ethical obligation.

The theory of a prime obligation is a description of the fire service's base ethical obligation. It is a touchstone for all other responsibilities and obligations associated with fire department activities. The obligations and responsibilities discussed later in this chapter can be justified on their own merit. However, they are rooted in their common relationship to the prime obligation.

This chapter will review several responsibilities and obligations associated with fire department operations. Some of those responsibilities may not seem directly connected to emergency response. Remember that each of those responsibilities is rooted in its compliance to the prime theory of obligation described previously.

Professional Responsibility

Ask fire fighters what the public can expect from their fire department, and you will likely get an answer such as, "They can expect our best efforts based on the staffing and equipment that we are provided." At face value, this seems like a reasonable answer. However, ask private citizens what they expect from their fire department and you will get a somewhat different answer. They expect a fire fighter to arrive promptly, and they expect the department to efficiently and competently do whatever is necessary to solve whatever problem that prompted their call for help. Again, these are reasonable expectations; however, the public also expects the fire fighter to risk his or her life to save theirs, the lives of their families, and even their possessions.

Clarification

The public expects the fire service to:

- Respond quickly.
- Perform effectively.
- Be willing to risk fire fighters' safety to save them, their family, and their possessions.

The public assumes that there are enough fire fighters to do the job expected. They assume the department has all the equipment it may need, and they expect that everyone on the crew is well trained and physically up to whatever challenges are presented. These are very high expectations, but, after all, isn't that why they pay their taxes? Many people may take exception to the notion that fire fighters should be willing to risk their lives for mere possessions. After all, doing so is not in keeping with the Everyone Goes Home Life Safety Initiatives (National Fallen Fire Fighters Foundation, 2017). Still, those expectations are in keeping with the commonly held image of the fire service. For the most part, the fire service has been complicit in instilling that image, and when not complicit, we certainly have not made any great effort to dispel the image of the superhero fire fighter. The public expects heroic efforts and expert results. So what exactly does the fire service really *owe* them? What are the actual obligations of a fire department?

Does a fire department have an ethical obligation to provide service to the citizens of its jurisdiction? Yes. That responsibility starts with our mission statement. We exist to "Save Lives and Protect Property." This is a promise of service, and people depend on it. They have come to expect heroic efforts because nearly every portrayal of the fire service promises it.

BOX 6-1 Everyone Goes Home Life Safety

The Everyone Goes Home program was founded by the National Fallen Fire Fighters Foundation in 2004. It provides free training and resources to implement the 16 Fire Fighter Life Safety Initiatives. The goal of the Everyone Goes Home program is to reduce the number of preventable fire fighter line-of-duty deaths and injuries. For more information go to https://www.everyonegoeshome.com.

Not only are they banking on the promise of extraordinary service, they assume they have paid for it. Whether it be from municipal taxes, fire district taxes, or donations at the firehouse fish fry, members of the public financially support the fire service, and in return it is either directly or implicitly promised that *their* fire fighters will show up and save them from disaster. It is not far from an exaggeration to say that it is almost a sacred bond. This heavy responsibility permeates the fire service and affects every department in every community.

Herein lies the first ethical responsibility of the fire service, fire departments, and individual fire fighters: *to always act in the public's best interest.* In the case of the fire service, public interest is served through the provision of emergency response services.

In 1922 Walter Lippmann asserted that acting in the public interest was a rational response (Figure 6-1). He wrote, "The public interest is what men would choose if they saw clearly, thought rationally, acted disinterestedly and benevolently" (Lippmann, 1922). The fire service and, by extension, individual fire departments exist specifically to provide a public service. A fire department's mission and functions clearly assert responsibilities and obligations with life-and-death consequences. The ethical obligations inherent in these responsibilities are complex and unique to our profession.

All of those responsibilities spring from or are directly connected to our primary ethical obligation to serve the public interest. This obligation exists because the consequences of our performance have a significant impact on others. We have a duty to

Figure 6-1 Walter Lippmann, 1889–1974. American author and social commentator.

Library of Congress Prints and Photographs Division Washington, D.C. 20540 USA http://hdl.loc.gov/loc.pnp/pp.print

perform based on our mission statement and an obligation of competency based on our social contract with the public to provide service for which they pay.

The Ethical Responsibilities of a Fire Department

Effective Response

A fire department exists to respond to emergencies. The public assumes that the agency is prepared for those responses. The public assumes that competent help is available based on an actual or implied promise made by the department to come to their aid. Based on that promise of service, each local fire department has an ethical obligation of competency. As every fire fighter knows, the delivery of quality emergency services is a complex process. The first requirement of a fire department is to identify and recruit qualified personnel. Acquiring qualified personnel requires unbiased recruitment and selection processes, and competency-based testing to identify individuals who are physically, emotionally, and intellectually suited for the work. Further, to meet its ethical obligations to both the public and the fire fighters, the department has the responsibility to train and equip fire fighters adequately. It is also incumbent on the fire service to implement departmental leadership capable of the responsible supervision of its personnel. Effective leadership is dependent on the provision of adequate training and educational opportunities for potential administrators. Beyond human resources capabilities, administrators must seek adequate resources to provide for public safety. Given the amount of resources needed, leadership must deploy those resources as efficiently as possible to provide for an effective response.

Most fire service students will quickly recognize that the responsibilities listed previously represent a fairly simplistic and common view of fire department operations. They represent the bare essentials necessary for the implementation of competent fire protection. Beyond the practical necessity of accomplishing fundamental principles of fire protection, there rests an ethical obligation to do so. Do

Clarification

The bare essentials necessary for the implementation of competent fire protection include:

- Identify and recruit qualified personnel.
- Train and equip fire fighters adequately.
- Implement departmental leadership capable of the responsible supervision of its personnel.
- Provide adequate training and educational opportunities for potential administrators.
- Seek and deploy resources as efficiently as possible to provide for an effective response.

not underestimate the implications of that statement. If fire protection services are to be provided as a publicly funded service, and the public is led to believe that their safety and that of their property is adequately protected, then there exists an ethical obligation for the department to effectively and efficiently provide that protection. Funding, logistics, and time constraints may be legitimate barriers to performance, but they cannot be an ultimate excuse for not meeting ethical obligations. A local fire department or fire district has a contract with the citizens it serves. The provision of effective and competent respondents represents the minimum level of responsibility in meeting the terms of the contract. Anything less is a breach of trust.

Every day across the country, fire departments are struggling to meet the aforementioned minimum responsibilities. Tight budgets, staffing issues, and ever-increasing service demands are conspiring to make the provision of public safety more and more difficult. We have an ethical obligation to try to provide for public safety, but we are confined by the resources available to us. All too often the public has an unrealistic notion of its local department's capabilities.

Most citizens assume that the local fire department has adequate staffing and resources. They also assume it will arrive in minutes and be fully capable of meeting any demands made on it. If those expectations are inaccurate, it is incumbent on the local fire department or district not to facilitate false expectations. The public has a right to know what the department's capabilities are.

The American fire service has invested a great deal of time and effort into fire fighter safety. The Everyone Goes Home initiative (National Fallen Fire Fighters Foundation, 2017) is focused on changing fire fighters' mind-sets regarding safety. Fire fighters' health and welfare are critically important and even ethically mandated. However, the inclusion of risk benefit and risk management is not exactly in line with the common belief of the "heroic" fire fighter who will risk all odds and dangers to save people and their property. We have a responsibility to educate the public that there are limits to our capabilities and to how far fire fighters will push the envelope Figure 6-2.

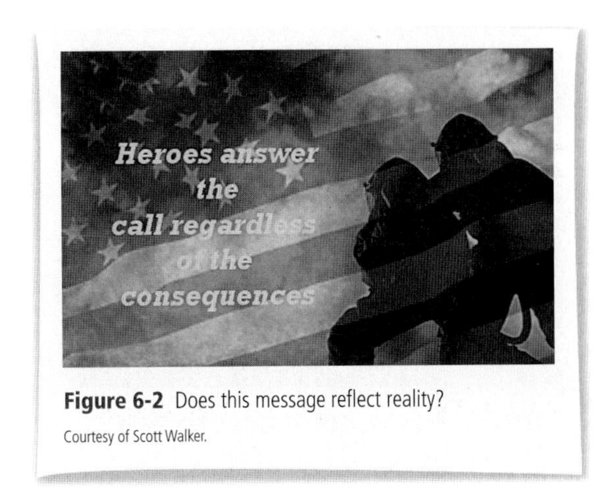

Figure 6-2 Does this message reflect reality?

Courtesy of Scott Walker.

An Obligation to Fire Fighter Safety

It is no secret that the fire fighter's work can be risky. Just as fire fighters act in the best interest of the public, they have a right to assume that the department is looking out for their best interests. Fire departments have an obligation to provide adequate staffing to undertake operations safely and, conversely, to limit operations deemed excessively dangerous due to a lack of resources. Fire fighters can also expect well-trained officers, making intelligent risk-benefit decisions. It is reasonable to assume that the fire fighter will undertake reasonable risk to protect the well-being of people he or she serves. It is also reasonable to assume that the department has an obligation to limit that risk as much as is reasonably possible. Even in today's fire service culture of heightened safety awareness, fire fighters are still conducting interior attacks in buildings of limited or no economic value. Too often fire fighters are being asked to conduct interior fire operations alone, or without a Rapid Intervention Team (RIT) team standing by, or without proper ventilation. If a fire department has an ethical obligation to protect its fire fighters' safety, then risking fire fighters' health and safety in needless or under-resourced operations is inherently unethical Figure 6-3.

Figure 6-3 Fire fighter safety is a prime responsibility.

It has been clearly demonstrated that the conditions and environments in which fire fighters work have significant long-term health consequences. If we are to ask our fire fighters to risk their safety, health, and well-being, then we must take all reasonable steps to protect them. Those steps include the implementation of annual physicals, as well as access to mental health resources and wellness programs.

Failure by governmental and departmental administrators to provide reasonable safety measures for fire fighters is tantamount to a breach of trust. We ask fire fighters to risk much because much is at stake. Yet it is an absolute truth that a fire fighter's life is as valuable as any citizen he or she may serve. Financial considerations cannot be an ethical justification for failure to provide reasonable health and safety provisions to fire fighters. Where resources are inadequate to provide for safety, high-risk operations must be scaled back to minimize risk to an acceptable level. From an ethical point of view, asking people who are risking their lives to do more with less is not only dangerous, it is irresponsible.

Please note, the ethical attachment of resource allocation only applies specifically to high-risk operations. While reduced staffing may place hardships on fire fighters performing nonhazardous duties, including fire prevention activities, those hardships are not necessarily unethical. Certainly, arguments regarding fairness, working conditions, and efficiency may have merit, but those arguments do not necessarily demonstrate a breach of ethical responsibility when life is not at stake.

Manage Resources Responsibly

A **fiduciary** is a person or an organization that holds a position of trust or confidentiality (Lemke and Lins, 2014). Fiduciary responsibility usually involves the handling of money. As an agency that utilizes public funds, fire departments and fire districts have a fiduciary responsibility. With that responsibility comes several attendant ethical obligations to the public. Certainly, fire departments must follow federal and state ethics laws regarding bookkeeping, expenditures, purchasing practices, and transparency requirements. However, our obligations to the public go far beyond those minimum legal requirements.

Clarification

Financial considerations are not an ethical justification for failure to provide reasonable health and safety provisions to fire fighters. From an ethical point of view, asking people who are risking their lives to do more with less is not only dangerous, it is irresponsible.

Fire department operations are expensive. They represent a significant portion of municipal and county budgets. In many jurisdictions public safety, including both fire protection and law enforcement, can represent up to one-third of a community's budget. It is well known that city administrators are often frustrated with emergency services' costs, including the fire service. Every dollar spent on fire protection is a dollar that cannot be used to build infrastructure, attract new businesses, and supply other services such as trash collection. Still, cities and counties budget large amounts of money for the provision of fire protection because it is critical to the quality of life in the community.

Clarification

Fire departments, including administrators and division chiefs, have an ethical responsibility to produce honest budgets based on justifiable need.

Regardless of their level of commitment to public safety, administrators must constantly balance the cost of ideal protection versus the economic realities of a community. In that pursuit, the fire service has an obligation to be a responsible partner in that decision-making process. The ideal would be to have a fire station in every neighborhood, and a relatively new fire apparatus fully staffed according to NFPA standards. The sad truth is that most cities cannot afford that, and so it becomes incumbent on the fire service to responsibly assist local government in planning for the most cost-efficient level of service that it can possibly afford.

Unfortunately, some departments approach budgeting almost like the child's game of Hungry Hungry Hippos. The philosophy of "Get as much as you can while you can" can lead to irresponsible, and even unethical, financial practices. Fire departments, including their administrators and division chiefs, have an ethical responsibility to produce honest budgets based on justifiable needs. We must constantly assess the difference between what we need and what we want.

Certainly, departments must budget for unforeseen emergencies. Unfortunately, as budget years approach their close, some departments go on spending sprees assuming that if they "don't use it they will lose it." This is an example of confused priorities. As department members it is easy, and even necessary, to develop a sense of responsibility for the organization's welfare. While protecting the organization is appropriate, it should not supplant the primary obligation of the fire department, which is to act in the public's best interest. Capriciously spending down budgets for the sole purpose of protecting future allotments is unfair to taxpayers and represents a breach of trust.

Hand in hand with responsible budgeting is the obligation for proper maintenance of assets. It is the responsibility of every department member at every level to maintain buildings, vehicles, and equipment to maximize their life span. The obligation to "take care of" department assets applies to everything the department owns. Most fire fighters take good care of their bunker gear, their apparatus, and their tools. Unfortunately, the same respect is not always extended to furniture, workout equipment, and office equipment. Regardless of its perceived importance, everything a fire department "owns" is really other people's stuff. How often are fire axes left at fire scenes, or firehouse furniture damaged from misuse? How often are things replaced rather than repaired? As inconvenient as it may be, the public has a legitimate right to expect a reasonable amount of frugality from publicly funded departments and employees.

Keeping Current

One of the fire service's true strengths is its respect for tradition. Our traditions remind us of who we are; where we came from; and, most importantly, why we are here. Our traditions give us roots and, similar to roots, they provide us with a foundation for growth. Our traditions must not be a hindrance to growth and change. If used properly, new ideas, methods, and technology can make us more proficient, safer, and more cost efficient. The fire service has an ethical obligation to provide quality service and to utilize resources responsibly. It follows then that we likewise have an ethical obligation to employ methodology and technology that enhances those capabilities.

These responsibilities go beyond the employment of the latest high-tech "gizmos." As users of taxpayers' money, departments should be willing to explore all aspects of their methodology. That exploration includes, but is not limited to, emergency response and business practices. To most efficiently serve the public, a department must be open to change. Concepts such as private/public partnerships, mutual aide, training consortiums, and even departmental reorganizations should be approached with an orientation toward the public welfare. Certainly, fire department consolidations and reorganizations can be threatening to a department's self-interest, as well as to that of the individual fire fighter. Remember, however, that our mission statement is to save lives and protect property. It is not to save lives and protect property and protect the fire department's interests. Harking back to Lippmann, doing that which is in the public's best interest is not only rational but an ethical requirement. Strategic planning must focus on the mission statement, and so we must approach it with an open mind regarding the implementation of technology and innovations in tactics, management, and administration.

Clarification

Strategic planning must focus on the mission statement, and so we must approach it with an open mind regarding the implementation of technology, and innovations in tactics, management, and administration.

Texture: Eky Studio/ShutterStock, Inc.; Steel: © Sharpshot/Dreamstime.com

Change can be scary, unpopular, and tough to manage. However, departmental administrators have the obligation to lead. Leadership often means taking people where they do not necessarily want to go. If innovation produces greater levels of service or efficiency, it is leadership's duty to utilize it and the rank and file's duty to support it.

Leading Community Risk Reduction

The fire service's core mission is fire protection. It is well known that the most efficient way to protect the community from fire is to prevent it from occurring. Few if any fire professionals would argue that fire prevention activities are the most efficient and effective methodology employed by fire departments to protect the community.

If we have an ethical responsibility to provide fire protection and if fire prevention activities are a more efficient way of doing so, then we have an obligation to pursue fire prevention activities aggressively. Ironically, even though it is the most efficient way to meet our mission, this obligation has considerably less support than fire suppression. Lack of public support is rooted in a lack of understanding. At an elementary level, the public understands that if their house is on fire, they need someone to put it out. However, the immediate benefit of code enforcement and inspections are somewhat less apparent. The public often does not entirely connect risk-reduction activities with their personal safety, nor does it emotionally embrace risk-reduction activities as it does emergency response. This is consistent with human behavior. Going to the hospital when having chest pains is of central importance to someone having a heart attack, yet preventing a heart attack through diet and exercise has less urgency because it lacks immediacy. A lack of immediacy often makes fire prevention tough to justify to elected officials and the public because it involves placing a value on a nonevent. Unfortunately, too many departments make the same mistake. Certainly, they outwardly extol the virtues of fire prevention activities, yet when budgets force tough choices, fire prevention activities are often some of the first programs sacrificed by departments Figure 6-4.

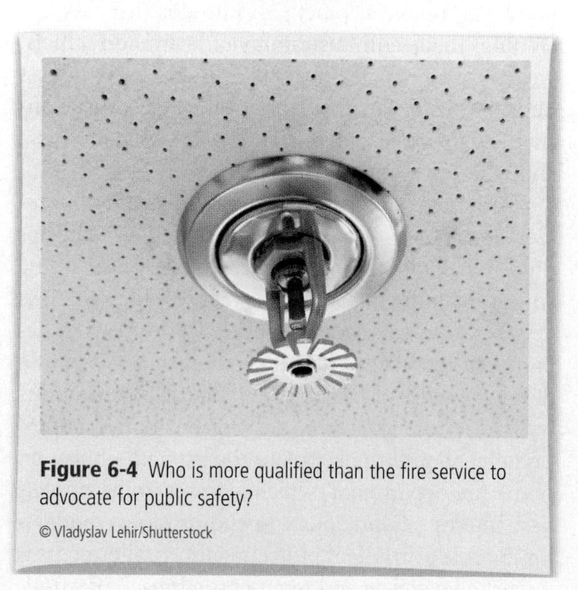

Figure 6-4 Who is more qualified than the fire service to advocate for public safety?

© Vladyslav Lehir/Shutterstock

Lack of support for fire prevention activities by fire suppression personnel is usually not about a lack of understanding. It is rooted in self-interest. Fire fighters like fighting fire. It is challenging, it satisfies ego, and its results are instantaneously understandable. Fire inspection and code enforcement lacks glamor and can be tedious, and the benefits of effort are not readily apparent. In short, well-done inspections do not make the evening news, and no one makes hero posters for fire marshals. The attitude is understandable, but is it consistent with our obligations?

A department exists to protect others. Over 300,000 fire fighters are paid to protect others. Departments have an ethical obligation to protect others. The barriers to initiating robust risk-reduction programs require vigorous leadership. It is a reasonable assertion that fire departments not only have an obligation to perform fire prevention activities, but they also have an objective obligation to be visible advocates for fire safety and the implementation of fire-safe building practices. As fire sprinkler systems represent the single most effective life-safety feature in the building, we have a special obligation to be vigorous advocates for their mandatory installation. If a local fire department does not lead the initiative to make a safer community, who else will?

While fire protection is our core undertaking, it is not our only undertaking. The last 30 years have seen a shift in fire department operations to an all hazards response platform. The emergence of EMS as a fire department function, as well as hazardous materials teams and technical rescue teams, has broadened the scope of fire department responsibilities. With these responsibilities come obligations to educate and advocate for public safety initiatives beyond fire safety. Many fire departments are undertaking advocacy programs for child safety seats, seatbelts, vaccinations, safe driving, and public first aid Figure 6-5. While these programs go beyond public expectations, they are consistent with the modern fire service's role in the community as a benevolent agent.

Figure 6-5 The fire service exists to serve the public.

Being a Positive Influence in the Community

If you accept the premise that being a fire fighter is something more than just a job, then clearly additional ethical obligations exist. And if you accept that a fire department represents something more to a community than just a public service akin to trash collection or street maintenance, then the following question arises: Does the fire service have an obligation to be a positive influence in the community? As repeatedly stated throughout this text, both fire fighters and the public jointly embrace the image of the fire service as a cultural icon. If we support that image, does that not constitute a promise? If so, then does that promise constitute an obligation? Working under this assumption, what does this obligation entail?

Diversity

You can argue that the community's sense of ownership in its fire department creates a departmental responsibility to represent equally all members of the citizenry. As such, it is reasonable to assert that a fire department has an obligation to provide equal access to departmental membership. Beyond legal requirements, a fire department has an ethical obligation to utilize fair and objective criteria for employment or membership. It is also arguable that a department has an obligation to actively recruit representative members of the community. However, remember that the fire service has a primary ethical obligation to save lives and protect property. As a result, recruitment and selection standards must not interfere with the department's ability to perform.

Diversity initiatives are rooted in the department's obligation to interact fairly with the community it serves. The same requirement mandates transparency—not only in selection practices but in all facets of operations. We act on the public's behalf, and the public should have input on the types and levels of service provided. The department also has an ethical obligation to provide equal access to services, and service levels should be fairly distributed throughout the community based on need.

Figure 6-6 Alan Brunacini, 1937–2017, was a retired fire chief, author, and lecturer.

Courtesy of Nick Brunacini.

Community Service

In large part, due to the work of former fire chief Alan Brunacini, the phrase "customer service" has become pervasive in the fire service vocabulary. Throughout the 1990s, articles, conference presentations, and leadership seminars all stressed its importance. In Chief Brunacini's book, *Essentials in Fire Department Customer Service* (Brunacini, 1996), he stressed the importance of the fire service embracing some concepts of customer service used in the private sector. These concepts include professional excellence; politeness; and respect for the dignity, privacy, and property of our customers (Figure 6-6).

Beyond the idea of excellent customer service, many departments have expended a great deal of energy into becoming integral parts of the community's social fabric through **community-based initiatives**. Community-based initiatives are programs that are not directly related to public safety but rather are in support of improving quality of life within the community.

A fire department may undertake these initiatives alone or in partnership with other agencies. Examples of these initiatives include participation in Big Brother/Big Sister programs, literacy programs, partnerships with charitable organizations, and various community events and festivals. The rationale for these activities runs along two general schools of thought. The practical justification is that the greater the affection the public has for the fire service, the more influence it has at city hall. The second school of thought argues that the fire service is essentially a public service agency and that, if it has a capability to provide a service, it should—it is just the right thing to do.

The adoption of community-based initiatives has sometimes been controversial. Its supporters have argued that it is really just an extension of what fire fighters have been doing for years. Individual fire fighters and the International Association of Fire Fighters have a long history of community service and charitable work. Common examples include financial support for burn camps, Meals on Wheels, and the St. Baldrick's Foundation. Most notable is the Fill the Boot campaign in support of the Muscular Dystrophy Association every Labor Day weekend. Supporters ask, Why shouldn't fire departments officially become involved in what fire fighters are already doing privately? Some feel that community-based initiatives are simply in keeping with the long tradition of the fire service. They can accurately point out that the history of the fire service began first as citizens coming together to serve

Figure 6-7 The Fill the Boot campaign is representative of the long history of the fire service's community service programs.

and protect their community, not as a profession but out of a sense of compassion and duty **Figure 6-7**.

Critics argue that community-based initiatives should remain in the hands of off-duty fire fighters and unions. They assert public community initiatives should not become official fire department policy because they are not consistent with its fundamental mission statement. As one fire fighter put it, "You do not see accountants and real estate agents standing on corners collecting money on Labor Day weekend." Others suggest that the "Saint Fire Fighter" image sets a nearly impossible behavior standard. As a fire chief once said, "The public begins to see us as guardian angels, not as people doing the job. Sooner or later we're going to let them down and there's going to be hell to pay."

Some have argued that customer service initiatives can detract from response capabilities by diverting time and resources away from the department's core mission. As a result, community-enrichment programs are not only not ethically required but may be actually unethical because they distract from a preexisting ethical obligation. The argument is certainly a compelling one if customer service and response efficiency are mutually exclusive. Whether the two initiatives can coexist must be judged independently from department to department based on its resources.

Logically, you can draw a straight line from a fire department's duty to respond efficiently and its ethical obligation to do so. The negative consequences of failure, along with the public's expectations, make a fairly clear connection between performance and ethics. The same can be said for the concepts of fire fighter safety, fiduciary responsibilities, and risk-reduction initiatives. The connection between ethical responsibilities and community-based customer service initiatives is not so clear. Customer service and community-enhancement programs can be a good idea; they enhance public image, raise morale, and even have political benefits. But is there an ethical obligation to do "good" in the community?

The theory of care ethics (see Chapter 4, *The Philosophy of Ethics*) would imply that any agent or agency that is in a position to do good has a responsibility to do so. Social contract theory would suggest that public agencies have an inherent responsibility to improve the lives of its citizens. It is the comfortable fit of the fire department's mission with that of community service that has urged so many departments to focus on community-enrichment programs.

However, you could argue that deontology would support the notion that if no specific duty is attached, and others are capable of providing the service, then perhaps no obligation exists. This notion would support the idea that customer service initiatives may be a good idea but not necessarily an ethical or moral imperative.

Certainly, the public has a right to form expectations and, if the public wishes something, the department may feel compelled to provide it. However, this does not necessarily constitute an ethical obligation. The fire service does not promise customer service as a condition of its existence, nor will the public likely suffer direct injury by its absence.

While the attachment of ethical obligation to customer service programs remains subject to debate, it would seem that the public and its representatives have a right to define what they expect for their investment. As representatives of the taxpayers, fire chiefs have a right to set service standards to meet those expectations. Collective bargaining agreements aside (they represent a binding agreement between fire fighters in the department regarding labor conditions), it would seem fire fighters may have an obligation to meet those standards if legally imposed by the department's chief.

What Do You Think?

Are customer service initiatives just a good idea or are they an obligation?

The Fire Fighter's Responsibilities and Obligations

The Role of Roles

As we have discovered, an important concept in personal and professional ethics is that of responsibility. Our sense of personal responsibility tends to be internalized, attempting to meet self-imposed standards of behavior arising from either our sense of virtue or a subjectively described greater good. Whereas professional ethics are duty based, either as a condition of professional membership or out of a personal sense of obligation. The ethical responsibilities of a fire fighter are rooted in our obligations to the public. However, fire fighters are more than just public servants.

Like most people, fire fighters lead complex lives. They have a myriad of obligations and responsibilities as a condition of their employment, but they also have responsibilities to family and friends, and as members of society.

Social scientists explain that our many obligations and responsibilities are grouped into what they call **social roles**. A role is defined as a form of personal identity associated with particular responsibilities and obligations. Examples of roles include: spouse; parent; friend; and, of special interest to this discussion, fire fighter. Each of these roles comes with expectations of behavior that impart specific obligations or responsibilities. As we learned in Chapter 3, *Influencing Behavior*, responsibilities can either be imposed externally (objectively) or internally (subjectively) (Cooper, 2012).

One of the fundamental ethics problems that fire fighters deal with is the competing demands of their various roles. At times fire fighters may likely feel as though they're serving many masters. The demands of home life, career, fellow fire fighters, and their own personal interests all seem to compete for attention.

Two fundamental causes of ethical dilemmas and questionable behavior are, first, an inconsistency between the responsibilities associated with our assumed roles and, second, an inconsistency between obligations and our personal values, beliefs, and needs.

An excellent example of conflict between roles and personal values garnered national attention in 2015. Kimberly Davis, a county clerk in Rowen County, Kentucky, defied a federal court order to issue marriage licenses to same-sex couples (Blinder, 2015). She contended that her religious beliefs prohibited her from sanctioning same-sex marriage. While many supported the relative validity of her *moral* stance, the *ethics* of the situation are relatively clear. She accepted a position in county government. As a condition of that appointment she agreed to uphold that county's laws and regulations. This creates an obligation to perform her duties as prescribed by law. As a county clerk (like a fire fighter) she has an ethical obligation to fulfill the duties that she willingly undertook. If her conscience prevents her from carrying out those duties, she can either seek to legally address the issue or resign. She does not have the option of accepting payment for doing a job that she refuses to do, regardless of how well-intentioned her objections may be.

This is not to say that moral objections are not valid or ethically relative. The most common defense argued by Nazi party officials at the **Nuremberg trials** was that they had an obligation to obey legal orders. As a result, they were not morally culpable for the results of laws implemented by the Nazi party. There is a long judicial and ethical history recognizing that following immoral orders is not a viable defense for committing atrocities. Hence, we see a dilemma. Moral obligation can and often does conflict with the responsibilities associated with social roles.

A second source of dilemma is between competing roles. An example of this is a fire fighter who must choose between an opportunity to attend the National Fire Academy for specialized training or attending a friend's wedding. The problem with role conflicts is that there is no objective way to prioritize competing roles. Professional development and friendship are both important facets of someone's life. The responsibilities associated with each role are likely viewed as equally important. Often, we are forced to prioritize responsibilities using criteria that are completely unrelated to our personal values, such as cost or convenience. These decisions can cause a great deal of internal emotional turmoil.

For public servants, understanding the responsibilities of their position (roles) becomes critical, as does an understanding of their own personal values and beliefs. By understanding both facets of our ethical selves, we can avoid imagined conflicts and better deal with real ones.

Objective Responsibilities

Objective responsibilities are those that are imposed on us from an outside source. Most exist as a condition of our employment, or membership in a department. As such, objective responsibilities are closely associated with duty-based ethics or **deontology**. Ethic obligations are attached because we willingly agreed to the department expectations when we voluntarily joined it. We form a contract with the department and, by extension, with the citizens it serves. Hence, we have a duty to meet expectations. We also share the ethical responsibilities of the department itself, which arise from the department's promise contained in its mission statement. The fire fighter's objective responsibilities related to the department's mission are numerous. The following are some key objective responsibilities.

The Obligation of Competency

Fire fighters routinely deal with life and death situations. Whether at the fire scene, an EMS call, or technical rescue, citizens count on our competency. As fire fighters, we have made a tacit promise to the public to respond efficiently and effectively. The import of the consequences of our actions coupled with our agreement to undertake these responsibilities imparts upon us a significant ethical obligation to perform competently. Being unprepared or unwilling to meet our competency obligation is a breach of trust not only to the public but to our fellow fire fighters who likewise count on us. The obligation for emergency response competency insists that we treat every classroom session and training session as if lives depend on it Figure 6-8. Because so many of our nonresponse duties affect our ability to perform competently, even mundane duties take on significance. Every morning inventory and maintenance assignment must be given the respect that each deserves. With so much on the line, the smallest details matter.

Considering the nature of our work, and the consequences of our actions, it is fair to say that "good enough" is probably not good enough. A department has a right to expect the fire fighter to be a

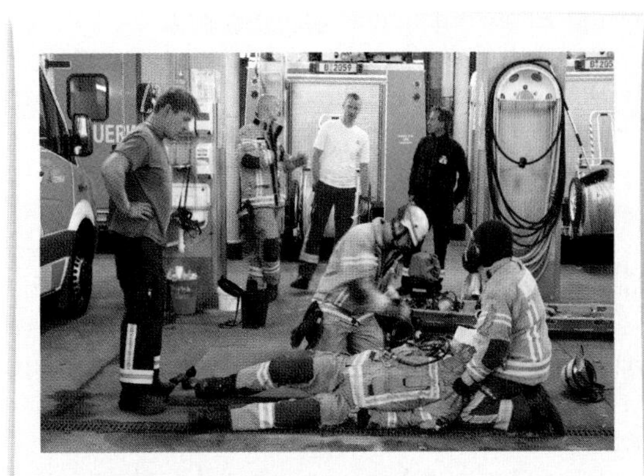

Figure 6-8 The obligation for emergency response competency insists that we treat every classroom and training session as if lives depend on it.

lifelong learner, and to grow and develop skills and knowledge continually. Again, people depend on fire fighters, and fire fighters willingly agreed to answer the call. As such they have an obligation to bring their best game every day.

Competency goes beyond emergency response. All disciplines within the fire service matter. Whether it is inspections, code interpretation, hydrant inspections, prefire planning or public education details, all tasks that fire fighters do have significant potential consequence. At times nonemergency assignments can seem mundane and unrewarding. It is important to understand that doing them is not only a mark of professionalism but an ethical obligation under the same terms and conditions as is emergency response.

An Obligation to Safety

Earlier in this chapter, the obligation of fire fighter safety was explored as a departmental responsibility. It is important to understand that this same ethical obligation applies to every fire fighter as well. The department owes the fire fighter as safe a working environment as can be reasonably provided. The obligation to fire fighter safety is both a moral and ethical obligation as much rooted in humanity as responsibility. Likewise, the fire fighter must accept responsibility for his or her safety; this obligation arises from a responsibility to others, including family and fellow fire fighters Figure 6-9.

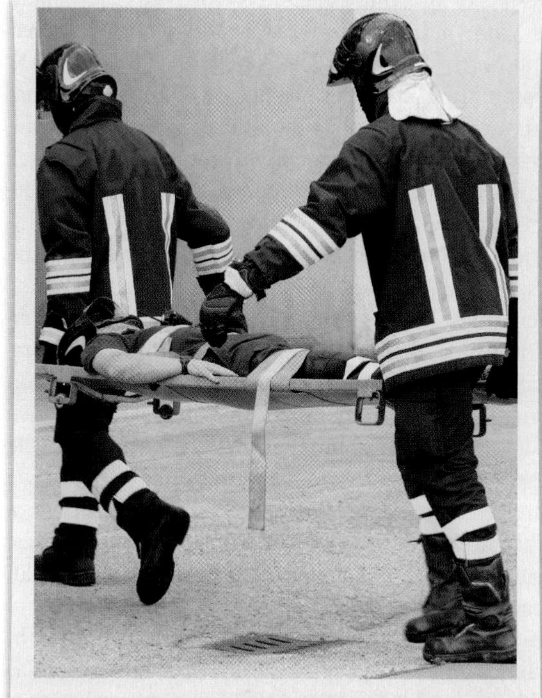

Figure 6-9 A fire fighter must accept responsibility for his or her own safety.

Injuries inhibit fire fighters from performing their duties. Firefighting is difficult work. Injuries either compromise or even prohibit fire fighters from doing their job effectively. Their injuries can and usually do have negative impacts on their fellow workers.

An injury-related absence can cause reduced staffing or backfilling by someone unfamiliar to the rest of the crew. An injury can inhibit team cohesiveness and reduce team productivity. Reduced staffing can also pose a significant risk to the crew's remaining fire fighters.

Injuries can have an emotional and financial impact on the fire fighter's family. Extended absences from work can, in some cases, have dire financial consequences. Further, extended convalescence redistributes family duties to spouses and children. For young children, a parent is a soul source of security. Seeing a mother or father significantly injured can have long-lasting traumatic effects. For many reasons, fire fighters have an obligation to protect themselves from injury for their family's sake.

Injuries are expensive. Fire fighter injuries can, directly and indirectly, cost a department significant money. Beyond medical bills, on-duty injuries can increase workers' compensation rates, insurance rates, and overtime. In part, these are funds that otherwise could be used for equipment and training.

It is more than a feel-good platitude to say that every fire fighter counts. Fire fighters' families depend on them, their departments depend on them, fellow fire fighters depend on them, and the public depends on them being there. A fire fighter's obligation for his or her own personal safety goes beyond self-interest.

An Obligation to Policy and Authority

A department's ability to perform its functions depends on organization, clear lines of authority, and unity of command. It is the department's responsibility to issue rules and policies that support the department's mission and strategic priorities. Therefore, it is the obligation of the individual fire fighter to follow those rules and policies. This obligation rests on two principles. First, the fire fighter voluntarily enters a department and, in doing so, accepts implicitly or explicitly the conditions of membership in the department. By joining a fire department or any other organization, the member agrees to its organizing principles, its mission, and by extension the policies that support that mission. Should a policy prove illegitimate because it is contrary to the department's mission, is illegal, or is inherently unfair, a fire fighter has an opportunity for procedural redress. Barring that, rules are most definitely *not* made to be broken. Fire fighters are typically sworn personnel—this involves giving the oath to uphold rules.

Second, if a person accepts the ethical obligation of the department to protect public safety and welfare, then, by extension, the members of the department have a likewise obligation. Any action by a fire fighter that disrupts the department's ability to provide safety or service is tantamount to a dereliction of that same duty that binds the department. The department's mission of saving lives and protecting property is a tremendous responsibility, and so is supporting the officers tasked with meeting those responsibilities. Just as a commitment to the department's mission requires fire fighter support and loyalty, there exists a similar obligation to support the department's leadership, which is engaged in accomplishing that mission.

Unfortunately, rules and orders can conflict. Even the best-intentioned fire fighters may find themselves in a situation where policy and values seem to be at odds with each other. Consider the following example.

Fire fighter Smith works for the city of Anywhere, USA. Smith self-identifies as a member of the Christian Right. He strongly believes America is a Christian nation and that Islam is at war with America and Christianity. On reporting for duty one morning, Smith learns that his company has been assigned a public education detail at the local Islamic center. Smith is greatly upset as he believes that the orders conflict with his personal values, his religious beliefs, and his sense of patriotism. As a result, he refuses to participate in the assignment.

The correctness or incorrectness of Smith's views are irrelevant to his objective obligations. Like the Kimberly Davis case, Smith is in error by placing his personal views ahead of his professional duties. While he may feel a religious or moral obligation, those views remain subjective. He clearly has an objective responsibility to perform the duties for which he is paid, and, as a fire fighter, he has an obligation to support the department in its core mission of promoting public safety.

Conversely, while inspecting a comedy club Lieutenant Smith finds a dysfunctional sprinkler system. The building is owned by a friend of the fire marshal, and he tells Smith to disregard the apparent violation. In this case, Smith's obligation to public safety (remember the prime obligation?) is more important than his obligation to obedience to the fire marshal's authority.

An Obligation for Physical Fitness

Similar to competency, fire fighters have an obligation to take those steps necessary to meet the requirements of the career they chose to pursue. This obligation is essentially true of most professions, whether a person is a bank president or a checkout person at the local grocery store. Employees have an obligation to perform the duties for which they are paid. This obligation is particularly heightened for emergency workers. Beyond the quid pro quo of work for payment, fire fighters deal with life and death consequences and so have a heightened obligation to the citizenry and to their fellow fire fighters.

In order to meet those obligations, fire fighters must prepare themselves for the incredible physical demands of the profession **Figure 6-10**. Many fire fighters mistakenly believe that firefighting is an exercise in brute force. However, studies have shown physical demands of the fire fighter require a combination of cardiovascular health, strength, and

Figure 6-10 Fire fighters have an ethical obligation to be physically capable of meeting the responsibilities of their profession.

flexibility. This requires a regular exercise regimen, proper diet, abstinence from excessive alcohol and drugs, and proper sleep.

The public and a fire fighter's coworkers have every right to expect that a fire fighter will show up for duty physically capable of fulfilling his or her obligations and responsibilities. A team is as strong as its weakest member, and so every fire fighter has a responsibility to contribute her or his best efforts. This obligates the fire fighter to have body, spirit, and mind prepared for each shift.

Subjective Responsibilities

Subjective responsibilities are somewhat more difficult to define and prioritize. Unlike objective responsibilities, which are based on the expectations of stakeholders, subjective responsibilities tend to be internalized or self-imposed. As such they are also open to self-interpretation.

As subjective responsibilities are rooted in values, they can be fairly described as associated with the ethics theories of consequentialism and virtue.

A fire fighter's feelings about loyalty and tradition often differ from that of his or her coworkers. Nonetheless, responsibilities arising from a personal sense of values and priorities are as important as those based on others' values and expectations. While less definitive, there is some general consensus regarding subjective responsibilities for fire fighters; these are based on shared values within the profession.

Respecting Tradition

As we discovered in Chapter 5, *Professionalism on the Line*, traditions are important to the fire service. They give a fire department, and its members, a sense of continuity and identity. Traditions can also have a positive effect in reinforcing important shared values within a team. Traditions represent and reinforce the department's legacy and its shared culture. They should not, however, be confused with routines, habit, or organizational inertia. Work schedules, hose loads, and fire tactics are *not* traditions, but those things that celebrate the department's identity, its past, and its values are. Respect for seniority, awards banquets, shift celebrations, and swearing-in ceremonies are all examples of traditions. Many senior fire fighters bemoan a perceived growing indifference to fire service traditions. Their concern begs the questions, Is the fire service losing touch with its traditions? and If it is, is it because this generation of fire fighters has a different set of personal priorities, or is there a failure of leadership in passing on the importance of tradition?

Almost 40 years ago, on the last day of his probation, a young fire fighter was handed a piece of chalk and told to climb to the top of the central station hose tower where he was instructed to write his name and date of hire on the upper walls. There on the wall were the names of hundreds of fire fighters along with their hire and retirement dates spanning many decades. As you may have guessed, this author was the fire fighter in that hose tower. At the time, I thought the wall was interesting but not particularly important. I can promise the reader that upon my retirement date 30 years later I found myself back on top of that tower with chalk in hand. Over the years I came to appreciate the incredible importance of having a living history within a department. It would be disappointing if that tradition were not still being carried out. Shared history implies ownership, and a sense of ownership is elemental to responsibility Figure 6-11.

Figure 6-11 Traditions give department members a sense of community and pride.

There are articles that tell of more and more fire fighters no longer sharing meals together. Rather, each brings in his or her own particular food for the day and eats when the mood strikes, or when there's time. A meal may not seem important in the grand scheme of a fire department's mission, yet breaking bread together builds cohesiveness and a sense of community. Similar to ownership, a sense of belonging is also fundamental to instilling a sense of responsibility

Most fire fighters agree, in principle, that members of the department have an obligation to respect tradition. Unfortunately, traditions can become victims to competing priorities. Senior members and company officers have an obligation to carry on traditions. Cohesiveness, unity, and shared experience builds trust. Trust is essential to a highly functional fire company.

The Obligation of Cohesiveness

One of the most significant differences in fire department culture, as compared to other career fields, is that fire fighters literally live together. As a result, the concepts of team, loyalty, and cooperation are pervasive in fire department culture. Most fire fighters would agree that these "virtues" are essential to effective fire ground operations.

Maintaining a positive work environment can be challenging. A single disruptive personality can jeopardize an entire engine house's cohesiveness and even morale. Bitterness, argumentativeness, and cynicism can create a toxic work environment and destroy initiative and overall effectiveness. Certainly, fire fighters have bad days. However, negative social behavior should be the outlier, not the norm. A spirit of cooperation and sociability are nearly imperative in firehouse culture, and every fire fighter has a responsibility to uphold her or his part of team unity. Fire fighters have a right to expect a nonhostile work environment. In a culture that promotes teamwork, mutual support among team members is a reasonable assumption. Certainly, a friendly and cheerful work environment is pleasant and thus is its own reward. However, there is a tangible obligation to maintain a productive work environment where individuals continually work in close quarters.

The Obligation to Learn

In Chapter 5, *Professionalism on the Line,* we mentioned intellectual curiosity as an important virtue to fire fighters. The fire fighter's tasks are of extreme importance and, because of that, proficiency is critical. That responsibility goes beyond basic competency. Open-mindedness and curiosity are essential for personal growth and development. They allow for continual improvement and professional development.

Earlier in this chapter, the need for a fire department to remain contemporary was reviewed. There is an obligation to explore the possibilities of new technologies and methodologies. The fire fighter's attitude toward innovation is critical in facilitating his or her department's commitment to modernization. The fire fighter's willingness to explore possibilities and be open to change can facilitate the department's efforts in achieving that goal. Conversely, unreasonable resistance to change by a majority of members can entrap a department in **cultural inertia**. The ability of a department, and its individual fire fighters, to grow professionally demonstrates a commitment to placing professional growth ahead of comfort with the status quo. Fire fighters would do well to remember that everything they know was new information at one time.

Balancing Work and Private Life

Being a fire fighter is more of a way of life than an actual occupation. Certainly, for many, being a fire fighter becomes ingrained in self-identity. Being a fire fighter can make great demands on time and commitment. In addition to the normal shift, there are constant overtime opportunities, off-duty training sessions, and community events. Additionally, fire fighters often socialize together, work second jobs together, and attend department functions and meetings together. For some, the fire service can become the focal point of the fire fighter's professional and social life, even to the exclusion of family, non–fire department friends, and hobbies.

For others, the demands of family and the attraction of personal pursuits can make meeting the demands of the fire service difficult or uncomfortable. They must ask themselves, Do I attend the paramedic Continuing Education Class or do I take my wife to a movie? Do I go to the department's training meeting or the neighborhood block party? Personal life and department activities both matter. In a sense both should be priorities. It is here that the challenge lies.

Being able to balance career and family demands is difficult for almost everyone. However, the fire service can be a "demanding mistress." Fire fighters who do not have a balance can become burned out or suffer marital problems. Fire fighters who ignore professional responsibilities in lieu of personal pursuits risk failing to meet their obligations as fire fighters. Fire fighters have an obligation to maintain awareness of personal and professional demands and actively seek to balance those important responsibilities. One of the fundamental reasons we study ethics is to gain a better understanding of competing responsibilities and obligations, and hopefully to develop appropriate coping mechanisms.

The Obligation to Give Back

The importance of leadership cannot be overstated in the fire service. A fire fighter who comes to actualize his or her potential fully has learned much along the way and has much to offer. However, many of our best and brightest turn away from the responsibility of promotion. This is may be understandable in that the financial rewards of promotion rarely compensate for the added responsibilities. Most high-ranking officers will tell you that it is more "fun" to be a fire fighter than an officer.

Does a fire fighter have an obligation to "step up and give back" through leadership positions? A principle of ethics is the relationship between rights and responsibilities. Having the right to refuse something does not necessarily negate the existence of a responsibility to do it. While certainly individuals have a right to choose their own career paths and the level of responsibilities they are willing to undertake, that right does not necessarily imply that there is not an obligation to undertake leadership roles.

If the best and brightest of our fire fighters turned their backs on promotional pursuits, the department they serve would suffer from their lack of contribution; as such, so does its ability to function effectively. Assuming that we acknowledge that the fire department and, by extension, its members have an obligation to provide for its citizens' safety and welfare, then it follows that qualified fire fighters have an obligation to provide leadership when needed.

If a fire fighter's fire department, battalion, and company are efficient and professional, it is due to the leadership of those who came before him or her. They benefit from those efforts and, as such, there exists an obligation to contribute to continued future success.

Further, if a fire fighter's department needs improvement, and the fire fighter is capable of doing so, doesn't that person have an obligation to contribute to its improvement? Some departments have histories going back more than a century; they are steeped in tradition. Those traditions are the result of one generation of fire fighters after another shouldering the responsibilities of leadership. As a fire fighter benefits from the efforts of those generations, he or she has an obligation to contribute to the department's future success.

Fire fighters gain skills, knowledge, and experience because of the effort and commitment of others to teach them. You can argue that acceptance of leadership and mentoring, without being willing to do the same for others, is self-serving. If a fire fighter personally benefits from leadership by those who were willing to accept that responsibility, there is a tangible obligation to pay it forward.

Finally, a fire crew depends on the competency of its leaders to make intelligent decisions at emergency scenes. An officer's skills, knowledge, and experience can mean the difference between a fire fighter going home after a shift or going to the hospital. If a fire fighter has the qualifications and requisite skills to be an officer, and those qualities can contribute to the safety of her or his fellow fire fighters, there is an obligation to render those services.

Loyalty

Loyalty is a virtue and a highly prized one in the fire service. However, it is so ingrained in our culture that it is not only a virtue to aspire to but an expectation as well. Fire fighters work in close quarters with each other, and they spend a significant amount of time with each other. The nature of their working conditions and the dangers that fire fighters face together create strong bonds between crew members. Loyalty is important in the fire service and, if properly focused, it promotes a sense of belonging and trust. These are key elements in teambuilding, and firefighting is and must be, a team effort. The virtue of loyalty imposes certain obligations on fire fighters, including the willingness to commit to competency. Fire fighters trust each other to perform their duties at emergency scenes. That trust is essential, making the commitment to excellence an act of loyalty. Loyalty also requires having each other's back, helping when asked, listening when necessary, and stepping in when needed. Loyalty requires fire fighters to protect each other's safety, health, and personal interests, as well as their own. However, loyalty can never be an excuse to cover up others' unethical acts. To help conceal others' unethical acts is to enable them. By its very definition, coverup is unethical. As repeatedly stated throughout this chapter, the first and most important obligation of the individual fire fighter is to act in the public's interest. This obligation supersedes loyalty to the department, or to individual fire fighters if they are acting contrary to the public good.

The Responsibility of Image Protection

Imagine six castaways floating in a large rubber life raft in shark-infested waters. Typical of people confined in a small space, each of the castaways has eventually "claimed" part of the boat as his or her space. After a few days, castaway "Bob" begins amusing himself by poking the rubber raft with a sharp stick. The other castaways object to the behavior because possibly puncturing the raft puts them all at risk. Bob retorts that it is his section of the raft that he is poking and he has a right to do it if he wants. Does Bob have a right to "poke the raft," or is there an obligation to refrain from actions that may negatively affect others? Most readers would recognize a social responsibility not to imperil others. You can apply the same theory to an obligation for reputation management.

Earlier, this text explored the contribution that a positive public image makes to the fire service's overall success. The responsibility of individual fire fighters in contributing to that public image should be obvious. If the fire service has an obligation to fulfill its mission, then individual members have an obligation to engage in actions that contribute to the success of that mission, including maintenance of a positive public image.

The inherent responsibilities of reputation management go beyond the practical obligation of supporting the fire service's mission. An argument can be made that individual fire fighters have a debt to the legacy of their individual departments Figure 6-12.

Most fire fighters are proud of being fire fighters. That pride is partly due to the nature of the work they undertake. Fire fighters also take pride in the respect and affection the public has for the profession. Many fire fighters also take pride in the values and traditions associated with the fire service and their individual departments. This pride is based on a long history of service, bravery, and self-sacrifice. That source of pride is the result of the honorable efforts of

Figure 6-12 Fire fighters have a responsibility to contribute to and protect the public image of their department and fellow fire fighters.

millions of past and contemporary fire fighters. The action of every fire fighter contributes to or detracts from that legacy. Every fire fighter honored for bravery, or recognized for an act of kindness, helps ennoble all fire fighters. Conversely, the foolish, selfish, or unethical actions of some fire fighters tarnish the reputation that so many have worked to build. Like it or not, the actions of any given fire fighter reflect on the whole profession, especially in this day of social media and an increasingly cynical population.

If our actions can have a negative impact on the entire profession and the legacy that others built, is there an obligation to protect that reputation and legacy? Similar to Castaway Bob, if a fire fighter's actions can harm others, then that fire fighter has an obligation to refrain from those actions.

The Special Case of Union Leadership

This author spent many years as a member of the International Association of Fire Fighters and served as an executive board member of one of the local chapters. Fire fighter unions, like all unions, play an important role in protecting the safety, working conditions, and rights of fire fighters. With that said union leadership can face significant conflicts of responsibility, particularly in the fire service.

Union leaders are often fire officers. The role as representatives of fire fighter rights can, at times, become uncomfortably close to conflicting with obligations of maintaining organizational discipline. Further, union leaders negotiating contracts may have to balance their obligation to represent the personal and financial interests of fire fighters against what they may professionally feel are legitimate administrative initiatives for new services. Imagine a fire chief wishing to initiate or expand EMS services. Captain Smith may feel that the extension of EMS services is good for the public and supports it. On the other hand, Captain Smith as a union president may likely feel an obligation to file a grievance against the initiation of a new EMS program as it represents a change in working conditions, resulting in an increased workload without compensatory pay increases.

As union officials, union officers have an ethical obligation to represent fire fighters' interests. However, as fire fighters, they have obligations and responsibilities to the well-being of the department and the success of its mission. Both responsibilities are real, both responsibilities are compelling, and failure to meet those responsibilities can have significant negative consequences. Fortunately, enlightened leaders in both the International Association of Fire Fighters (IAFF) and the International Association of Fire Chiefs (IAFC) have worked together to educate union leaders and fire officers about how to ethically manage both interests.

It is also fortunate that in most cases fire administrators and union officials likely share the same values. Both should be committed to fire fighter safety. Both should want fair pay. Both should want quality working conditions. As long as labor and management deal with each other honestly and respect the obligations that each have, ethics conflicts should be avoidable in most cases. Still, it is imperative that union members have a clear understanding of their ethical obligations both to the union and to their individual departments. The attending legal and ethical dilemmas represented within both positions should be anticipated and well understood.

Respecting Privacy

Respecting and protecting the privacy of those we serve is both an objective and subjective responsibility.

Objectively, there are restrictions that limit a fire department's prerogative to release personal information. An example of this is the **Health Insurance Portability and Accountability Act of 1996 (HIPAA)** ("Summary of the HIPAA Privacy Rule," n.d.). HIPAA legislation restricts the release of a patient's medical and personal information within very narrow parameters. Any department that provides emergency medical services is bound by the HIPAA regulations. Other regulations exist that protect financial information, Social Security numbers, and other types of personal information as well.

Subjectively, we have a humanistic responsibility to protect the privacy of those we serve. Fire fighters routinely interact with people who are having the worst day of their lives. We often enter homes when people are not expecting company. On medical calls people routinely must share personal information regarding the conditions that would be highly embarrassing if publicly known. A fire fighter with any amount of experience can tell many stories relating the amazing, comical, and tragic ways that people live their lives. It is important to remember, however, that the fire service's customers are trusting fire fighters with nearly unlimited access to their lives. They do this because they must, not because they want to. They depend on us for help. They rightfully assume that we will cause them no harm. They have every right to feel betrayed if we cause embarrassment or worse by breaking that confidence.

Because of the nature of our work, people are constantly interested in what we do. Fire fighters are routinely asked to share details about the calls we run, and it's only natural to want to share a good story with interested friends and family. However, sharing the stories can be destructive to our ability to do our job. If the public does not feel that they can trust us to keep their confidences, they are more likely not to seek help. Imagine a mother who calls an ambulance to care for her sick child. Further imagine that her home is in a complete state of disarray—dirty dishes everywhere, the floor full of clothes, toys, and everything else imaginable. For a fire fighter, this is not hard to imagine. They see it fairly routinely. Now, imagine a week later that same mother comes across a shared Facebook picture of her very messy home accompanied by several humorous and judgmental comments. She is embarrassed and angry. Is it not reasonable that she will be reluctant to call for help the next time her child is sick or injured? In this likely scenario, a child suffers because of the fire fighter's insensitivity.

Fire fighters must avoid gossip. At its face, it is wrong because it takes advantage of those who are in need. Beyond that, it can impair a department's ability to perform its duties. Finally, it displays a lack of personal integrity and empathy. The more calls fire fighters answer, the more likely they are to be desensitized to the elements of humanity they are dealing with. In other words, homes become tactical problems, patients become a set of vital signs, and personal tragedies can become nothing more than interesting calls. A loss of empathy can cause even well-intentioned fire fighters to lose touch with the consequences of an unfortunate remark.

Not long before this book was written, two medics became entangled in a scandal, resulting from a "friendly" contest. They began taking pictures of themselves with unconscious patients. It is likely that these medics considered themselves good people and probably never thought about how taking pictures of "selfies" with patients would look to others. Still, consider how you would feel if a medic took a picture with your unconscious daughter, son, or spouse. You would feel outraged and betrayed. The point of the story is that these medics may not have been "bad" people. They may have just lost perspective on the responsibility to respect the privacy and dignity of the people they serve.

Clarification

A sense of obligation to the well-being of others, to their humanity, and to their privacy reflects on the personal integrity of the fire fighter.

WRAP-UP

Chapter Summary

- This chapter focuses on responsibility as it applies to ethics, opening with a discussion on public expectations of the fire service and its local fire departments.
- A comparison was introduced between the fire service's reputation and the reality of the promised service levels. A review was conducted of those activities the fire service is actually obligated to provide the public, and from where those obligations spring. Particular attention is paid to the fire service's express mission statement, "To Save Lives and Protect Property."
- Fire departments have specific responsibilities to their local jurisdictions. The first and foremost responsibility is to serve the public interest through competency in its core mission.
- Other obligations include the provision for fire fighter safety, the responsible management of public resources, adaptability, a commitment to community risk reduction, and a culture of public service.
- Fire fighters, like other members of the community, have a myriad of responsibilities to numerous tasks and individuals. Those responsibilities can be grouped into "roles."
- Fire fighters have many roles beyond that of being a fire fighter. Those roles include family member, spouse, parent, friend, community leader, and self-care provider.
- The many obligations and responsibilities that fire fighters face can create situations in which competing priorities demand hard choices.
- Fire fighters must seek balance in their professional and personal lives to remain productive, as well as physically and emotionally healthy.
- Individual fire fighters have a responsibility first to the public they serve, and second to the department, including its other members. Those responsibilities can be classified as objective or subjective.
- Objective responsibilities are those that are imposed externally. They may include conditions of employment, legal obligations, or performance expectations by stakeholders.
- Subjective responsibilities tend to be more internalized. They are those feelings of obligation fire fighters impose on themselves as a result of their personal values, beliefs, and sense of virtue.
- Fire fighters' objective responsibilities include an obligation for competency in the essential skills and methodologies of their profession. This obligation is rooted in both expectations by the public and the department.
- A fire fighter's job can deal with life and death. The community counts on fire fighters, and fire fighters willingly accept that responsibility.
- That acceptance of responsibility creates a social contract and an ethical obligation. Other obligations include a commitment to their own personal safety, a commitment to maintaining their physical and emotional capacity to perform their duties, an obligation to respect the privacy and dignity of the public they serve, and an obligation to uphold departmental rules and policies.
- Fire fighters also have subjective responsibilities. Although subjective responsibilities tend to be based on personal values and obligations, there are several that are consistent throughout the fire service and are typically represented in fire service culture.
- Common subjective responsibilities include a sense of obligation to uphold departmental

traditions, a commitment to loyalty and the team concept, commitment to professional development through continued lifelong learning, and an obligation to contribute to the department's legacy through the acceptance of a leadership role.

- The unique challenges of conflicting responsibilities by union leadership were explored.
- The chapter closed with a discussion of the fire fighter's obligation to protect the reputation of his or her department in the fire service by avoiding unethical actions.

Key Terms

community-based initiatives programs in support of improving quality of life within the community.

cultural inertia the tendency for a group of people to cling to traditions and ways of thinking that have outlived their usefulness even when better ways are presented.

deontology the ethics of an action should be judged relative to its compliance with a code of conduct.

fiduciary a person or an organization that holds a position of trust or confidentiality.

Health Insurance Portability and Accountability Act of 1996 (HIPAA) legislation that restricts the release of a patient's medical and personal information within very narrow parameters.

Nuremberg trials a series of military tribunals judging Nazi officials accused of war crimes. They were conducted from 1945 to 1949 in Nuremberg, Germany.

objective responsibilities obligations imposed from an outside source and are associated with duty.

prime theory of obligation a description of the base ethical obligation of the fire service; it is a touchstone for all other responsibilities and obligations associated with fire department activities.

social role a form of personal identity associated with particular responsibilities and obligations.

subjective responsibilities self-imposed obligations that tend to be rooted in personal values associated with virtue ethics and consequentialism.

Challenging Questions

To check your understanding of this chapter's material, answer the following questions. It is highly recommended that you discuss your viewpoints with fellow students, peers, coworkers, and friends to discover their opinions as well.

- The public assumes that fire fighters are willing and capable of meeting any challenge for which they are called. Do individual departments have an obligation to inform the public of the realities that reduced staffing and lightweight construction are placing on interior firefighting?

WRAP-UP

- This chapter reviews several of the objective obligations and responsibilities of fire fighters. Of those listed, what do you believe is the most important responsibility of the fire fighter?
- Should a fire fighter be subject to discipline for posting on social media humorous stories about calls, which citizens may find embarrassing? Does the citizen's right to privacy outweigh the fire fighter's First Amendment rights to free speech?
- Either from the Internet or from magazines, research some recent accounts of fire fighters' inappropriate behavior. Briefly explain what happened, and then answer the following question: What professional responsibilities did the fire fighter fail to honor? Discuss what ethical issues are associated with the failure to meet those responsibilities.
- Identify a tradition within your local fire department that you feel upholds a productive value or virtue. Explain the relationship between the tradition and the virtues that are benefited by it.

Case Study Conclusion

Revisit the case study at the beginning of the chapter. Spend a few minutes considering the questions posed at the end of the case study. In light of the information shared in this chapter, have any of your original observations changed?

In this case study, we see several of the elements of objective and subjective responsibilities. Further we see an example of the potential conflict between responsibilities. Smith is caught between several ethical obligations. First, as Jones's company officer, he has an obligation to fire fighter Jones's potential career. If, however, he places credence in training chief Campbell's opinion that Jones's issues are significant enough to warrant termination, he has an obligation to the department and to the rest of his crew to assist in Jones's removal. Smith also has a loyalty obligation to both of the superior officers, who seem to be working at cross purposes. Whether or not they are acting ethically may or may not be ethically relevant to Smith.

It is Smith's desire to balance his perceived subjective obligations to the other individuals involved in the case study that is causing him to mistakenly believe that he has an ethical dilemma. In reality, this is a straightforward ethical problem.

Smith's first objective obligation is to the demands of his position, or competency. He has an ethical obligation to honestly evaluate probationary fire fighter Jones's performance. This represents a straightforward duty, supported by his voluntary acceptance of his position's responsibilities. Altering the evaluation in any way would be a breach of his duty. The consequences of his actions are neutral. Jones's fate lies in the judgments of his superior officers, and their ability to meet the obligations of their positions. In other words, Smith's honest evaluation of Jones does not endanger the department, nor does it harm Jones. His obligation is to provide an unbiased assessment to those who will have the responsibility of seeking the best outcome.

Smith's best ethical path is to inform both superior officers in writing of his impressions of Jones's current and potential performance, and then extricate himself from the process.

Chapter Review Questions

1. As a tax-supported public service agency, what is the primary responsibility of the fire department?
2. What is an objective responsibility?
3. What is a subjective responsibility?
4. What is HIPAA?
5. Explain the term "social roles."

References

Blinder, A. 2015. "Kentucky Clerk Denies Same-Sex Marriage License." *New York Times*. Published September 1, 2015.

Brunacini, A. 1996. *Essentials of Fire Department Customer Service*. Oklahoma City, OK: International Fire Service Training Association.

Cooper, T. L. 2012. *The Responsible Administrator: An Approach to Ethics for the Administrative Role*. San Francisco, CA: Jossey-Bass.

Dylan, B. 1979. "Gotta Serve Somebody." *Slow Train Coming*. Los Angeles: Columbia Records.

Lemke, T. P., and G. T. Lins.2013. *ERISA for Money Managers*. Egan, MN: Thomson Reuters.

Lippmann, W. 1922. *Public Opinion*. New York: Harcourt, Brace and Company.

National Fallen Fire Fighters Foundation. 2017. "Everyone Goes Home." Accessed June 8, 2018. https://www.everyonegoeshome.com.

"Summary of the HIPAA Privacy Rule." n.d. *Health Information Privacy*. U.S. Department of Health and Human Services. Accessed June 8, 2018. https://www.hhs.gov/hipaa/for-professionals/privacy/laws-regulations/index.html.

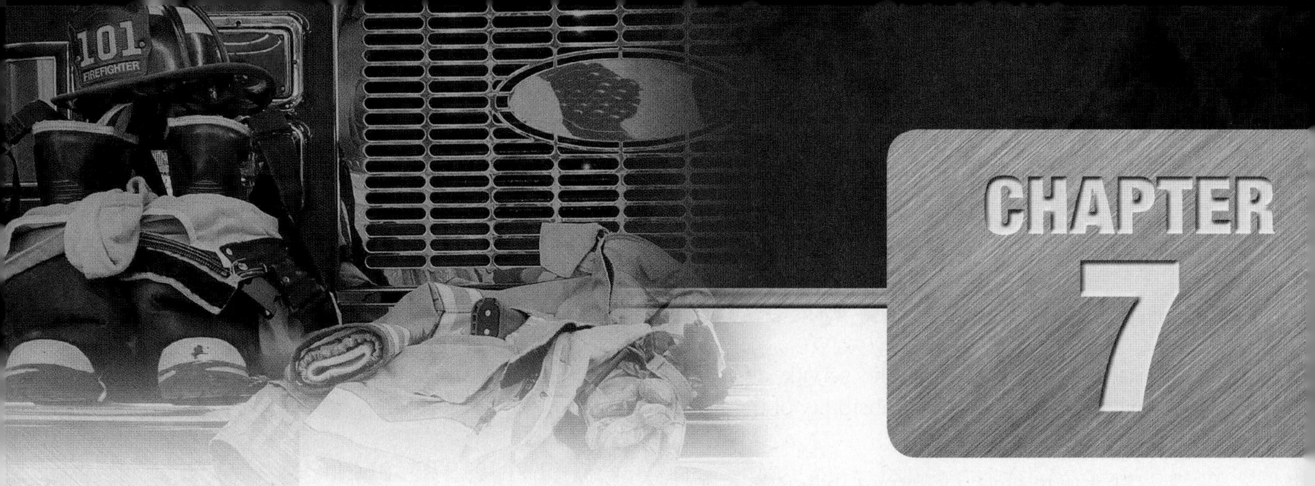

CHAPTER 7

Ethics and Leadership

"The only thing necessary for the triumph of evil is that good men do nothing."

—Edmund Burke

OBJECTIVES

After studying this chapter, you should be able to:

- Depict the elements of ethical leadership.
- Discern the importance of ethical leadership to a department.
- Characterize the theory of transformational leadership.
- Interpret the theory of servant leadership.
- Assess the leadership responsibilities of fire officers.
- Understand the elements that influence ethical leadership.

Case Study

Lieutenant Jones is in charge of station seven. It is a quiet station in the suburbs that handles only a few calls a day. Like the fire fighters he supervises, Lieutenant Jones is a departmental veteran with more than 15 years on the job. He has been a fire officer for seven years.

Lieutenant Jones has a fairly simple theory of leadership. He believes a happy shift is a good shift. He wants his fire fighters to like him, and he looks for a path of least resistance in accomplishing those things that his superiors absolutely require of him. Jones believes in not sweating the small stuff; as an example, he sees no particular value in departmental policy regarding uniform conformance or work schedules. He has been known to cover for fire fighters who report late to work, and he does not enforce restrictions on "free time" versus work time. He frequently allows his fire fighters to take naps during the middle of the day and has even occasionally let fire fighters sign training sheets for in-company drills that did not occur. He rationalizes his behavior as "building morale by avoiding busywork." He repeatedly reminds his crews that "what happens at station seven stays at station seven."

Although Lieutenant Jones's crew does not complain about their lieutenant's cavalier attitude toward responsibilities, other fire fighters resist being stationed with Jones. The other officers and fire fighters on the shift commonly call station seven the "retirement home." Less kind epitaphs for the station house are the "land of misfit toys" and the "hospice house for careers."

- In what ways is Jones letting down his subordinates?
- What, if any, ethical obligation to his superiors does Jones have as a fire officer that he is not meeting?
- What sort of attitudes would you expect Jones's crew to have about their house culture and specifically about him as a leader?

Introduction

It is 7:15 AM. The engine company crew has just finished their morning inventory and radio check. As one of the crewmembers puts on a pot of coffee, the company officer picks up a set of keys, and says, "It's time to take a walk across the street." He puts a portable radio in his hip pocket and, along with one of his crewmembers, walks across the street. They knock twice and let themselves inside. Their purpose is to help Mr. Smith out of his bed and into his wheelchair. Mr. Smith's mother is his caretaker, and she is unable to be there until 10 AM, and so the fire fighters lend a hand. There is a return visit at 9 PM to help Mr. Smith back into bed. This routine occurs every morning and every evening for over two years. There is no call from Mrs. Smith for help, and no incident report is written. Smith has no relatives on the department; the routine began because at one time the Smiths were "frequent flyers." Collectively, the crews who staff the engine house decided it would be simpler just to take care of it off the books.

Impressively, this routine goes on without complaint from fire fighters, and it is continued even if the normal company officers are not present. The individual crewmembers take it upon themselves to "take a walk across the street."

This story is true and, happily, it is not an isolated incident. Similar occurrences happen every day across the country. Why? Helping a stranger with a mundane daily task is not part of the job description. There is no ethical obligation to assist folks with daily life, but the station house's officers and crew collectively decided it was the right thing to do. The fact that it is done without complaint, and even in the absence of an authority figure requiring it, speaks greatly to the house's culture.

Most important, their actions speak to the leadership quality of the organization. These actions occurred long before fire service authors were writing about customer service or servant leadership, yet almost instinctively these individuals understood the concepts of instilling values and being oriented toward service. The officers were modeling positive behavior.

A fair question may be, How does this story apply to the ethics of leadership? As it turns out, a lot. Both leadership and ethics are important undertakings by department officers. As you will learn, fire fighters' sense of obligation to something other than themselves has a significant impact on the occurrence of ethical violations in an organization. The importance of a fire officer's ability to shape attitudes and instill virtues cannot be overstated. An organization's ethics is directly related to its institutional values and its approaches to delivering service. Ethics is a cultural condition; anything that affects culture affects ethics. Leadership, most important, ethical leadership, likely has the greatest impact on a fire department's culture than any other influence.

Consider for a moment the leaders and mentors that personally affected your life. Ask yourself what it was about them that so affected you. Was it their technical expertise, or was it their ability to inspire through their personal values and steady example?

Leadership may be one of the most frequently discussed topics in the fire service. There are many articles discussing the need for leadership, what leadership looks like, and even if there is a leadership crisis in today's fire service. While many of these articles list the virtues and qualities of a leader, they rarely discuss the ethics of leadership. This chapter will focus specifically on the relationship between ethics and leadership.

We will explore both the concepts of **ethical leadership**, and **leadership in ethics**. While the previous sentence may sound like wordplay, the concepts of ethical leadership, and leadership in ethics, are quite different. The former refers to the ethical quality of leadership provided by fire officers, while the latter focuses on those behaviors that instill ethics in others.

Specifically, this chapter will focus on numerous facets of ethical leadership, including leadership orientations. This chapter will also review the ethical obligations associated with leadership. For clarity, it will categorize those responsibilities as obligations owed to an officer's superiors, and obligations owed to an officer's subordinates. Finally, the virtues and personal qualities inherent in ethical leaders will be reviewed.

As you move through this chapter, you should keep in mind that the topics generally are addressed relative to "line" officers—those men and women who supervise emergency response personnel. We will explore in depth the topic of ethical leadership at the administrative level in Section 3.

Ethics and Leadership

People want leadership. Both groups and individuals complain when they feel it is absent or impoverished. One of the most common criticisms of disillusioned fire fighters is the perceived absence of competent leadership. In groups where there is an absence of leadership, almost invariably an unofficial leader will emerge. Whether by acclamation or self-proclamation, a leader who possesses qualities perceived as worthy by a group will find followers. This is true within any organization but even more so in the fire service.

The fire service operates in a rather unique culture. Fire fighters perform hazardous duties on a routine basis, and they work in prolonged close quarters with their peers and their supervisors. These factors provide a catalyst in forming significant relationships with each other, even in cases where they are not particularly warm relationships. The fire service has a culture that places a high value on paramilitary structure, competency, teamwork, and personal virtues. These cultural aspects conspire to create a high expectation of leadership performance especially related to these concepts.

Leadership classes typicaly stress the need for vision, competency, charisma, and experience. It

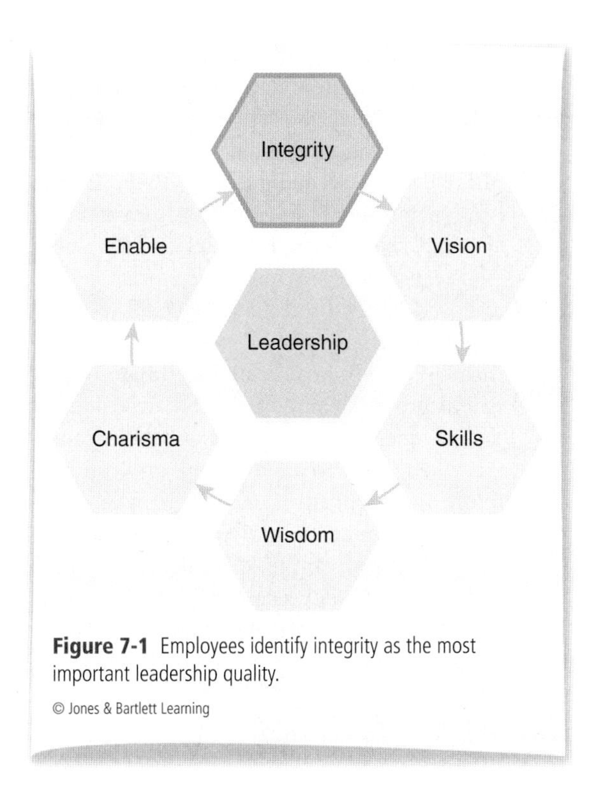

Figure 7-1 Employees identify integrity as the most important leadership quality.

may be surprising to find that several studies have indicated that of all the expectations placed on leadership, the most desired qualities of a leader are honesty and integrity Figure 7-1 (Keith, 2013). Both are strongly connected to ethics. Surprisingly, ethics is rarely discussed within fire service articles about leadership. Whereas the individual traits of integrity and honesty are typically identified as important personal traits when discussing leadership, a discussion of ethics as a prerequisite for leadership is rarely included in the majority of professional articles on the subject.

Is Ethical Leadership Important to a Department?

If the individual virtues associated with ethical behavior are important to employees, then it seems obvious that ethical behavior in department leaders is of paramount importance. Does ethical leadership matter? The answer seems to be an emphatic yes!

Research has indicated that people who believe their leaders are ethical have higher degrees of job satisfaction than those who do not, and they are less likely to leave their places of employment. There is also evidence that employees are more likely to emulate role models who demonstrate high ethical standards (Mayer, Kuenzi, Greenbaum, et al., 2009).

Most important, there are strong indications that organizations populated by ethical leaders have less incidence of employee misconduct. This appears to arise from a social learning theory called **modeling**. The theory asserts that followers take behavior cues from trusted figures in authority positions. Fire fighters bring with them their own sense of values and beliefs as to right and wrong. However, there is credible evidence to suggest that within a workplace, the behavior expectations of leadership can have a strong influence on work behavior and, depending on the individual, greatly affect how that person applies his or her personal beliefs to work behavior (Bandura, 1985). Leaders modeling ethical behavior have a strong positive influence on the behavior of followers. Conversely, in organizations where leadership is not trusted, a siege mentality can form. Employees are more likely to act in their own self-interest either out of a perceived necessity for self-preservation or out of an attempt to achieve some form of social justice. Negative work cultures can clearly motivate unethical behavior and, as such, are usually dysfunctional in accomplishing necessary tasks.

The absence of ethical leadership not only forfeits whatever potential benefits might be gained from it, but it also will likely induce cascading negative effects in productivity, initiative, and morale.

Specific Benefits of Ethical Leadership

A leader's ability to influence behavior by modeling ethical leadership is important, but it it does not represent the extent of the benefits derived from ethical leadership. Ethical leadership influences both individuals and the organizational culture pervasively.

The following are several important benefits the fire department derives from ethical leadership.

- **Ethical leadership instills trust.** While it may sound trite, it is true that one cannot lead if no one will follow. People only follow leaders they trust. The essential elements of ethical behavior are rooted in virtue, an attempt to seek just outcomes, and consistent respect for duty and obligation. These same elements are essential in the development of trust.
- **Ethical leadership brings credibility.** Related to trust is the concept of credibility. Followers must believe that a leader believes in the values and priorities that they espouse. Most fire fighters have heard the expression "You have to be able to walk the talk." The expression reflects an expectation of credibility. A lack of ethics reflects poor character and hence the lack of credibility.
- **Ethical leadership promotes collaboration.** One of the common principles of inspired leadership is making those around you stronger and more productive. It is a well-accepted theory that collaboration is more effective than the use of coercive power. To lead is to facilitate others' success. Collaboration is difficult if not impossible between individuals who lack respect or trust for each other.
- **Ethical leadership promotes morale.** Followers who believe that their leaders have integrity and an interest in others' welfare typically feel "safe." A sense of safety is a fundamental requirement for good morale and personal growth.
- **Ethical leadership allows for the occupation of the moral high ground.** A leader who is ethically challenged cannot reasonably insist on high levels of behavior from subordinates. Unethical leaders cannot expect to be followed in situations where personal integrity is important. The notion of "Do as I say and not as I do" has never been an effective motivational method. Hence, the unethical officer marginalizes his or her ability to lead.
- **Ethical leadership affords self-respect.** Leadership can require tough decisions, and it can force unpopular choices. Confidence in your ethical orientation and values is necessary when the inevitable second-guessing arises. Leadership is an exercise in confidence; it cannot be accomplished effectively if you lack self-respect. Self-respect is nearly impossible for an unethical person, and those unethical people possessing it are exercising self-deceit.
- **Ethical leadership creates a climate of personal growth.** Employees who are confident in the integrity and honesty of their leaders are more likely to attempt new tasks and explore new challenges. A culture where fairness is assumed reduces the perceived risks associated with failure. Confidence in leadership's fairness allows an organization to more easily create a culture of learning.

Joining Ethics and Leadership

The value of ethical leadership is apparent in the connection between the success of the leader and followers' perceptions that the leader is essentially ethical. Ethical leadership also contributes to employees' success. Ethical leadership is desirable, but where and how do leaders learn to lead ethically?

We learn ethics and values from parents, teachers, and religious leaders, but those lessons do not necessarily relate directly to the responsibilities of leadership. Surprisingly, most literature regarding leadership is relatively mute on the subject of ethics. Typically, discussions on leadership focus on how to motivate people to do things. Absent are in-depth considerations of the ethical implications of intended outcomes or methodology. History is rife with powerful and influential leaders who sought unethical goals or used highly questionable methods in the pursuit of benign objectives. Clearly, the concepts of ethics and leadership are intertwined but not necessarily mutually supporting. As such it is not enough to study leadership absent understanding its ethical implications.

It would seem logical that an individual would learn ethical leadership the same way that he or she learns personal ethics—through the example of role models. Just as the child learns the principles of honesty and integrity from teachers and parents, so too the future leader learns the fundamentals of leading with integrity and honesty from leaders he or she can emulate. They model behavior that they observed. Still, the example of leadership may not necessarily be completely instructive in developing ethical leadership skills. Seeing ethical leadership does not necessarily facilitate being able to practice ethical leadership. For instance, watching a professional golfer swing the club can be instructive, but it does not necessarily enable you to perform the task.

One theory suggests that the development of **emotional intelligence (EI)** is necessary to becoming a successful ethical leader. Emotional intelligence is the ability to understand and manage the emotional aspects of human interaction. This understanding must include both self-awareness and an empathetic understanding of others' emotions (Goleman, 2005). Daniel Goleman is an American psychologist who popularized the notion of EI Figure 7-2. According to Goleman, there are five main elements necessary for emotional intelligence to manifest into ethical leadership.

Figure 7-2 Daniel Goleman. Developed the popular understanding of EI.

© Astrid Stawiarz/Getty Images North America/Getty

The first requirement of an effective ethical leader is **self-awareness**. Understanding your preferences, prejudices, and emotional predispositions is important in facilitating fair and honest interactions with followers. Self-awareness facilitates understanding of personal motivation and is very useful in avoiding moral disengagement (see Chapter 3, *Influencing Behavior*). Self-understanding is fundamental to an ethical being.

To be aware of your strengths, weaknesses, wants, and needs is only half the equation. The ethical leader must exercise **self-control**. Making decisions based on emotion, or preconceptions, often leads to unethical actions. Just as important is that emotional self-control allows for the creation of open and honest relationships with followers through responses rooted in fact and appropriate to the circumstance.

Understanding **motivation** is the third characteristic of an effective ethical leader. A leader who understands his or her personal motivations finds it much easier to avoid self-serving actions. Leaders who evaluate their motivations are more likely to act in the interest of their followers. Leaders who are motivated by an internal **locus of control** are more likely to accept personal responsibility, which is important to ethical leadership.

The fourth quality of EI associated with effective, ethical leadership is **empathy**. Leaders who are empathetic or outwardly focused are aware of others' needs and wants. Empathy requires understanding. Empathetic individuals are much less likely to act in a self-serving fashion. Conversely, egotistical or self-centered individuals are more likely to act in ways that are destructive to others even if they have no ill will toward them. The failures of nonethical leaders are often rooted in ignorance of others' needs. Ethical leaders must possess outward focus.

The fifth and final quality of EI possessed by successful ethical leaders is that of **social skills**. It is not enough to be well intentioned, properly motivated, or outwardly focused if a leader is not able to effectively interact with followers. The ability to effectively communicate, mediate, and facilitate must be incorporated into effective leadership, ethical or otherwise.

Clarification

The five characteristics of EI include:

- Self-awareness
- Self-control
- Motivation
- Empathy
- Social skills

Texture: Eky Studio/ShutterStock, Inc.; Steel: © Sharpshot/Dreamstime.com

Another suggested tactic in becoming an ethical leader is to integrate facets of ethical conduct into daily behavior (Thorton, 2013). It is theorized that ethical behavior becomes habit and so is integrated into leadership style. The following are several habits recognized as essential to developing ethical leadership.

- Actively resolve complex ethical problems and take responsibility for making ethical decisions carefully.
- Incorporate ethics in day-to-day duties.
- Cultivate a respectful environment, build trust, and demand open communication.
- Do not confuse policy with ethics; legal does not necessarily mean right.
- Expect ethical behavior from everyone.
- Embrace ethics as an ongoing learning journey. Ethics is not something a person has or does not have—it is something a person cultivates.

Effective ethical leaders focus on what is right. Their methods are not only justifiable within the principles of ethics, but they are also justifiable for their contribution to successful organization management. We expressed that one of the primary benefits of ethical leadership is that it invokes ethical behavior in subordinates. While modeling ethical behavior is important, alone it does not support the development of behavior standards. Developing ethics in an organization requires effort that is both deliberate and sustained (Nortz, 2010). To influence ethical behavior in others, leaders should:

- Articulate a clear version of good within their realm of influence. Defining expected behaviors is the first essential step in encouraging those behaviors. It is important for fire officers of any rank to understand they are responsible for the development of behavior standards. It is a mistake to assume that defining expectations lies only with administrative officers.
- Invest the time and effort as necessary to pursue expected behaviors. Diligence can be taxing, but an important concept in ethics is that it is easier to correct attitudes than bad habits. Another important element is to understand that the reinforcement of behavior standards requires consistency. If a fire officer is ethical 90 percent of the time, he or she likely will have a reputation as being unethical. The fire officer that requires appropriate standards of behavior from his or her fire fighters 90 percent of the time confuses his or her subordinates' expectations nearly 100 percent of the time. Consider the toddler who throws a tantrum when denied candy. Nine out of 10 times the parent admonishes the poor behavior. However, every now and then the parent gives in to a path of least resistance and allows the candy. Even though the tantrum only works 10% of the time, the chance of success is almost certainly enough to reinforce the behavior. Although firefighters are not children, the power of positive reinforcement of poor behavior should not be ignored.
- Officers should hold themselves personally accountable for the behavior of their subordinates. This can be difficult and at times even seem unfair. However, the unhappy truth is

that part of leadership's responsibility is the articulation of behavior standards and the reinforcement of values. The effective ethical leader accepts the challenge of "the buck stops here."

Approaches to Ethical Leadership

The fire service is a service-oriented organization. We exist to help other people. In that pursuit, we have a culture of teamwork, personal development, and learning. Of the many leadership theories, two seem highly appropriate in the development of ethical leadership within the fire service. As leadership theories go, both are relatively new. James Burns introduced **transformational leadership** in 1978, and Robert K. Greenleaf introduced **servant leadership** in 1970 (Figure 7-3). It is the orientation of both theories that makes them highly appropriate for the fire service in general but, in particular, as they pertain to ethics in fire service leadership.

Transformational Leadership

Transformational leadership seeks outcomes by shaping both individual attitudes and the general work culture. Ultimately it seeks to align individual employee needs with desired organizational outcomes. Transformational leadership utilizes employee empowerment and development both as a motivational technique and as an ethically justifiable priority. Transformational leadership enhances motivation, morale, and performance through several mechanisms (Burns, 2003). What makes transformational leadership particularly effective for the fire service is that it focuses on reinforcing the fire fighter's sense of identity within a team, which is accomplished by positive reinforcement, the provision of growth opportunities, and empathy. Through empowerment, the transformational leader instills a sense of potential accomplishment. Through a commitment to the team's welfare, the transformational leader imparts to followers a feeling of being a part of something bigger than yourself. To any fire fighter, these should be familiar concepts. The transformational leader is a role model for followers and inspires them by challenging them to take greater ownership in their work, understanding their strengths, and aligning their personal values with those of the department to optimize performance. In a practical sense, these precepts match well with the fire service's ongoing culture of emphasizing team performance, commitment to training, and commitment to internal promotion.

Figure 7-3 James M. Burns, 1918–2014. A Pulitzer Prize–winning historian and political scientist who first introduced transformational leadership.

Transformational leadership is built on four elements, sometimes referred to as the four I's (Reggio, 2009).

The first element of a transformational leader is *individualized focus*. Transformational leaders focus on the individual needs of their followers. They act as coaches and mentors and are genuinely interested in the development and growth of their subordinates.

Second, the transformational leader is committed to *intellectual growth*. For a leader, learning is a value not only for its practical application but also for its positive effect on the follower. Transformational leaders seek teachable moments and consider it their responsibility to advance the careers of their followers (mentoring and succession planning).

A third transformational element is to motivate *inspirationally*. A transformational leader articulates a vision of success and optimism; she or he constantly strives for excellence. No one is more influential in setting the firehouse's mood than is the company officer. A pervasive feeling of optimism and belief in the worth of effort is infectious. Fire fighters' attitudes toward their work is a direct reflection of their officers' attitudes.

Finally, and most important, transformational leaders exert *idealized influence*. They provide a role model of high ethical behavior. They instill pride in their followers and both demand and freely give respect and trust. They set high standards of integrity for themselves and their followers.

Clarification

The four elements of transformational leadership include:

1. Individualized consideration
2. Intellectual growth
3. Inspirational motivation
4. Idealized influence

Transformational leadership is based more on a leader's attitude than methods. The transformational leader feels a sense of responsibility for the quality of the workplace as well as for the quality of the work being done. This form of leadership requires personal investment in employees' well-being and, more important, in the growth and development of their employees at a humanistic level.

Servant Leadership

Servant leadership is a philosophical approach to leadership oriented to putting the well-being of people and communities first.

Unlike leadership styles that incorporate authoritative, top-down management of personnel, servant leadership focuses on collaboration, communication, trust, and ethical use of power. The notions of public service and the public servant have a very long history. In 600 BC the influential Chinese philosopher Lao Tzu wrote, "The greatest leader forgets himself and tends to the development of others" Figure 7-4 (Tzu, 1868). The concept of servant leadership is also embedded in Christian ethics. In Mark 10:43–44, Jesus said, "Whoever wants to become great among you must be your servant, and whoever wants to be first must be made servant to all."

Figure 7-4 Lao Tzu was a contemporary of Confucius in the sixth century BC. He is generally credited with establishing the philosophic movement of Taoism.

However, it was not until 1970 that a specific leadership philosophy based on a servant first orientation was articulated by Robert K. Greenleaf Figure 7-5 (Green Leaf Center, 2016). The majority of leadership theories tend to focus on the acquisition of power, the legitimization of authority, and the use of authority to accomplish ends. Servant leadership differs in that its focus is almost entirely rooted in ethics and morality. The theory suggests that the only legitimate use of authority is for the betterment of those being served (the led). Greenleaf's theory not only applies to the individual leader but

Figure 7-5 Robert Greenleaf, 1904–1990. Founder of the modern servant leadership theory.

Courtesy of The Greenleaf Center for Servant Leadership.

to an institution as a servant as well. Again, servant leadership asserts that the only legitimate exercise of authority by an institution (fire department) is in the provision of service (Green Leaf Center, 2016).

As we have frequently repeated in this text, fire departments exist to provide a public service. It is inherent in our mission statement and culture. The connection between the fire service and servant leadership should be readily apparent. The fact that servant leadership is rooted in an orientation that both responds to a duty to serve and protect, as well as a consequentialist desire to seek a positive outcome, makes this form of leadership inherently ethical.

Several studies have concluded that the defining element of servant leadership is morality. A servant leader models ethical behavior and encourages integrity and moral reasoning in colleagues. These concepts are embedded as guiding principles of servant leadership (Keith, 2013). The founding principles of servant leadership include:

- **Serve people:** It is important not to confuse servant leadership with assuming a submissive or subordinate role to those being led. The servant leader does not take orders from followers or, in the extreme, run errands for them. Rather, the concept of service is that the leader assumes an orientation of responsibility for the followers' welfare. The servant leader places people and their welfare ahead of short-term goals. The same thoughts are expressed in care ethics and to some extent in the theory of social contract ethics. Servant leadership differs in that these principles are applied to motivation and organizational accomplishment rather than personal interaction.
- **Facilitate personal development:** This principle is an expression of respect for each person's potential and recognition of the importance of personal growth and self-actualization. It is consistent with **Maslow's hierarchy of needs** (see Chapter 3, *Influencing Behavior*) in that it suggests that self-actualized individuals have a greater capacity for achievement. As an employee's capacity grows so does the organization's capacity.
- **Exercise foresight:** Vision, and a sense of purpose, are necessary for any leader. This principle is not unique to servant leadership; rather it expresses the common understanding that to lead requires an understanding of current and future needs. What is unique about servant leadership's approach to foresight is its relationship to an obligation as a caretaker. Servant leadership places an ethical responsibility on protecting the interests of the organization and its followers; this requires a degree of vigilance.
- **Ethically interact with stakeholders:** The orientation of service expressed in servant leadership extends not only to followers but to all stakeholders within an organization. It mandates ethical dealings with superiors, customers (the public), fire fighters, contractors, vendors, and any other individual or company dealing with the department.

Similar to transformational leadership, servant leadership is an expression of orientation as opposed to methodology. Both leadership styles are humanistic in application. Further, both leadership styles focus on the demands of ethics as well as the

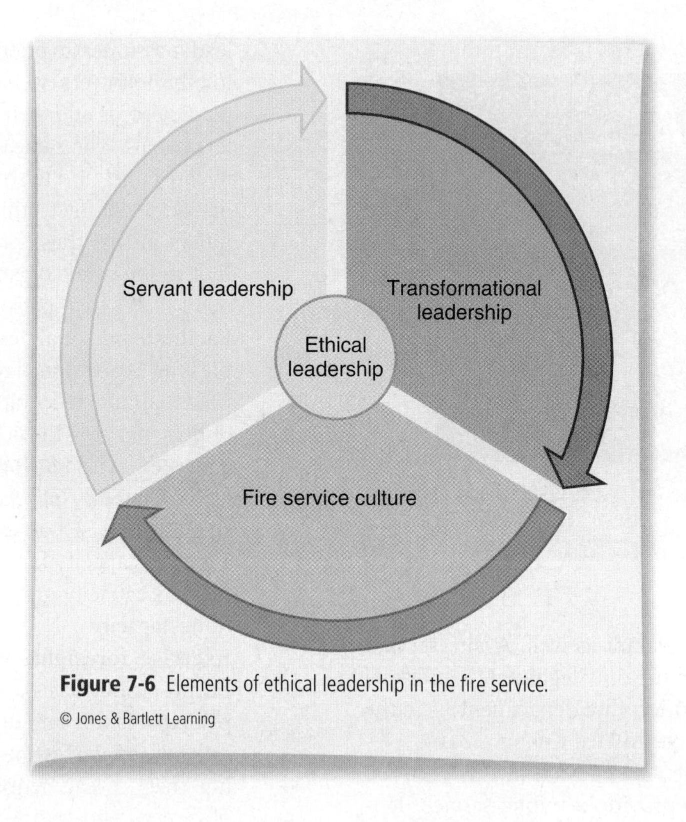

Figure 7-6 Elements of ethical leadership in the fire service.

unique challenges of managing fire fighters. The fire service's culture is oriented toward the development of employee skills, as demonstrated by our robust training schedules, and the fire service's long tradition of internal promotions. A prominent element of fire service culture is the concept of "taking care of our own." The attitude is expressed not only in department culture but in policy. Few industries have as an intense focus on employee safety as the fire service. Fire departments typically have very liberal employee wellness programs, as well as generous insurance, pension, and leave policies. Those traditions are highly consistent with the attitude of servant leadership Figure 7-6.

The fire service is a paramilitary organization with a clearly identifiable chain of command and respect for authority. However, company officers, up through the ranks to chief officers, have strong personal connections with fire fighters in addition to their managerial relationships with them. More than most professions, fire service officers have an almost parental relationship with subordinates in that the relationship is authoritative but also personally connected.

The fire service's culture is already invested in the founding concepts of servant leadership. There is a high degree of organizational accountability for the welfare of subordinates, investments in and personal growth of subordinates, and managerial relationships built on personal accountability of leaders to followers.

Ethical Leadership and Influence

There is evidence that the effects of ethical leadership trickle down and influence lower levels in the department (Mayer, Kuenzi, Greenbaum, et al., 2009). The absence of ethics at the top of an organization can have a negative cascading effect on the organization's culture. Proactively, administrative officers influence department ethics by setting department priorities and policies. They guide the department's culture by setting tone and example.

While chief officers influence culture, it can be argued that company officers have the greatest impact on *building* and reinforcing a department's culture, especially as it relates to ethics. Ethical culture is shaped on apparatus floors, in dayrooms, and on training grounds. Culture is reinforced every day. Company officers interact with fire fighters directly; whether consciously or subconsciously, they lead by example. Although administrative officers have a greater degree of **positional authority**, company officers tend to have greater **referent authority**. Their access and proximity to fire fighters strengthen their influence. Conversely, the ability to exert direct influence on ethical behavior diminishes as layers of positional authority are added.

A fire chief's lack of proximity can be particularly exacerbated by the size of the department. In small departments, fire chiefs have greater access to direct interaction with their fire fighters than in very large departments.

Clarification

The role of the company office in departmental ethics cannot be overstated.

- Ethical influence is proportional to proximity.
- Standards of behavior and policy compliance are subject to the company officer's willingness to enforce them.

In general, the influence of ethical leadership in the fire service is directly related to proximity. While ethical leadership is important at all department levels, the role of the company officer cannot be overstated. Company officers are not only the eyes and ears of departmental leadership, they are also its voice. Company officers are the department's representative on the ground. Policies, procedures, and departmental initiatives must pass through the company officers on their way to the fire fighters. Standards of behavior and compliance to policy are subject to the company officer's interpretation and willingness to enforce. The success of a department's commitment to ethics is, in a very large degree, dependent on the company officers' support of it.

Leadership's Obligation to Fire Service Ethics

Most would agree that ethical leadership is desirable in any organization. Most of the fire service would certainly agree that ethical leadership is particularly important in an organization empowered with the immense responsibilities of the fire service. We should also recognize that the imperative of ethical leadership is magnified by a local fire department's reliance on public image.

The case for the necessity of fire service leaders to behave ethically is self-explanatory. However, does modeling appropriate behavior satisfy the fire officer's obligations when it comes to ethics? Should a fire officer be held responsible for the ethical behavior of subordinates? There is a theory of culpability in ethics management that extends responsibility to supervisory officers. President Harry Truman may have expressed it best by the popular expression "The buck stops here." More than a few fire officers have found themselves answering some very difficult questions regarding their subordinates' behavior. Some have even found their careers derailed by their perceived inability to control their fire fighters' behavior. Is this fair? Is there an inherent leadership obligation to shape values and facilitate the personal and ethical growth of subordinates? Is leadership in ethics inherent in the job of the company officer, or is the exercise of adequate management skills enough to meet departmental obligations? As one officer puts it, "My crew gets its work done, they train on what they are supposed to train on, and they obey the rules. You can't ask for more than that." Perhaps we should ask for more than that.

Anyone familiar with the study of fire service leadership will recognize that these questions are a

What Do You Think?

Should fire officers be held accountable for the attitudes of their subordinates? If so why? If not, why not?

Texture: Eky Studio/ShutterStock, Inc.; Steel: © Sharpshot/Dreamstime.com

common element in the distinction between leadership and management. There is almost universal agreement in fire service literature that fire officers have an obligation to the department and their fire fighters beyond simple management of time and resources. It is typically argued that the success of the department is dependent on the fire officer's ability to motivate, educate, and mentor. The department needs its officers to lead Figure 7-7.

If you accept the popular premise that leadership is elemental to a fire department's success and that leadership, particularly ethical leadership, is consistent with fire department values and traditions, then you can only conclude that leadership is an obligation for company officers. It follows that if leadership is a responsibility, then ethical leadership is as well.

Just as the fire service needs to constantly teach and reinforce core skills, we need to teach and reinforce the ethical values associated with being a fire fighter. A fire department cannot afford to have incompetent fire fighters, nor can it afford to have fire fighters who are self-serving, dishonest, disinterested in team success, or uncommitted to public service.

Figure 7-7 Fire departments can only perform as well as the officers who lead them.

© TheImageArea/E+/Getty

Fire officers, particularly company officers, have the greatest influence on the quality of a department's culture (particularly in ethics). If it is also true that maintaining the department's culture is essential to the department's welfare, then it would make sense that fire officers have a responsibility to care for that culture. The following is a list of the fire officer's general responsibilities. Each is strongly related to leadership ethics.

- **Delivery of emergency services.** Excellence of service not only depends on a fire fighter's ability to perform a task but also on his or her commitment to doing it well. Employee commitment is a reflection of the officer's ability to instill values consistent with the department's mission.
- **Fire fighter's safety and well-being.** A fire fighter's safety depends first on the officer's competency. Safety further depends on the officer's commitment to safety and ability to teach fire fighters. Those responsibilities are an ethical obligation rooted in the acceptance of responsibility and in humanistic concern for the welfare of others.
- **Fire fighters' prosperity.** Within the fire service, promotion is not only based on educational qualifications gained by the fire fighter but also the fire fighter's access to quality mentoring. Because the fire service relies almost entirely on internal promotions, the individual company officer has a significant effect on the future career trajectory of his or her fire fighters. Ethically, an officer must accept responsibility for the future success of the department by facilitating the success of those the officer is charged with leading.
- **Reputation management.** A fire department's reputation depends on its members' behavior. The fire officer's ability and willingness to enforce policy is only a first step. Ethical

> leadership shapes attitudes and the behavior habits of fire fighters. These are key elements in obtaining high standards of behavior.

Imagine that you just hired a painter to repaint your home. The painter has an ethical obligation not only to put paint on your home but to also do all of those preparatory and ancillary tasks needed to successfully complete the job. You are paying a professional to do professional work. You and the painter have essentially entered into a contract. The terms of that contract insist that you pay money and that the painter performs all necessary services for adequate completion. You both are ethically obligated to meet those obligations. This same theory applies to fire officers. They are entrusted with an obligation not only to manage fire fighters' time but also to do all the preparatory and ancillary tasks needed to effectively fulfill the department's mission. In this case, the ancillary tasks are those associated with leadership, and a company officer has an ethical obligation to meet those responsibilities.

Hallmarks of Effective Ethical Leadership

Achieving "effective, ethical leadership" encompasses two distinct goals. Effectiveness is an expression of accomplishment. A leader is tasked with numerous responsibilities. The officer's relative effectiveness depends on how well those responsibilities are met and under what conditions. In a loose sense, the effectiveness of leadership is measured quantitatively.

The other half of the equation refers to ethics. Ethics can be thought of as measured qualitatively. The quality of ethics in leadership is likewise an accomplishment, but it differs in orientation. Ethics is a reflection of values. Ethics is not a measurement of what is done but how it is done.

Fortunately, effectiveness and ethics are not mutually exclusive, but they are mutually supportive. It is well demonstrated that practicing ethics builds leadership credibility, motivates followers, and elevates commitment levels. Effectiveness by its nature requires a commitment to responsibilities. The effective leader must be ethical, just as an ethical leader must be more effective than a nonethical leader. If effectiveness and ethics are desirable traits in fire service leaders, the logical question is, How do you become both effective and ethical in your leadership style?

First, the fire officer must recognize and accept that he or she has responsibilities that go beyond basic management. Then, the officer must be committed to accepting responsibility for the welfare and needs of others.

Next comes understanding the principles of ethical leadership. The orientations of both transformational and servant leadership require an internalization of their precepts. In other words, the effective ethical leader must understand and buy into these concepts.

Last comes practice. As Aristotle suggested more than 2,000 years ago, ethical behavior must be habituated. Habit springs from repeated intentional behavior. The following are some hallmarks of ethical leadership:

- **A commitment to excellence:** The development of skills is a hallmark of both transformational and servant leadership. It fulfills the ethical obligation to perform and grow the capabilities of subordinates.
- **Modesty:** The ethical leader understands his or her limitations. It is only through self-understanding that personal growth occurs. Hubris is counterintuitive to ethics as it promotes self-interest.
- **Mentorship:** The effective leader recognizes the responsibility to teach. This responsibility comes out of the practical need to create competency, as well as an ethical responsibility to enrich the capacities of your fire fighters.
- **Generosity:** The effective and ethical leader gives his or her time, attention, and knowledge freely. He or she is committed to the betterment of others, not only as a sense of obligation but also as a product of virtue.
- **Compassion:** People sometimes fail. Individuals most prone to failure are those who are willing to risk failure to achieve. The effective

and ethical leader recognizes that growth comes from addressing shortcomings with the intention to teach rather than punish.

- **Integrity:** The ethical leader is effective because people trust him or her. The ethical leader possesses moral authority because he or she models moral behavior. An officer who lacks respect is severely limited in the ability to lead.
- **Optimism:** Optimism is an expression of confidence and hope. Subordinates do not emotionally invest in a leader if they do not believe that their efforts will bear fruit. Followers insist that leaders have a vision of a better state. This applies to company officers as well as fire chiefs. Conversely, pessimism is an expression of fear. It may be perceived as a fear of failure, a fear of trying, or a fear of the future. Leadership cannot be effective in the presence of fear.
- **A heightened sense of responsibility:** The ethical and effective leader must demonstrate a sense of obligation to his or her responsibilities. Responsibilities and obligations are a hallmark of integrity, commitment, and service. Neither ethics nor effectiveness can be born out of irresponsibility.

Final Thoughts

Ethics is a cultural condition. It not only requires members oriented to ethical behavior, it also necessitates ethical leadership. Ethics must permeate every level of an organization and must influence all decisions. Ethical cultures can be achieved only if leaders are ethically intelligent. They must be simultaneously committed to ethical outcomes and ethical methodology. Effective ethical leadership requires a fundamental shift in commonly accepted ideas about the achievement and use of authority.

Clarification

- Ethics is a cultural condition; it permeates all levels of an organization.
- That which affects culture affects ethics.
- Leadership greatly affects culture.

Texture: Eky Studio/ShutterStock, Inc.; Steel: © Sharpshot/Dreamstime.com

WRAP-UP

Chapter Summary

- We first explored what employees expect from leaders, as well key elements they identified as worthy of respect.
- Studies have indicated employees most desire integrity and honesty in their leaders. They value this more than competency or charisma. It was found that fire fighters who believe that their leaders are ethical tend to report higher job satisfaction and commitment to the organization and its goals.
- Trusting that leaders have employees' best interests at heart is a key element in commitment to follow an individual. Job satisfaction and job performance are both related to trust in leadership.
- Beyond building trust, ethical leaders tend to inspire ethical behavior. This concept is known as modeling. To the extent that modeling inspires ethical behavior in employees, it is clear that an absence of an ethical model has a direct correlation to aberrant behavior in employees.
- In organizations where employees do not trust leadership, a siege mentality can form leading to employees acting in their own self-interest. The absence of ethical leadership not only fails to inspire ethical behavior but may directly motivate unethical behavior.
- There are several benefits of ethical leadership: trust building, the acquisition of moral high ground, enhanced collaboration, and increased confidence in organizational support.
- Enhanced employee confidence in leadership has a positive impact on employees' willingness to take chances with new methodologies and ideas.
- Three concepts associated with ethical leadership were explored: the acquisition of emotional intelligence, as well as the leadership theories of servant leadership and transformational leadership; the compatibility of the fire service's organizing principles, in its traditional culture, with that of servant leadership; and the relative influence of ethical leadership based on perceived authority, influence, and proximity.
- Proximity has the most significant effect on a fire officer's ability to influence behavior.
- The company officer may be the most important resource the department has for building an ethical culture. The fire officer's obligation to provide ethical leadership was explored.
- Several elements of responsibility were identified and explanations given as to why there is an ethical imperative for the fire officer to model ethical behaviors. Beyond that, it was demonstrated that there is an ethical obligation to provide for the safety, well-being, and development of his or her subordinates.
- Ethical behaviors are facilitated through continued learning, seeking excellence, mentoring, and providing opportunities for personal development.
- The chapter closed with a review of those qualities most associated with ethical leadership.

Key Terms

emotional intelligence (EI) the ability to understand and manage emotional aspects of human interaction.

empathy the action of understanding, being aware of, being sensitive to, and vicariously experiencing the feelings, thoughts, and experiences of another.

ethical leadership the act of leading individuals using ethical methods and seeking ethical outcomes.

leadership in ethics leadership methods intended to raise the ethical standards of followers.

locus of control the extent to which people believe they have power over the events in their lives.

Maslow's hierarchy of needs a theory that suggests, as humans, we have needs that can be categorized as physiological, safety, social, esteem, and self-actualization. These needs are listed in their order of primacy.

modeling a theory suggesting behavior cues from trusted individuals influence levels of ethical behavior within organizations.

motivation the act or process (such as a need or desire) that causes a person to act.

positional authority authority bestowed as a condition of rank.

referent authority authority related to respect and trust.

self-awareness an awareness of your personality or individuality.

self-control restraint exercised over your impulses, emotions, or desires.

servant leadership leadership theory developed by Robert Greenleaf oriented to putting the well-being of people and communities first as a management priority.

social skills the tools that enable people to communicate, learn, ask for help, and get needs met in appropriate ways.

transformational leadership leadership theory by James M. Burns that stresses employee empowerment and development.

Challenging Questions

To check your understanding of this chapter's material, answer the following questions. It is highly recommended that you discuss your viewpoints with fellow students, peers, coworkers, and friends to discover their opinions as well.

- Critics of servant leadership theorized that it diminishes the authority of the leader. Considering the fire service's authority structure, do you see this is a problem? If so, how can it be avoided?
- What, if any, leadership challenges do you believe are uniquely present in the fire service?
- What is the most important trait that you assume fire fighters expect from their officers?
- If you have not done so already, read the quote by Edmund Burke at the beginning of this chapter. Burke is expressing the idea that we are all responsible for our culture's morality. Give two historical examples whereby unethical leaders came to power because of the apathy of followers. How can an individual resist unethical leadership?
- Identify a leader who was ultimately found to be unethical. What actions did the leader take that were found to be unethical, and what do those actions say about the leader's character?

Case Study Conclusion

Revisit the case study at the beginning of the chapter. Spend a few minutes considering the questions posed at the end of the case study. In light of the information shared in this chapter, have any of your original observations changed?

Lieutenant Jones is a classic example of impoverished leadership. Beyond his obvious dereliction of responsibility, there are significant indications that Jones lacks integrity. Although his primary intent is to be "liked," that intention should not be construed as being charitable toward his fire fighters. He is acting to achieve a self-serving goal. He is showing a complete disregard for the welfare, self-respect, and prosperity of his fire fighters. He is also showing a complete disregard for the ethical responsibilities assumed as a trusted fire officer. Jones's attitude demonstrates disloyalty to his supervisors by not enforcing policy. His dishonesty in documenting activities is modeling an attitude that rules do not matter. Jones is also failing to meet his responsibilities to protect the public; even if his crew managed to maintain some level of competency, the culture of his house can only encourage sloppy attention to detail and a lack of enthusiasm for quality services. Finally, and just as important, he is ignoring any responsibilities that he may have toward his crew's welfare. He is placing them in an environment that creates a career dead end. He is not only failing to develop their skills, he is also, directly and indirectly, creating a culture unsupportive of initiative or ambition.

The Lieutenant Joneses of the world are not that unusual. Common experience indicates that fire fighters working with this type of officer may be content, but they are rarely happy. They are almost always aware that their team is not held in high regard, and they often report a sense of boredom from a lack of accomplishment.

Surprisingly, there is no correlation between an officer's popularity and a willingness to disregard policy enforcement. Fire fighters, similar to most people, make decisions about liking or disliking an individual based on personal habits and compatibility, not on perceived benefit from an association. Fire fighters may or may not like an officer for a number of reasons, but rarely is likeability based solely on job performance. In almost all cases the Jones model officer does not hold the respect of the fire fighters he or she manages. As a result, even though Jones asked very little of his crews, officers similar to Jones typically find fire fighters are more likely to be insubordinate.

Chapter Review Questions

1. What is referent authority?
2. Describe the hallmarks of servant leadership.
3. What is transformational leadership?
4. List the qualities of an ethical leader.
5. List three benefits of ethical leadership.

References

Bandura, A. 1985. *Social Foundations of Thought and Action: A Social Cognitive Theory.* Upper Saddle River, NJ: Prentice-Hall.

Burns, J. M. 2003. *Transforming Leadership.* New York: Atlantic Monthly Press.

Goleman, D. 2005. *Emotional I.* New York: Bantam Books.

Green Leaf Center. 2016. "The Servant Leader: What Is Servant Leadership." Accessed June 11, 2018. https://www.greenleaf.org/what-is-servant-leadership.

Keith, K. M. 2013. *The Ethical Advantages of Servant Leadership.* Accessed June 11, 2018. http://www.toservefirst.com/pdfs/The-Ethical-Advantage-of-Servant-Leadership.pdf.

Mayer, D. M., M. Kuenzi, R. Greenbaum, et al. 2009. "How Low Does Ethical Leadership Flow? Test of a Trickle-Down Model." *Organizational Behavior and Human Decision Processes* 108 no. 1.

Nortz, J. 2010. "It Takes More Than Good Intentions to Be an Ethical Leader." *Rochester Business Journal.* Published March 5, 2010.

Reggio, R. 2009. "Are You a Transformational Leader." *Psychology Today*. Accessed June 11, 2018. https://www.psychologytoday.com/blog/cutting-edge-leadership/200903/are-you-transformational-leader.

Thorton, L. F. 2013. *7 Lenses: Learning the Principles and Practices of Ethical Leadership.* Richmond, VA: Leading in Context.

Tzu, L. 1868. *The Tao Te Ching.* London: Trubner and Co.

CHAPTER 8

Ethics and Diversity

"To see what is right and not to do it is want of courage, or of principle."

—Confucius

OBJECTIVES

After studying this chapter, you should be able to:

- Explain the meaning of diversity.
- Understand the history of workplace diversity in the United States.
- Determine the general benefits of diversity in the workplace.
- Recognize the common barriers to attaining a diverse culture and workplace.
- Discuss the status of diversity in the American fire service.
- Assess the ethics of diversity.
- Discern the methodology for achieving diversity.
- Decipher the complexities of managing fire service diversity.
- Explain how diversity is a reflection of professional ethics.

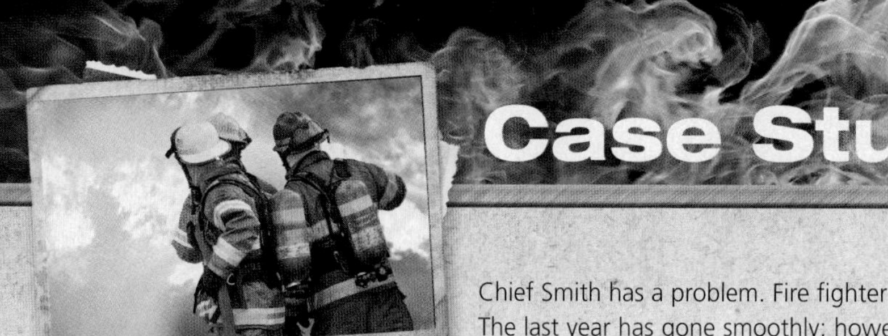

Case Study

Chief Smith has a problem. Fire fighter Jones has recently completed her probation. The last year has gone smoothly; however, now he has a harassment complaint sitting on his desk. He also has an entire shift at odds with each other. The majority of the shift is angry at fire fighter Jones and want nothing to do with her. Sadly, they have been systematically shunning her. Incidents have included walking out of the room when she walks in, refusing to speak with her, and placing anonymous notes in her locker urging her to either "grow up" or go home. Jones and a few supporters feel as though she is being treated unfairly and are considering legal action.

It all started the day Jones completed her probationary year. In keeping with a long tradition of the department, Jones's fellow fire fighters attempted to "initiate" her to the department with what they call a "trial by fire." The trial involves pulling the newly initiated fire fighter from the bunk in the middle of the night, strapping the person to a backboard face-down, and receiving a group spanking by the rest of the station house. Jones was aware that a hazing tradition existed, but details were cryptically kept from her.

On the night of the initiation, six of her fellow fire fighters drug Jones out of her bunk in the middle of the night. They attempted to force her onto the backboard while she kicked and screamed desperately trying to fend them off. Although no physical harm came to her, several fire fighters essentially manhandled Jones in an attempt to restrain her. During the struggle, Jones's clothing became disheveled "accidentally" exposing her undergarments. After several minutes the company officer realized that fire fighter Jones was nearly hysterical with fear and ordered the proceedings to stop.

The next day Jones reported the incident to her shift commander. She said she did not want to file formal charges, but she wanted to make sure that the rest of the fire fighters understood that such behavior was not welcome and she would file charges if it happened again. The shift commander informed Jones that she was overreacting, he assured her that the incident was actually part of an initiation that has been going on for over 20 years, and she should be proud that she was being accepted by the shift. However, if she insisted, he would talk to the fire fighters and warn them that she was offended and they would have to desist from their tradition.

After a meeting between the fire fighters and the shift commander, the fire fighters began their shunning of fire fighter Jones. They made it clear that if she wanted to be treated differently, they would treat her as an outsider. They also made it clear that they resented her "snitching" to a superior officer and her disrespecting their traditions.

- Was this incident actually harassment or just a misunderstanding?
- Do you feel that the shift commander handled the issue properly?
- If you were fire chief, how would you respond to this problem?

Introduction

To say that diversity is an important issue in America is an exercise in understatement. Few topics elicit so much emotion as does diversity and its related issues. It is fair to say that the very history of the United States is a study in diversity. Consider that the very first written document produced by the founding fathers stated their intention to form a government based on a simple first principle, "We hold these truths to be self-evident, that all men are created equal, and that they are endowed by their Creator with certain unalienable Rights . . . ." For purposes of this text, it is important to note that this is not only a declaration of intent but also an important statement of ethics. The fact that the founding fathers justified their intent to create an independent state based on an ethical argument in support of diversity is remarkable. The pursuit of that lofty principle has shaped elections, forged laws, and even started

Figure 8-1 Depending on with whom one speaks, fire service diversity is celebrated as an accomplishment or derided as an abject failure.

© Thinkstock Images/Stockbyte/Getty

wars. The Constitution's Bill of Rights is designed to protect the individual's right to be different, both in expression, and in thought **Figure 8-1**.

From the moment the country was formed, there's been an internal battle between the majority's wish for conformity and individuals' personal freedoms. The displacement of American native tribes, slavery and the Civil War, Reconstruction, the Jim Crow laws, the great European migrations of the 20th century, and Japanese internment during World War II are all examples of turmoil related to diversity rights. Politically, those battles continue today and continue to shape our politics and our future history. Race relations, women's issues, immigration, and LGBTQ issues are in many ways legacy arguments dating back to the founding of our country. You cannot understand the American experiment without understanding that the struggle with diversity in all its forms has had a consistent and important influence on American life.

Through most of that tumultuous American history, the fire service remained more or less unaffected. The issue of diversity in the fire service is relatively new and has only become a mainstream topic of debate in the last 40 years. In that time, it has become a significant issue for fire service personnel across the country and is becoming more important with each passing year.

Depending on with whom one speaks, fire service diversity is celebrated as an accomplishment or derided as an abject failure. It is often discussed as a moral and practical necessity, and also dismissed as a dangerous experiment in political correctness. For many, discussions of fire service diversity elicit frustration, fear, anger, and uncertainty. Ironically, for many, the topic is met with indifference.

This chapter will explore the fundamentals of diversity. It will introduce background information as to what diversity is and attempt to dispel its misconceptions. The chapter will explore the fire service's relationship with diversity, focusing both on its current status, common initiatives in attaining diversity, and barriers that prevent diversity from being implemented in many departments. The chapter will review the ethical implications of diversity initiatives and the ethical implications of fairness and justice regarding the treatment of minorities in the firehouse.

What Is Diversity?

If you search for the word **workplace diversity** in the dictionary, you will find the following definition, "the condition of having or being composed of differing elements: variety; especially: the inclusion of different types of people (as people of different races or cultures) in a group or organization" ("Diversity," 2015). Certainly, the definition is accurate; unfortunately, it's not particularly enlightening. The definition of "workplace diversity" is somewhat more helpful, "Similarities and differences among employees in terms of age, cultural background, physical abilities and disabilities, race, religion, sex, and sexual orientation" (Web Finance, 2018). While this definition is more on point, it still brings us no closer to understanding the complexities of the subject.

One of the challenges of diversity is that the term means different things to different people. Even within the academic community, discussions of diversity can range from employment initiatives to social interactions among diverse groups. Diversity is a complex subject; it encompasses many interrelated concepts. In some cases, confusion over the subject of diversity is rooted in a misconception of intent, or even terminology. For example, two fire fighters may be debating the value of diversity. While one fire fighter may be expressing opposition to the use of quotas in promotions, the other may be expounding the general value of a diverse work culture. They are both talking about diversity, and they may believe that they are talking about the same issue, but, in reality, they are talking about two separate elements within a larger realm of diversity. In general, the subject of diversity covers three important concepts:

1. A description of human traits, customs, ideas, and beliefs that are common to a class of individuals or are unique to specific individuals.
2. Employment practices intended to attain a balance of representation of identified "classes" of minorities.
3. A management/leadership theory intended to utilize the unique skills, talents, and viewpoints of diverse populations.

The first meaning of diversity listed tends to be relevant mostly to social scientists. For purposes of this text, it is the second and third diversity concepts that are of most interest because they reflect those elements with which organizational leaders and members routinely wrestle.

To gain an understanding of the literal and cultural meaning of diversity, it is instructive to review the thought processes that have shaped contemporary views on the subject. Our modern understanding of workplace diversity begins with the implementation of the Civil Rights Act of 1964, which forbade discrimination based on race, color, religion, sex, or national origin. The act set the stage for decades of court actions that came to shape the current legal impetus for nondiscriminatory hiring practices. Frequently referred to as a **compliance-based diversity strategy**, early diversity initiatives represented the original model for diversity, and were rooted in affirmative action and supported by the necessity to comply with equal opportunity employment objectives (Jewson and Mason, 1986). Compliance-based strategies have been a lightning rod for diversity critics since their inception. Typically, complaints coalesce around the general notion of **tokenism**. Critics charge that diversity initiatives promote reverse discrimination, and the employment and promotion of unqualified candidates. Proponents of compliance-based strategies argue that diversity in the workplace cannot be achieved voluntarily, and the social justice gains of diversity hiring justify the use of mandates to encourage minority hiring.

A refined approach to diversity, typically called **equal opportunity**, soon gained mainstream acceptance. It focused on the creation of a "level playing field." Rather than promoting affirmative action or quota programs, this approach to diversity sought to legally restrict unfair hiring and promotional practices that either facilitated unfair treatment in hiring practices, or policies that had an indirect negative impact on minority employment candidates (Cox, 1991; Cockburn, 1989). Proponents believed that by providing a level playing field, minorities could successfully compete in a fair market. The main thrust of their argument was that diversity initiatives should not impose unfair conditions on any group, including majority groups. They believe that quotas and affirmative actions constituted *a second wrong in an attempt to make a right*. Opponents were divided into two camps: those who believed that a level playing field could not be created by solely restricting unfair practices, and those who felt that the restrictions gave minorities an unfair advantage.

The third and most recent philosophical approach to diversity gained mainstream attention in the 1990s. It is sometimes referred to as a **transformational approach to diversity**. It asserts that a logical case can be made for an inclusive workforce based on organizational necessity. Proponents argue that organizations who lack a diverse workforce or fail to utilize the individual talents presented by a diverse workforce place themselves at a competitive disadvantage in problem solving and serving customer needs (Cockburn, 1989).

General Views on Diversity

Diversity has been and will likely continue to be a hot-button topic in the fire service for the foreseeable future. One of the causes of consternation is that proponents of diversity initiatives are usually referencing the transformational understanding of diversity, while opponents are usually focused on the compliance-based meaning of diversity. The differing viewpoints are not the result of miscommunication so much as both definitions are highly relevant to fire departments dealing with the issue of diversity. They are different concepts, but they are deeply interconnected. Strong proponents of diversity initiatives tend to focus on the perceived benefits of diversity and tend to minimize the organizational stresses that diversity initiatives can present. The most severe critics of diversity initiatives tend to argue that implementation of diversity initiatives can create injustices and have negative impacts on workforce quality while producing only marginal benefits to the organization.

In most cases, opposing views tend to argue from the extremes in any given issue, with reality lying somewhere in the middle. However, diversity issues remain contentious because even the extreme elements of both sides have some kernel of truth within them. Diversity can be difficult to implement, and there have been cases where well-intentioned diversity initiatives have caused individuals to be disadvantaged. Likewise, there are significant expected benefits from the development of a diverse workforce.

Let's be frank: Achieving diversity can be hard. Common experience supported by an overwhelming amount of literature discussing the challenges of diversity clearly demonstrate that diversity can be hard to attain and harder still to manage Figure 8-2. Research has shown that social diversity in a group can cause discomfort, tense interaction, a lack of trust, and the perception of conflict. So what's the upside?

Benefits of Diversity

The general consensus of literature supporting diversity tends to identify several likely benefits in having a diverse workforce. These include:

- **Increased adaptability:** Having a diverse workforce with a wide variety of experiences and talents helps an organization to problem solve, as well as identify and exploit opportunities (Greenberg, 2013).

Figure 8-2 Diversity can be hard to attain and hard to manage.

© Graham M. Lawrence / Alamy Stock Photo

- **Increased service quality:** Having a diverse collection of skills and experience allows an organization to better relate to and serve a diverse population. There is a general belief that a homogenous organization cannot effectively identify and serve the needs of minority groups (Greenberg, 2013).
- **Avoidance of groupthink:** Diversity within an organization brings fresh viewpoints. Homogenous groups tend to lack loyal opposition and as a result can collectively make poor decisions based on shared misconceptions and preconceptions (Greenberg, 2013).
- **Better workforce engagement:** Organizations that encourage diversity tend to better utilize the talents, ideas, and experiences of their employees to better effect. This translates to better performance and a higher degree of employee satisfaction (Greenberg, 2013).

The diversity benefits listed are straightforward. That individuals with additional experience, knowledge, and viewpoints can add to an organization's knowledge is easily recognized. There is, however, an increasing body of knowledge suggesting that diversity benefits may go well beyond the addition of cognitive resources. Years of research by practitioners of psychology and sociology suggest that diverse groups are more innovative and benefit from increased communication skills (Phillips, 2014).

Diverse workgroups are more likely to seek consensus than are homogenous groups. The presence of "different" individuals within a group dynamic tends to change decision processes, as mixed groups tend to recognize a need (not often present in homogenous groups) for accommodation. It was likewise found that diverse workgroups tended to share greater amounts of information and thoughts than did homogenous groups, likely because members of homogenous groups assume agreement and a shared frame of reference. Conversely, members of diverse groups tend to feel more obligated to share information, assuming that there likely may be differing references and experiences Figure 8-3 (Phillips, 2014).

There also seems to be significant evidence that as individuals we tend to pay closer attention to information shared by coworkers outside our own group identity. When someone who does not look like us has a differing opinion, we tend to listen more closely. This probably occurs because we are more likely to expect different viewpoints (Phillips, 2014).

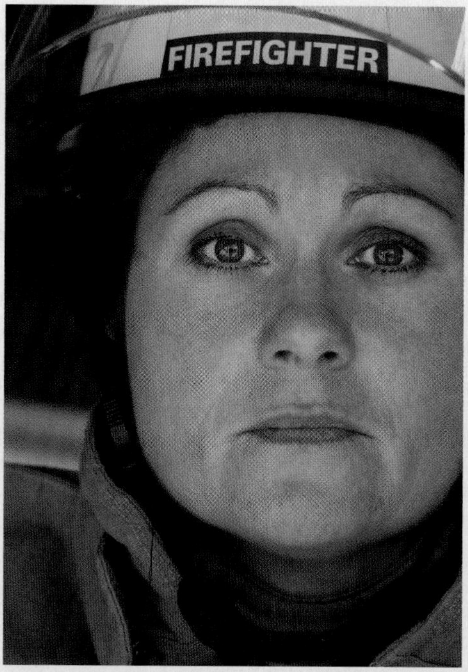

Last, individual members of diverse groups tend to prepare better for team tasks and exercises. They are also more likely to articulate their thoughts precisely, having collected their thoughts in advance and with more consideration. Studies have shown that members of diverse groups assume dissenting opinions and prepare for them, whereas members of homogenous groups are more likely to assume agreement and as a result place less effort into the decision-making process (Cox, 1991; Phillips, 2014; Page, 2007).

There are many advantages to a diverse workplace. Still, many organizations have a difficult time attaining and managing diversity. There is a plethora of authors who suggest that by and large diversity efforts typically fail. Statistical evidence backs up this belief. The last 10 years have seen no great increases in minority representation in almost all employment sectors, including the fire service (Fox, Hornick, and Hardin, 2006). Many organizations also report that even when minority representation has increased, many have not derived measurable changes in organization performance or culture. In spite of its likely benefits, diversity has not seen sustained growth over the last few decades, and organizations that have managed to increase diversity representation are often not reporting subsequent benefits (Davidson, 2014). There is a growing body of literature addressing barriers to diversity.

Figure 8-3 Diverse work groups tend to be more innovative and have better communication skills.

Barriers to Diversity

The great American endeavor to create a more diverse and fair workplace has more than five decades of experience behind it. While progress has been made in opportunities for minorities and women, it has been slow. The progress that has been made has come at a high cost. There is a long history of harsh words, legal battles, and even violence. The topic of diversity initiatives remains controversial and

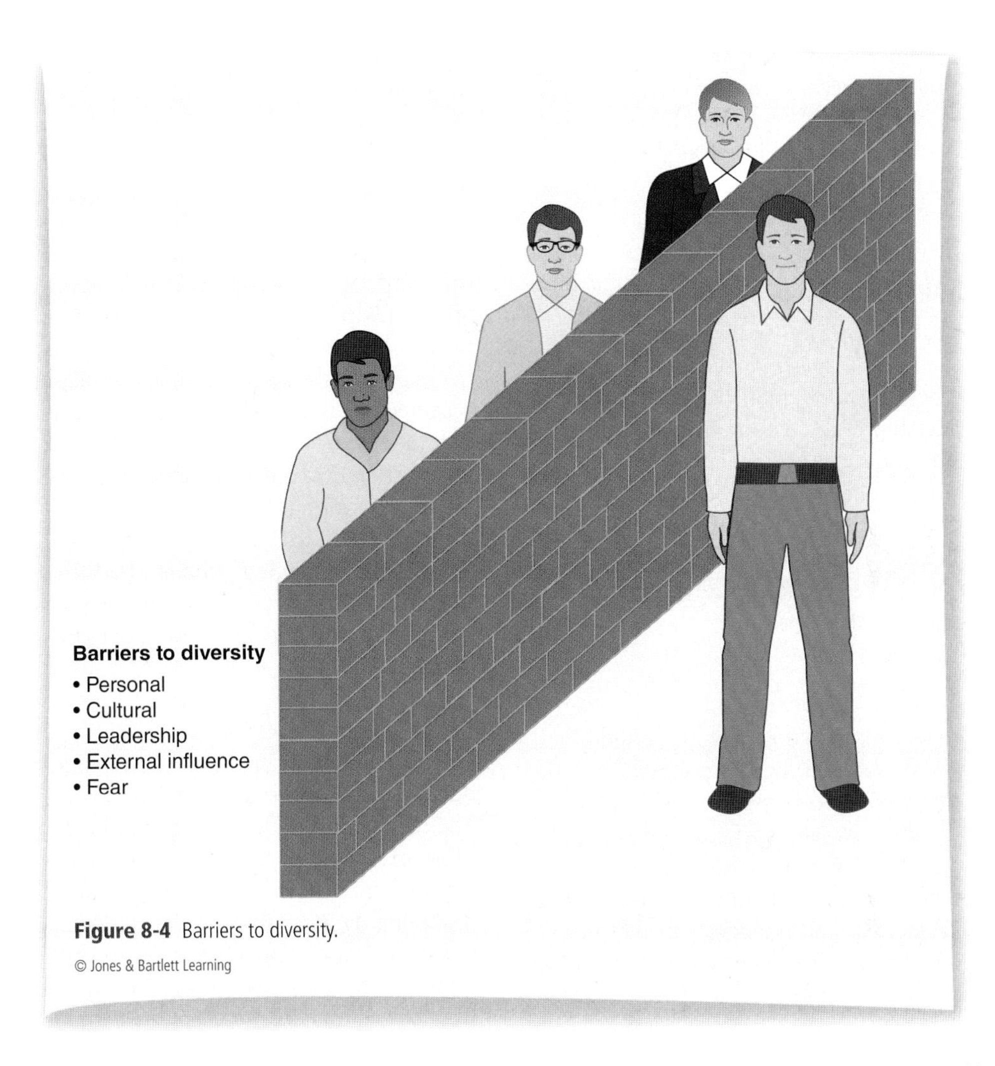

Figure 8-4 Barriers to diversity.

emotional. There are several elements at work resisting the diversity progress in the American workplace Figure 8-4.

Personal Inertia

There is a saying that variety is the spice of life. However, while we may enjoy variety in many things, there is evidence that people resist it in their relationships. We tend to associate with people who are similar to us, who look similar to us, who speak similar to how we do, and have similar values and opinions as we do. For example, if a young person walks into a room of older professionals, that person is likely to be very uncomfortable. If the young person sees someone of a similar age, he or she is very likely to gravitate toward that person under the assumption that they will likely have common bonds, thus making the social situation less awkward. The same phenomenon is true with race and gender. Humans are simply more comfortable socializing within similar groups. As such, we have a natural wariness of strangers. It is not uncommon for groups of similar individuals to exhibit hostility toward outsiders, especially toward those individuals who are substantially dissimilar in appearance or mannerisms. Sadly, differences in language, customs, appearance, sexual orientation, and nationality have been sources of violent conflict throughout history. Many social scientists observed that diversity tolerance and inclusion are higher social functions that must be learned in order to overcome the natural wariness of dissimilar individuals.

Inertia within Leadership

A common barrier to diversity is within an organization's leadership. Organization leaders are usually well served by the status quo. Even when organizational leadership recognizes values in diversity, they often find it difficult to implement meaningful changes in operational methods. This resistance is because they have a personal interest in the methodologies, policies, and cultures that they help create. Leadership often likes the idea of diversity but not necessarily its practice.

Cultural Inertia

In homogenous groups, the majority often does not see a need or value in diversity. Similar to their leadership, the status quo serves them well. There is comfort in the familiar, and they rarely accept that different is necessarily better. Well-intentioned individuals typically do not hold animosity toward the integration of women and minorities in the group; they just do not see a need to change methodology. This mind-set often results in indifference toward inclusion initiatives, with the resulting undermining of their potential benefits.

The Tyranny of the Majority

Many organizations that have attempted to integrate minorities and females into existing homogenous groups report frustration on the part of new members. It is common for majority groups to expect new members to assimilate into the existing culture and conform to its ideals and habits. Group members may openly welcome the addition of women and minorities into groups, yet thoroughly expect them to act as the majority group acts. Typically, the majority finds this to be a reasonable expectation. However, it can cause a great deal of stress on new hires who do not share the majority's cultural tendencies. Even in groups where conformance is not necessarily expected, there could be great pressure on new members to conform and fit in. Unintentional social ostracizing or marginalizing can result from nonconformance. Beginning in the 19th century, America has embraced a "melting pot" approach to immigration. It is still embraced by many. The melting pot analogy describes the expectation by the majority group that newcomers adapt to socially accepted behaviors, while simultaneously their presence somehow enhances the overall culture. This approach has two effects: it tends to homogenize the group, thus negating any value that might've been realized by diversifying the workforce, and, more important, the melting pot often leads to dissension, frustration, and conflict. We will discuss alternative approaches to the melting pot later in this chapter.

Fear

Often, members of majority groups find the introduction of "diverse workers" to be intimidating. The recognition of the necessity to be diversity sensitive or *politically correct* causes some individuals discomfort. Stories of harassment suits, or internal discipline as a result of unintended inappropriate behavior, cause some employees to avoid fully engaging with women and minority employees out of fear of repercussion. Some employees report that the extra pressure associated with being around people who are *different* is simply not worth the effort. As a result, walls go up, and *cliques* are formed.

Cultural and Social Barriers

Even among the best intentioned individuals, barriers associated with language, nonverbal communication, and customs can be challenging. Differences and experience can cause unexpected reactions to social interactions, and communication can become more difficult in the absence of cultural references. This issue is particularly exaggerated in the millennial generation who, more than most, rely on contextual speech references. Sadly, those inexperienced with intercultural exchanges often mistake the absence of understanding cultural references with a lack of intelligence. Diversity candidates often find assimilation and promotion challenged by barriers completely unrelated to job performance. For example, many older fire fighters may recognize the term "fubard" as a common description for a situation that is unacceptable. The term is a World War II military reference that has little meaning to millennials, women, or minorities.

External Influences

No workplace, including a firehouse, exists in a vacuum. Political and social strife outside the workplace shapes opinions of both minority and majority group members. Stereotypes, racism, sexism, and ultranationalism exist and can wreak havoc on the work environment. While organizations may make policies mandating socially accommodating behavior, it is impractical to assume that people will check their feelings and preconceptions at the door when they come to work. This is particularly problematic in small work groups and work environments where employees work closely together for long periods of time, such as a firehouse.

Emerging Group Patterns

Research has indicated that certain group patterns will emerge when new members are introduced into work groups that are dissimilar from the majority. Typically, the relationship between the group and the dissimilar member can be characterized by one or more **normalizing behavior patterns**. First, dissimilar members may attract a disproportionate amount of attention and scrutiny. This behavior has the effect of creating significant performance pressure. Second, majority members tend to develop exaggerated impressions of minority members. They may become hypersensitive to habits, especially those that reinforce preconceived notions. Last, minority members will either adapt behavior to conform to preconceived assumptions held by the majority or modify behavior to assimilate into the larger group (Kanter, 1977).

Benefits Outweigh the Challenges

It is evident that there are many challenges associated with diversity. It is likewise evident that there are many benefits associated with both employing a diverse workforce and utilizing diverse employees' experience and talents. Diversity initiatives require a lot of resources, time, and energy. They demand much from leadership and the employees. All the benefits and challenges discussed previously apply to the general workforce. It has been well established that the fire service is a unique culture, with its own traditions, vernacular, and customs.

The fire service has the unique issue of being a workplace that has much of the elements of a home and family. Fire fighters spend nearly a third of their lives together, not only working but sharing meals and chores, and in many cases just hanging out together. As has been previously explained, the fire service also has a unique relationship with the general public. Fire stations are often unofficial community centers, and, in many cases, local fire stations are part of the social fabric of the neighborhood. The question becomes, Do diversity experiences in the private sector translate to the fire service?

Diversity and the Fire Service

The face of America is changing. A report by the Pew Research Center indicates that minorities will become the majority by 2050. Asians are the fastest growing minority in the United States. The African-American population is projected to nearly double in the next 25 years. Millennials are on the verge of surpassing baby boomers as the largest segment of the U.S. population. Surprisingly, 43 percent of millennials identify as nonwhite. The population of Americans under the age of 40 has never been more diverse than it is now. Women are becoming a more dominant presence in the workforce. Women now represent 40 percent of the sole or primary family breadwinners. Although still underrepresented, women have experienced incremental yet steady growth into leadership positions (Cohn and Caumont, 2016).

The population composition of the United States is changing. However, the American fire service remains entrenched in its hiring practices and consequent demographics. In 2014 the United States had a total of 1,134,400 fire fighters. Of those, 69.5 percent were volunteers, and the remaining 30.5 percent were career fire fighters.

Ninety percent of volunteer fire fighters are white, and roughly 7 percent of volunteer fire fighters are female (National Fire Protection Association, 2012). These ratios have remained relatively static over the last decade. It is important to note that the overall number of volunteer fire fighters is diminishing. Since 2004 the number of volunteer fire fighters has dropped by almost 12,000 (National Fire Protection Association, 2012). A common concern expressed by volunteer fire chiefs is the inability to recruit and retain members. With a population that is becoming increasingly diverse, the inability to attract minorities and women becomes increasingly problematic for volunteer departments as they seek to maintain adequate staffing levels.

Likewise, career departments have struggled with diversity. In 2000, African Americans represented 9 percent of the fire service. Hispanics made up 5.2 percent, and women 3 percent. By 2012 African Americans were reported to comprise 7.2 percent of the American fire service, while Latinos represented 9.4 percent. As of 2015 women represented nearly 50 percent of the workforce, yet only 3.9 percent of fire fighters are women (National Fire Protection Association, 2012).

Despite its intended objective of diversifying, clearly the fire service has made little progress in changing its demographics. Although leadership in the fire service has repeatedly committed to diversity over the last two decades, many still ask the question, Should it?

Is it important for the fire service to actively pursue diversity? Certainly, there is a plethora of literature supporting diversity initiatives within the fire service, but the lack of meaningful progress suggests that it is clearly not a universally accepted principle among rank-and-file employees.

Resistance to Diversity

Some have argued that the fire service gains no particular benefit in diversity initiatives, and diversity initiatives may well do more harm than good. The argument goes that as a paramilitary organization with entrenched policies, procedures, and methodology, there is no real value in the new perspectives offered by a diverse population.

It has been asserted in fire service blogs and magazines that the command structure in the fire service precludes any meaningful advantage in diversity. They suggest that in a paramilitary structure diversity of opinion is irrelevant; they assert that we are here to do what we are told.

Others have expressed concerns that diversity initiatives "water down" the talent pool; they rationalize that because of the extreme demands of the profession, testing and hiring procedures should seek the very best of the best Figure 8-5. Any accommodation for diversity lessens the efficiency of the team and the life-and-death services it provides.

Although rarely argued publicly, there are those who suggest that diversity initiatives will only improve optics and that a mostly all-white, all-male department is at *least* as efficient in service delivery

Figure 8-5 Does the fire service really benefit from diversity?

as a diverse department. This argument is often accompanied by the observation that when a citizen's house is on fire, that person does not care whether a diverse workforce responds. More important, the fire does not care! In short, diversity does not have significant benefit, it is all about appearances, and the public really doesn't care that much about who works at their fire department.

Finally, and perhaps most passionately believed by opponents to diversity initiatives, is the idea that diversity accommodations unfairly punish qualified white male fire fighters. Not only has the issue of fairness been debated in firehouses, but it has also been subject to several legal challenges as well.

Are these critiques valid? Are diversity initiatives only an attempt at political correctness? Do they interfere with properly carrying out the mission of the fire service? Do diversity initiatives punish majority employees?

There are some fundamental flaws in the premise of all these critiques. The first flaw is the assumption that the fire service solely provides fire suppression. True, it is the fire service's core function; however, fire fighters spend very little time actually on fire scenes. Much more time is spent in fire prevention activities, inspections, EMS calls, rescue calls, and so on. The fire service is no longer a "one trick pony." Proponents of professional development in the fire service are quick to point out that it provides a myriad of services and that fire fighters need to be multitalented. It stands to reason that diversity of experience and talent would supplement that need.

There is also the assumption that the standard fire service testing process identifies the *best* candidates. Most fire fighters can attest to the fact that many highly successful fire fighters were not number one or two on their prospective hiring lists. High test scores do not necessarily translate into high performance. The standardized hiring process used by most departments is designed to identify *minimum* qualifications, not aptitude. Common sense, supported by case law, has demonstrated that the establishment of minimum qualifications is an imperative for emergency services, and few support the notion of hiring any candidate who does not meet minimum qualifications. However, there is no credible evidence to suggest that giving preference to *qualified* diversity candidates in any way negatively affects actual job performance.

Clarification

The standardized hiring process used by most departments is designed to identify minimum qualifications, not aptitude.

The belief that the fire service's command structure negates the value of diverse perspective and experience is an opinion usually held by rank-and-file members, not by administrative personnel. Groupthink and organizational inertia have long been recognized as impediments to fire service professionalism. More than most, organizations with strict chains of command benefit most from loyal opposition.

The issue of fundamental fairness may be the most complex argument about diversity in the fire service. Traditionally, personal objections and legal challenges have been made in cases where organizations have changed qualifications, testing conditions, or scoring methods to achieve higher success rates among minority candidates. These types of initiatives are sometimes called quota systems or are mistakenly called affirmative action. Complaints about creating parallel testing conditions usually claim "reverse discrimination." In these cases, white male candidates are considered to be put at a substantial disadvantage in access to hiring and promotion. Moreover, in some court cases, judges have found these practices illegal. The confusion lies in the assumption that all diversity initiatives incorporate some form of "rule change" to promote the interests of minority candidates. The reality is that most fire departments limit diversity initiatives to the recruitment of candidates who then compete for positions under conditions identical to those of other candidates. Hiring preference is given to otherwise qualified candidates who have successfully completed a standardized test identical to that given to all other candidates. The argument for giving preference is that there is value in having diversity in the workplace and that giving preference points for diversity is justifiable in the same way as giving preference points to military veterans or individuals holding college degrees.

The argument that the fire service derives no benefit from fresh perspectives is likewise flawed. The assumption that the fire service has no need for fresh ideas or methodologies is not only inaccurate, it is dangerous. As firefighting becomes more technically challenging, and as ancillary services become a bigger part of the fire services, innovation and flexibility in approaches to service delivery will become increasingly important.

Incentives for Diversity

Earlier in this chapter, we discussed the general benefits of diversity, but are there elements of diversity that are especially valuable to local fire departments? In general, a diverse fire department with an inclusive culture is much better at engaging the public it serves. The following observations support this thinking.

- Diversity creates efficiency of operational service through a better understanding of the unique needs and challenges faced by groups within the community, including cultural habits that may pose fire risks, and creates efficiency.
- Organizational diversity can be instructive in the development of targeted fire prevention programs. Fire prevention activities become more relevant to individual minority groups and likely are more cost-effective.
- Fire departments that fully engage with their community's demographic groups are much more likely to be better prepared for disasters through the development of more accurate disaster plans.
- Fire departments are likely to be more efficient in handling emergency scenes where strong community relations already exist.
- Many fire departments lament the difficulty of recruiting minority candidates. An emerging realization is that the department's ability to recruit qualified candidates requires community engagement, which requires a diverse and inclusive department—diversity begets diversity.

Socially, fire departments that have diverse workforces and an inclusive culture are more able to engage and communicate effectively within all facets of the community they serve. These open lines of communication create opportunities for customer service programs, as well as an increase in the efficiency of fire prevention operations. A positive interaction with minorities within a community also increases the stature the department enjoys as a political entity. Conversely, fire departments that suffer from minority distrust can be targets for political activism, lawsuits, and in some cases even violence.

Internally, fire departments with an inclusive culture report that fire fighters tend to become more self-actualized. Further, they are more likely to seek promotion and generally have a more positive attitude toward their work. Inclusive departments also report decreased numbers of sickness and absence days, and less turnover. Diversity can increase monetary efficiency by reducing overtime costs, training costs, and recruitment costs (Sagen and Pini, 2008).

By recruiting and employing women and minorities, fire departments not only gain a better understanding of the public they serve, they also tap into a wider pool of skills, thus creating internal opportunities for improved management, program development, and administration.

By having a diverse workforce, fire departments become more attractive to a wider group of potential work candidates. This improves their ability to recruit the best talent, most of whom have options other than the fire service. Members of fire departments understand that a career in the fire service is highly rewarding; however, highly talented potential candidates may overlook the fire service as a potential employer simply because they do not identify with it. For example, cultural changes over the last 10 years have seen increased numbers of highly skilled and highly trained women coming out of the military. Their physicality, experiences in high-stress and dangerous situations, and experience in working within predominantly male work environments make them ideal candidates for the fire service. Yet most do not pursue the fire service as a career opportunity because they simply never considered it as an option. The more women and minorities working in the fire service, the more the fire service is seen as a career opportunity.

A significant incentive for fire departments to pursue diversity employment is the avoidance of potential legal ramifications. As a public employer, fire departments are under scrutiny for their hiring and testing practices. There is an expectation that fire departments will actively recruit minority and female candidates. The absence of progress in recruitment invites court-ordered intervention.

The Ethics of Diversity

To this point, this text has focused on the issue of diversity as a practical matter. The question becomes, What are the ethical implications of diversity initiatives?

The most obvious ethical issue supporting diversity initiatives is that of fairness. Intuitively human beings generally recognize that individuals should be treated fairly. At a very basic level, an ethical argument can be made that all people should have equal access to gainful employment and promotion.

There is another side to the issue of fairness. In a larger sense ethics would imply that minority groups are entitled to fair treatment, and that past injustices may necessitate present remedial actions. Others would argue that redressing past wrongs at the expense of "innocent bystanders" is inherently unethical. The past few decades have seen no shortage of frustrated white males who have felt they were cheated out of jobs or promotions as a result of diversity initiatives. At a strategic level, addressing gender and racial inequalities seems just, yet it may well seem inherently unfair at a personal level.

A driving force behind the diversity movement of the last 40 years has been recognition that women and minorities have been "shortchanged" in the workplace, including the fire service. Proponents argue that as a society we have an ethical obligation to create equal pathways for success for all of society's participants. Immanuel Kant's *Moral Imperative* as well as John Locke's *Natural Rights Theory* articulate this consideration. Consequentialism would suggest that the greater good is served by having all groups within a society prosper. As a result, initiatives to help those in need of assistance is an ethical obligation.

Conversely, egoism or enlightened self-interest would suggest that society does not benefit by enabling minority groups to succeed at the expense of dominant populations. They would argue that this ultimately weakens the minority group as well as the majority group, and hence diversity initiatives are inherently unethical. It should be noted that this is a minority opinion rarely held by contemporary philosophers or ethicists. Almost all moral and religious ethical systems recognize some degree of obligation to "take care of one's neighbor."

Courts have consistently recognized the ethical value of diversity initiatives and have even imposed mandatory remediation for organizations with a demonstrated pattern of discrimination. They also, however, have found that creating "biased" hiring and promotion policies that unfairly treat majority candidates (reverse discrimination) are unethical and illegal Figure 8-6.

As emergency service providers, fire departments routinely deal with life-and-death situations. As such, a commonsense argument can be made that it is inherently unethical to risk the lives of citizens or fellow fire fighters by hiring unqualified candidates or promoting unqualified candidates to positions of responsibility beyond their capabilities. This axiom is nearly universally agreed on both in firehouses and in courtrooms. However, while this supports the notion that it is unethical to hire unqualified minority candidates, even with good intention, there appears to be no ethical preclusion from giving preference to women and

Figure 8-6 Hiring practices must be free from bias.

minorities who are *otherwise qualified* to serve as fire fighters.

While fire departments are ethically required to avoid any activity that impedes their ability to protect the lives and property of others, they have the parallel ethical responsibility of pursuing those things that facilitate those pursuits. The fire service also has a fiduciary responsibility to manage public assets as efficiently as possible. If a logical case can be made that fire departments benefit from the presence of a diverse workforce, then there is an ethical obligation to have one.

There is a humanistic consideration in the creation of an inclusive work culture. Several ethical theories suggest that an organization has a responsibility for the humanistic needs of its employees. As such, a work environment that is hostile to minorities or women represents an ethical breach of duty by management.

Finally, fire departments are public agencies. They are supported by tax dollars, including those tax dollars paid by minority groups. As public agencies, fire departments have an ethical responsibility to equally protect and serve all members of their community. Minority populations have every right to expect equal access to services and benefits (including employment) as any other community group.

Achieving Diversity

Demographics

The term **demographics** refers to categorizing a population based on specific markers. When speaking about diversity hiring initiatives, the primary markers are race, gender, and nationality. Most recently sexual orientation has also become a primary group identifier. Typically, laws regarding employment practices, discrimination, and equal opportunity typically reference primary markers. Not surprisingly, most discussions regarding diversity initiatives and recruitment policies also tend to focus on the same primary markers.

While public policy focuses on primary markers, individuals are much more complex. Beyond the primary markers, we tend to self-identify by numerous characteristics, affiliations, and qualities. Secondary factors in demographics include, but are not limited to: religion, marital status, economic status, education, appearance, and political views. It is the existence of the secondary factors that can complicate the attainment of diversity through the sheer pursuit of demographic balance. A diversity initiative aimed at hiring a particular percentage of women, Hispanics, or African Americans may or may not actually achieve diversity as intended because these markers do not speak to experience, thought, attitude, or outlook. Still, the achievement of diversity is nearly impossible if an organization consists solely of one homogenous group. While relying solely on demographics only tells half the story, it seems to be an essential half Figure 8-7.

Demographic Factors

Several tactics have been employed to achieve diversity through hiring a demographically diverse group. Each presents opportunities and challenges. Each has its supporters and its detractors.

Achieving Demographic Representation

Quota systems represent the most simplistic form of diversity hiring initiatives. It is perhaps also the most controversial. Quota systems are used almost

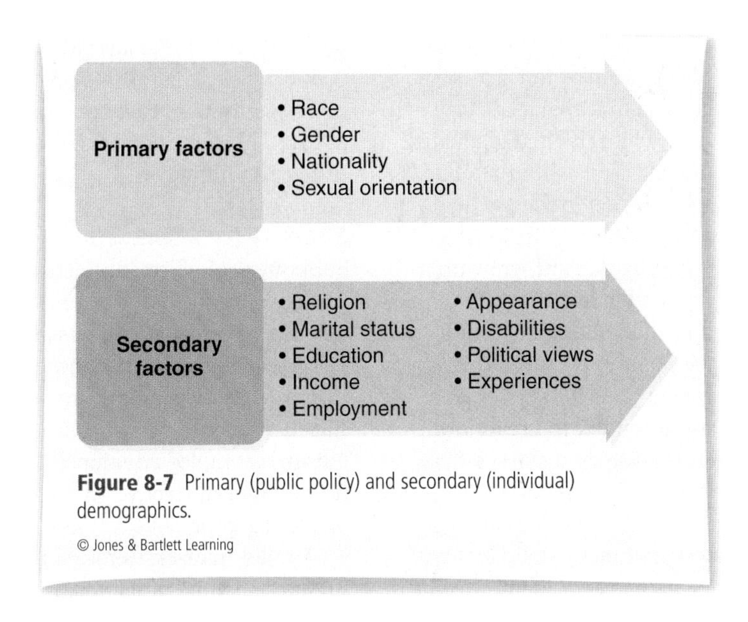

Figure 8-7 Primary (public policy) and secondary (individual) demographics.

exclusively by public agencies or as the result of a court-ordered mandate. It is a two-phase process. First, analysis is done determining the number of available minority workers within a population, then the data is used to create targeted hiring numbers. Second, an agency reviews its current workforce and determines whether or not there are underrepresented groups, typically based on racial and gender profiles.

The rationale for the use of quota systems is that some organizations that lack affirmative action diversity initiatives or that have been actively avoiding hiring minorities may need to be compelled to do so. Quota systems tend to be the most direct and pragmatic way to force organizations to diversify (EEOC, 2016). It is the qualities of simplicity and verifiability that attract public agencies and courts.

Quota systems are often maligned, and not without cause. They tend to be problematic for four reasons.

1. **They may set unrealistic goals by not taking into account the number of available qualified candidates within the hiring population.** As an example, females make up approximately 50 percent of the available workforce. At face value, they should theoretically hold 50 percent of the firefighting jobs in the country. Of course, the reality is that a much smaller percentage of women available to work for the fire department have an interest in doing so. In comparing the fire service to other nontraditional female job categories, a more realistic expectation would be around 17 percent.
2. **They discount merit-based hiring and promotion.** This can cause resentment within the workforce. This resentment not only negatively affects majority stakeholders but also tends to marginalize minority employees even if they are meeting or exceeding expected work goals.
3. **Quota systems can perpetuate stereotypes.** There is an assumption within the application of a quota system that by hiring an African American, or a woman, or a Latino, then diversity in thought, methodology, and experience is automatically gained. If a quota system is to accomplish anything beyond demographic representation, then there must be the assumption that all minorities are inherently different from white males in some meaningful

way. By extension, this implies that all minorities are to some degree similar to each other. As an example, let us assume that a law firm is completely staffed by white Harvard graduates from wealthy families, all of whom attended private prep schools. In order to achieve demographic diversity, they hire an African-American woman who went to Harvard, grew up in a wealthy family, and attended a private prep school. Is there a likelihood that she brings a particularly diverse point of view based solely on her race and gender?

4. **Quota systems are intended to create minimum hiring levels of protected classes.** They often have the unintended consequence of creating maximum standards as well. Many organizations who meet minimum standards will desist in diversity initiatives. Quota systems tend to create numerical targets rather than changing cultural attitudes toward inclusion.

Affirmative Action Plans

Affirmative action plans are often used synonymously with that of quotas, but they are different. Affirmative action is a governmentally required positive effort to diversify a workforce. Affirmative action programs set goals for progress in diversity, whereas quota systems set mandated specific numbers. Typically, affirmative action programs achieve those goals by active recruitment and by giving preference points to otherwise qualified candidates. Quota systems have no provision for preference points but simply mandate the hiring of specific numbers of protected groups regardless of qualification. In a quota system, employers are encouraged to hire qualified minority candidates when possible, but in the absence of qualified candidates, they may be required to hire minority candidates they feel come closest to qualified. Affirmative action programs are typically barred from fundamentally changing entry or promotion qualifications to accommodate diversity hires. Court cases have found that while opportunity initiatives are legal, they are barred from creating alternative qualifications for "protected groups." Landmark legal proceedings that shaped affirmative action law include *Gratz v. Bollinger* (2003) and *Regents of the University of California v. Bakke* (1978). Both of these cases set limits as to what constituted legal pursuits of diversity versus reverse discrimination in affirmative action initiatives.

Support for Affirmative Action

Supporters of affirmative action programs insist that proactive measures to increase diversity in the workplace are necessary to redress inequalities in the workplace. Their belief is that hiring practices have inertia; organizations (including fire departments) that are staffed primarily by white males tend to attract and ultimately employ additional white males. Active participation in affirmative action programs are needed to "break the cycle."

Unlike quota systems, supporters argue that affirmative action programs are not only race-based but also gender- and class-based. Affirmative action programs tend to benefit a wider array of underrepresented groups, including various nationalities, sexual orientations, and the economically challenged. The identification and hiring of economically challenged groups particularly has positive effects within the Latino community. Latinos are statistically more likely to be living in poverty than whites, yet in many equal opportunity employment initiatives Latinos are not specifically defined as hiring "targets."

There are detractors of affirmative action as well. Even though affirmative action does not typically reduce qualifications for entry or promotion, some still argue that preference points produce an equal and unfair access for minority groups to employment and promotion. Opponents also argue that, similar to quotas, affirmative action programs focus primarily on targeted groups and have no value for groups not identified as protected by law. Example groups include white immigrants, religious minorities, and the LGBTQ community.

Diversity Recruitment Programs

Diversity recruitment programs seek to encourage diversity hires by identifying and actively pursuing qualified candidates from minority pools. Often, these programs are continuing efforts by the organization and have no specific target goals for minority representation. These programs may or may not use preference points in the hiring process. Diversity recruitment programs are usually undertaken voluntarily.

A similar concept is used in the promotion of minority candidates who are already in the organization. These programs are sometimes called **mentoring programs** in which minority candidates are given access to specialized training programs to better prepare them for promotion opportunities.

The U.S. Equal Employment Opportunity Commission (EEOC) advocates targeted recruiting as a proactive method of achieving diversity. However, like quota systems and affirmative action programs, diversity recruitment programs may have limitations. Some states, such as California, preclude governmental agencies such as fire departments from recruitment campaigns that target specific groups (Fox, Hornick, and Hardin, 2006). Unfortunately, laws regarding recruitment initiatives can be vague. While most such laws express prohibition of recruitment practices that disadvantage any group, they do not typically define practices that are permitted or prohibited. As an example, placing a recruitment advertisement in *Stars and Stripes* targets veterans, but at face value, it does not disadvantage nonveterans. Conversely sending invitations to minority candidates inviting them to participate in testing programs has been found to be prohibited (Fox, Hornick, and Hardin, 2006). It is also important to note that specific recruitment practices that are allowed or disallowed may vary from state to state.

The IAFF published a white paper generally supporting diversity recruitment. In it, IAFF identified the best practices for achieving diversity through minority recruitment (Fox, Hornick, and Hardin, 2006). The report advocates use of specific language that indicates inclusiveness, the use of formal recruiting methods such as advertising that targets diversity groups, and the use of minorities and women as recruiters. It also advocates the use of mentoring groups that prepare candidates for qualification testing.

Any diversity hiring initiatives including recruitment are an exercise in futility if a fire department is utilizing biased employment, promotional, and human resources processes and policies. Additionally, recruitment efforts are ineffective if a fire department has a general culture hostile to inclusiveness. Even well-intentioned departments can create policies and procedures that are either disparate in their treatment of employees or disparate in impact. **Disparate treatment** is identified as a policy or procedure that by its nature discriminates against or unfairly marginalizes a particular group. As an example, requiring fire department job applicants to be male constitutes disparate treatment in that it excludes females.

Clarification

Disparate treatment and disparate impact are often confused. The first means to treat people differently. The second refers to treating people the same, but the action has an unfair result.

Disparate impact is a policy or procedure that at face value appears to treat all groups equally but may have an adverse effect on a particular group. Initiating a height requirement above six feet is an exercise in disparate impact. It differs from disparate treatment in that the height requirement is applied to all applicants equally, and theoretically any person from any group can meet the requirement. It is, however, a statistical fact that a significantly higher number of males are over six feet compared to females. Hence a six-foot height requirement has a disproportionate effect on females as opposed to males. While the example of height requirement seems like an obvious case of disparate impact, some common employment requirements may, in some cases, have unjustifiable disparate results. For instance, utilizing vocabulary testing may be a disadvantage for populations in which English is not a first language, and residency requirements by predominately white affluent communities may have a disparate impact on African Americans or Latinos.

What further complicates the issue of disparate impact is that, in some cases, practices that have disparate impact implications may be justifiable and legal. For instance, the candidate physical aptitude test (CPAT) has a higher failure rate for women as compared to men, yet it has been vetted as being specifically task related. As a result it has been found by several courts to be a valid test instrument.

Managing Diversity

Here's a thought experiment: Close your eyes and quickly describe yourself in five words or less. There is a high probability that your self-description did not reference your gender, race, or nationality. You very likely described yourself based on those qualities or affiliations that you feel, make you unique. Perhaps you described yourself as a fire fighter, parent, spouse, and football fan. Human beings are complex; those complexities go far beyond our primary or even secondary diversity factors. Each of us has unique backgrounds, opinions, values, and preferences.

There is an approach to diversity management that goes beyond the mere hiring of minority groups. The **transformational approach to diversity** asserts that diversity goes well beyond outward appearances and that simply hiring people of different color or gender or nationality does not make an organization more diverse. It is likely that an organization cannot attain diversity without hiring people from diverse backgrounds. However, it is also true that an organization can have a diverse workforce and still suffer from groupthink, organization inertia, and a culture of noninclusiveness.

Clarification

A fire department cannot attain diversity without hiring individuals from a diverse background. Having individuals from diverse backgrounds does not guarantee that a department will achieve cultural diversity.

The transformational view of diversity requires an organization not only to have representative groups but also to utilize and celebrate those differences they bring with them when they walk in the door.

The true challenge for an organization is not the acquisition of minorities or women into the department but rather accommodating their unique talents and skills. Many of the negatives experienced by departments in their diversity initiatives have resulted from a failure to grasp this concept. Not having a diverse workforce is only marginally less productive than hiring women and minorities only to marginalize them by compelling them to conform to the established culture. It both minimizes their contribution to the larger group and causes frustration and tension for the new employee.

There is an obligation for any new employees, including minority employees, to "get along." However, that obligation does not encompass the expectation of changing their personality, speech patterns. or cultural identity. It is unreasonable to hire a young woman into the fire service and expect her to be "one of the boys," just as it would be unreasonable to expect male employees to adopt a feminine persona. Yet the expectation for women and minority fire fighters to reshape themselves to fit into the appropriate pigeonhole happens far too often in firehouses across the country. This observation is supported by significant data based on female and minority fire fighters' experiences (Huwlett, Benedict, Thomas, and Moccio, 2008).

Several resources identify best practices in managing a diverse workforce within the fire service (Fox, Hornick, and Hardin, 2006; Huwlett, Benedict, Thomas, and Moccio, 2008; Willing, 2011).

Best Practices in Managing Diversity

- Utilize unbiased hiring and promotional selection methodology.
- Clearly articulate the department's commitment to diversity and the values of diversity to the organization.
- Review policies to make sure that they are not counterproductive to inclusiveness. For example, are policies for sick time, family leave,

personal days, and flextime unnecessarily restrictive to women?
- Avoid policies that may unintentionally segregate minorities and women. Certain accommodations must be made but avoid unnecessary policies and procedures that may inadvertently single out minority members of the department by identifying them as "special cases." As much as is possible, normalize the presence of minorities and especially women.
- Provide ethics and diversity training for all employees.
- Effective inclusive cultures link diversity initiatives to career paths. They identify skills and experience necessary for career success and ensure equal access to resources for individuals' professional development.
- Ensure that minorities and women within the department have a voice. This includes representation within the administration, on committees (such as safety committees, grievance committees), and within unions.
- Implement processes by which minorities and women have access to both formal and informal methods of redress in cases of perceived unfair treatment.
- Embrace a culture that seeks and fairly evaluates input and suggestions from all employees, at all levels of the organization.

A common challenge faced by officers in managing a diverse work culture is that of informal exclusion. This form of divisiveness can be intentional or unintentional; it can spring from fear, awkwardness, or genuine bias.

It is important for company officers to maintain an inclusive culture in the firehouse. This first means making it clear that the fire officer has an obligation to represent the interests and well-being of *all* fire fighters. Leadership and management actions must support team inclusiveness. Despite whatever differences in background may exist among employees, the fire company is a team, and every member is an equal part of it. Whether minorities or women are present in the firehouse or not, a basic culture of civility, respect, and accommodation for individual differences must be maintained. As one fire officer put it when asked how to treat a new female fire fighter, "Treat her exactly like you would like your sister treated; after all, she is your sister."

Clarification

Despite whatever differences in background may exist among employees, the fire company is a team, and every member is an equal part of it.

Texture: Eky Studio/ShutterStock, Inc.; Steel: © Sharpshot/Dreamstime.com

Women in the Fire Service

The ethical issues pertaining to women working in the fire service are relatively straightforward. As a matter of fairness and social justice, any qualified candidate should have equal access to employment of his or her choice. This fundamental right is consistently validated within the theories of virtue ethics, deontology, and consequentialism. Further, once gainfully employed a woman or any other employee should enjoy fair treatment, fundamental privacy, and an environment free of hostility. Again, these rights are consistent with every theory of ethics.

Clarification

As a matter of fairness and social justice, any qualified candidate should have equal access to employment of his or her choice.

Texture: Eky Studio/ShutterStock, Inc.; Steel: © Sharpshot/Dreamstime.com

As a profession, the fire service has given vocal support to the concept of equal access to employment by minorities and women. Yet data suggest that the fire service has significant issues in attaining anything close to diversity parity with other professions, including those professions identified as typically masculine, hazardous, strenuous, or dirty (Huwlett, Benedict, Thomas, and Moccio, 2008).

In 2008 the International Association of Women in Fire and Emergency Services (IAWFES) issued a report detailing the scope of the insufficiencies. Their findings include evidence that more than half of the fire departments in the United States have never had a female fire fighter. Women make up nearly 50 percent of the American workforce, yet less than 4 percent are fire fighters When comparing the fire service to other stereotypically male professions, such as construction, law enforcement, and auto repair, data suggest that roughly 17 percent of the fire services workforce could reasonably be expected to be female (Huwlett, Benedict, Thomas, and Moccio, 2008). This is significantly higher than present fire service female employment levels Figure 8-8.

Many fire chiefs that recognize the necessity to recruit and hire more women fire fighters lament that women simply do not view the fire service as a potential career choice and that is why women are so significantly underrepresented. Sadly, women may very well have a good reason not to view the fire service as a potential employer. Contrary to what we would like to believe, there is significant evidence that the fire service is hostile to minorities in general, and women in particular.

Figure 8-8 The fire service lags well behind other traditional male professions in recruiting women.

Again, the IAWFES report reveals unsettling data. On average 29 percent of fire fighters of color report some form of disparate treatment. Even more unsettling is that 84.7 percent of women report being treated differently because of their gender. More than 36 percent report that their gender has been a barrier to career advancement, and 50 percent have reported they have been ostracized or shunned in the firehouse. Forty-six percent of surveyed women report that their privacy has been invaded while showering or changing clothes, 30 percent report sexual advances, and 24 percent say that they have been denied training or were trained differently because of their gender (Huwlett, Benedict, Thomas, and Moccio, 2008).

The statistics clearly demonstrate that there is a significant problem in firehouse behavior toward women and minorities. The behavior described previously is unethical and represents a real problem to the fire service. Unethical treatment of female fire fighters by their departments creates a significant potential for legal action. Unethical behavior of any kind, but especially hostile treatment of female and minority fire fighters, can have a significant negative impact on fire department reputations. The perception that the fire service is hostile to females and minorities can impair its ability to recruit talented people, and not only within those populations. Talented, intelligent prospective employees (including white males) are often reluctant to join organizations seen as regressive or having hostile cultures. Negative attitudes toward minorities and women often spill over into other facets of the department culture. Impoverished leadership that fails to address diversity issues is also likely to tolerate bullying and harassing behavior among the department's general population. Regardless of the presence of minorities or women, incivility in the firehouse can have a cascading negative effect on morale and efficiency.

From an ethical as well as a practical point of view, the fire service must do better in treating its female employees. Leadership must take an active and decisive role in putting a stop to boorish behavior.

Fire departments are workplaces. No fire fighter has a right to expect the department to be an exclusive social club or a frat house.

The LGBTQ Community and the Fire Service

The last 10 years have seen a significant mainstreaming of the LGBTQ community. As members of the LGBTQ community have become more open about their sexual orientation, highly traditional workplaces have had to adjust accordingly. Certainly, the changes in social attitudes toward LGBTQ issues have not been without their share of controversy. While national fire service leadership and publications have been nearly universally in support of LGBTQ rights, reactions on individual apparatus floors have been somewhat more mixed.

From an ethical point of view, much of what has been discussed in this chapter regarding the role of diversity and women in the fire service pertains to the LGBTQ community as well. However, there are certain aspects of LGBTQ inclusion that are unique from those of people of color and women. One such difference is that of optics. Gender and race are easily viewable, whereas sexual orientation is not easily perceived. This lack of perception has, at least to some extent, impeded the national conversation regarding LGBTQ rights. Until the recent past the gay community was a silent minority. The treatment of LGBTQ people has only recently become a mainstream issue.

Another important difference is the religious connotation surrounding homosexuality. A significant portion of the country self-identifies as religious conservatives. Many religious conservatives oppose homosexuality as a matter of religious principle and moral obligation. Some are resistant to hiring LGBTQ people to high-profile public positions, and some express resistance to working with them.

Certainly, individual rights as guaranteed by the First Amendment of the Constitution support the individual's freedom to believe as he or she wishes. However, there is a large body of case law asserting that individual beliefs do not legitimize discriminatory action. Similarly, ethical theory nearly universally suggests that a person's beliefs do not justify hostility, violence, or the infringement of other people's rights.

Ethics suggest there are there are certain principles relevant to LGBTQ rights within the fire service.

- The fire service has an absolute obligation to equally protect and respond to the needs of all community members. Religious beliefs do not supersede the ethical obligation to the public inherent in being a fire fighter. To be blunt, if a fire fighter believes that his or her personal beliefs give that person the right to deny emergency services to any class of individuals within the community, he or she is in the wrong line of work.
- As a matter of fairness and law, persons who qualify to be fire fighters have a right to pursue it as a career. Besides the ethical principle at work, as a public institution, U.S. law guarantees equal access to government services and opportunities.
- All members of a fire department have a right to expect ethical treatment by both department leadership and the rank and file. This includes the right to work in a nonhostile environment, to be treated with respect, and to have equal access to promotion. Insults, harassment, disparate treatment, and shunning are not ethical.

When discussing the presence of LGBTQ persons in the firehouse, some fire fighters have expressed concerns regarding unwelcome sexual advances. Certainly, all fire fighters, gay or straight, have a right to expect not to be sexually harassed. This is not a "gay issue" any more than it is a heterosexual issue. Fire stations are workplaces. Unwelcome sexual advances should not be tolerated in whatever form they may take. There is, however, no evidence to suggest that gay fire fighters are any more likely to engage in inappropriate behavior than are heterosexual fire fighters. In short, the issue is a nonissue.

Diversity Is a Reflection of Professional Ethics

Issues surrounding diversity initiatives and the treatment of minorities, women, and members of the LGBTQ community in the fire service remain the subject of contentious debate and high emotion. The changes in American society and the push to change the culture of the fire service can be unsettling to some and even threatening to others.

There are ethical principles at work here that go beyond social science and political correctness. Much of this text so far has focused on the ethical obligations of the fire service not only to protect the public but to represent it as well. A fire department's hiring practices and treatment of its employees directly reflects on its relationship with the citizens who support it. The fire service is owned by the community—all members of the community. It has an ethical obligation to provide equal access by all community members to our services, and to the employment opportunities they provide with their tax dollars.

The fire service's culture is built on the premise of public service, and the virtues of courage and compassion. Those principles are rooted in ethics. As has repeatedly been stated in this text, the American fire service is by its very nature and its mission inherently rooted in ethics. How the fire service treats its citizens and employees is a direct reflection of its ethical character and how well it represents its founding principles.

WRAP-UP

Chapter Summary

- The basic elements of diversity were explored, with special attention paid to the ethical relevance of diversity as it applies to the fire service.
- The term "diversity" actually relates to several concepts including an academic description of various traits and customs of a particular group and employment practices associated with the development of a demographically representative workforce.
- Diversity also describes a management theory that focuses on the utilization of unique skills and traits possessed by individuals of various backgrounds.
- Various meanings of diversity make a single definition problematic; however, for purposes of study in this chapter, diversity was defined as "similarities and differences among employees in terms of age, cultural background, physical abilities and disabilities, race, religion, sex, and sexual orientation."
- American history is, in a sense, the story of integrating diverse peoples into a cohesive and unique culture. The great stories of our history are of immigration, settlement, slavery, and the struggle for equality.
- Our current struggles with diversity are in many ways a legacy of those struggles. Workplace diversity became an issue of prominence in the 1960s and became a mainstream political issue with the passage of the Civil Rights Act of 1964. Since that time workplace managers, social scientists, and lawmakers have wrestled with the meaning of diversity and how to achieve it.
- The fire service, similar to other private and public organizations, has invested much time and effort into understanding and implementing fair workplace policies. First attempts at workplace diversity developed in a time when racial tensions were high and discrimination was common.
- Mandatory hiring practices and racial quotas seemed to be the only effective method of integrating women and minorities into white male–dominated professions. Within a decade, opponents of hiring quotas argued that they were heavy-handed and not effective.
- Private- and public-sector organizations began investing in what they called equal opportunity programs such as affirmative action. The rationale behind equal opportunity programs was that laws should "level the playing field" for minority job candidates. This approach often rejected the promotion of minority hirings at the unfair expense of the majority.
- The third and most recent approach to diversity gained mainstream acceptance in the last few decades of the 20th century. Known as a transformational approach to diversity, it is based on making a business case to incentivize diversity initiatives.
- The approach argues that diversity employment benefits companies both economically and socially. This form of diversity initiative also recognizes an ethical obligation and a social responsibility for diversity hiring.
- The question was asked regarding whether or not diversity presented any real advantages for the fire service, and whether the difficulties in attaining and managing diversity offset those benefits. It was found that the mainstream consensus of diversity being important to the fire service was accurate.
- There are compelling arguments supporting diversity initiatives both from an ethical perspective and from a practical one. Several advantages of diversity in the fire service were explored.

WRAP-UP

- The hierarchal structure and strong traditions of the fire service were reviewed as possible barriers to inclusiveness, and it was determined that fire departments must ensure that minorities and females have a voice within the fire service organization. This not only includes having access to leadership but also through the promotion of minorities and women to positions of leadership.
- The fire service has an ethical and moral obligation to provide equal service to all members of the community and provide equal access to those services. As a publicly funded organization, the fire service has an ethical obligation to provide equal access to employment for the citizens who financially support it.
- All members of the department have a right to expect a nonhostile work environment and to be equally represented by fire department policy and leadership.
- There are numerous ethical principles that support the ethical imperative for fire fighters to treat all members of the department with respect and dignity.
- The chapter closed with discussions regarding current trends challenging fire departments managing diverse work cultures. Special attention was focused on the entry of women and the LGBTQ community into the fire service mainstream.

Key Terms

affirmative action a diversity effort intended to increase diversity representation through recruitment and employment from within a qualified pool of eligible minorities.

compliance-based diversity strategy diversity strategy targeted at complying with equal opportunity employment numerical objectives.

cultural inertia the tendency of groups of people to maintain the status quo.

demographics categorizing the population based on specific identifying markers.

disparate impact programs or policies that, at face value, treat all groups equally but have a disproportionate effect on one group over another.

disparate treatment programs and policies that facilitate an unequal treatment of a minority group.

equal opportunity a diversity initiative focusing on the creation of a "level playing field" by legally restricting unfair hiring and promotional practices that facilitate the disparate treatment of workers.

mentoring programs as applied to diversity, programs that identify, recruit, and prepare minority candidates for eventual employment.

normalizing behavior patterns reassertion of the status quo by encouraging group minorities to adopt the behaviors and patterns of the group while simultaneously marginalizing nonconformists.

personal inertia the tendency of individuals to be resistant to change.

quota systems mandated hiring targets of underrepresented minorities based on demographics.

tokenism a general term applied to the practice of selecting minorities simply for aesthetics or out of political correctness.

transformational approach to diversity the promotion of diversity through the making of a business case for social responsibility and the economic advantages of having a diverse workforce.

workplace diversity differences among employees in terms of age, cultural background, physical abilities and disabilities, race, religion, sex, and sexual orientation.

Challenging Questions

To check your understanding of this chapter's material, answer the following questions. It is highly recommended that you discuss your viewpoints with fellow students, peers, coworkers, and friends to discover their opinions as well.

- If you are asked to develop a recruitment program for female fire fighters, what sort of recommendations would you make?
- What cultural and social barriers seem to be inhibiting the fire service's ability to recruit and employ women and minorities?
- Does your department have openly gay fire fighters? If so, are they treated the same as straight department members? If not, what reaction would your department have if a member "came out"?
- Seek out three fire fighters of differing primary group identification (gender, race, age, or sexual orientation). First, ask them to describe themselves in five phrases. (For instance, "I am always in a hurry or "I am very laid back.")

  Next, ask them how their life experiences related to their group identifier has affected the qualities by which they describe themselves. As an example, if you approached a female fire fighter who described herself as "family oriented, career driven, intellectually curious, and deeply religious," ask how her gender has affected the development of those qualities.
- Develop a proposed recruitment program for your department targeted at an underrepresented minority.

Case Study Conclusion

Revisit the case study at the beginning of the chapter. Spend a few minutes considering the questions posed at the end of the case study. In light of the information shared in this chapter, have any of your original observations changed?

The case of fire fighter Jones's initiation is a classic example of a fire department that has a noninclusive culture, is poorly trained in diversity, and has leadership unprepared to handle diversity issues. The practice of hazing is at face value unprofessional, whether it be directed toward minorities, women, or anyone else. There are several positive traditions that can mark transitions and/or celebrations of achievement. They can include shift parties, special meals, and presentations of awards or plaques. Hazing serves no real purpose in building camaraderie. The fire fighters' actions in this particular case were egregious. Behavior by peer groups (men directed toward men) is likely to be perceived differently than by members of a minority group. To get a sense of the indignity of their actions, ask yourself how you would feel if your wife, sister, or daughter were treated this way. The incident was escalated from a complaint to a formal grievance by the subsequent shunning of fire fighter Jones as social punishment for reporting the incident. This is not only unethical, it is illegal. Jones had both an ethical and legal right to a nonhostile work environment. That the hostility was spurred by the actions of a superior officer demonstrates a systematic problem within the department.

WRAP-UP

A couple of points should be noted regarding the incident. First, the department should've outlawed hazing before the incident. Second, if the culture of the department were properly managed, the fire fighters would have realized that their actions were inappropriate. Third, even if the fire fighters were not aware of how their actions would be received by Jones, the company officer should have. Fourth, upon receiving the complaint, the shift commander should have aggressively put an end to the practice without placing any blame or responsibility on Jones. It should have been made clear that the actions were inappropriate at face value.

The chief now has three problems to deal with. First, if he does not aggressively deal with the issues of the complaint, he may put his department and himself at risk of further legal action. Second, he also has a culture problem to deal with. It is obvious there is some education needed within the rank and file regarding basic civility, empathy, and ethics. Finally, he has a fire fighter (Jones) who is in a difficult and emotionally stressful position. From her perspective, she has been attacked and then punished for objecting to it. She likely feels as though the department does not support her. Jones has a long career ahead of her, and these issues must be addressed.

Chapter Review Questions

1. Define disparate impact.
2. Define disparate treatment.
3. How does affirmative action differ from quota systems?
4. What is the Civil Rights Act of 1964?
5. Explain transformational approaches to diversity management.

References

Cockburn, C. 1989. "Equal Opportunities: The Short and Long Agenda." *Industrial Relations Journal* 20, no. 3, 213–225.

Cohn, D., and A. Caumont. 2016. "10 Demographic Trends That Are Shaping the U.S. and the World: Fact Tank." Pew Research Center. Accessed June 12, 2018. http://www.pewresearch.org/fact-tank/2016/03/31/10-demographic-trends-that-are-shaping-the-u-s-and-the-world.

Cox, T. 1991. "The Multicultural Organization." *Academy of Management Perspectives* 5, no. 2, 34–47.

Davidson, M. 2014. *The End of Diversity as We Know It.* Oakland, CA: Barrett-Koehler.

"Diversity." 2015. *Merriam-Webster's Collegiate Dictionary.* Martinsburg, WV: Merriam-Webster.

EEOC. 2016. "Equal Opportunity Terminology." Accessed June 12, 2018. https://www.eeoc.gov/employers/eeo1survey/terminology.cfm.

Fox, K., C. Hornick, and E. Hardin. 2006. *International Association of Fire Fighters Diversity Initiative.* Accessed June 12, 2018. http://www.iaff.org/HR/Media/IAFF_Diversity_Report.pdf. Published 2006.

Gratz v. Bollinger, 539 U.S. 244 (2003).

Greenberg, J. 2013. "Diversity in the Workplace: Benefits, Challenges and Solutions." Accessed June 12, 2018. http://resources.essentialpersonnelinc.com/i/80531273l1.

Huwlett, D., M. Benedict Jr., Sheila Thomas, and F. Moccio. 2008. *A National Report Card on Women in Firefighting.* New York: International Association of Women in Fire and Emergency Services.

Jewson, N., and D. Mason. 1986. "The Theory and Practice of Equal Opportunities Policies: Liberal and Radical Approaches." *Sociological Review* 34, no. 2, 307–334.

Kanter, R. M. 1977. "Some Effects of Proportions on Group Life: Skewed Sex Ratios and Responses to Token Women." *American Journal of Sociology* 82, no. 5, 965–990.

National Fire Protection Association. 2012. "Fire Statistics and Reports." Accessed June 12, 2018. http://www.nfpa.org/news-and-research/fire-statistics-and-reports/fire-statistics.

Page, S. 2007. *The Difference: How the Power of Diversity Creates Better Groups, Firms, Schools, and Societies.* Princeton, NJ: Princeton University Press.

Phillips, C. W. 2014. "How Diversity Works." *Scientific American* 311, 42–47.

Regents of the University of California v. Bakke. 438 U.S. 265 (1978).

Sagen, L., and T. Pini. 2008. "Diversity in the Fire Service—Why Does It Matter?" Accesed June 12, 2018. http://www.firehouse.com/article/10494144/diversity-in-the-fire-service-why-does-it-matter.

Web Finance. 2018. "Workforce Diversity." Accessed June 12, 2018. http://www.businessdictionary.com.

Willing, L. 2011. "Diversity in the Fire Service: The Fear Factor." Accessed June 12, 2018. https://www.firerescue1.com/cod-company-officer-development/articles/1204254-Diversity-in-the-fire-service-The-fear-factor.

FIRE & RESCUE
ADKINS
FAIRFAX COUNTY

SECTION

3

Ethics for Administrative Personnel

Building an Ethical Culture

"Integrity is doing the right thing, even if nobody is watching."
—C. S. Lewis (likely)

OBJECTIVES

After studying this chapter, you should be able to:

- Describe an ethical culture.
- Describe the hallmarks of an ethical culture.
- Understand the effects of culture on behavior.
- Understand compliance-based ethics.
- Understand integrity-based ethics systems.
- Identify common barriers to an ethical culture.

Case Study

Robert Smith is a 51-year-old fire fighter with 24 years of experience in the fire service. He has been the chief of the Anytown Fire Department for five years. In the brief time that he has been fire chief, he has come to believe that bureaucratic red tape and overregulation are a hindrance to efficient management.

Chief Smith is contemplating how to solve a problem. The fire fighter's union has been complaining that the dorm-style bunk room in station seven is old and dingy, and that the absence of separation between sleeping areas is a health hazard. He obtained an estimate for remodeling the bunk room and does not have adequate funds to cover the expense. Because of budget constraints, he is confident that next year's budget will not include funding earmarked for the building renovation.

The city of Anytown has specific rules regarding public building construction and renovation. All work is to be done by licensed contractors, who must utilize union workers paid the prevailing wage. It is this requirement that is making the remodeling of the station seven bunk room cost prohibitive.

To provide the safe work environment that he believes his fire fighters deserve and to save taxpayer money, Smith decides to solve the problem through some creative practices. Chief Smith "cuts a deal" with several fire fighters who agree to do the construction work either on duty or for "comp time" for off-duty hours if necessary. In exchange, the department will purchase all building material. The work progresses smoothly until the fire fighters discover black mold and asbestos while removing partition walls. Chief Smith assures the fire fighters that dust masks and bunker gear are adequate protection to remove the asbestos and moldy material. He further instructs them to place the material in garbage bags and "make it disappear."

The project is completed in three months. A few weeks later a fire fighter is in a bar and brags to one of his construction worker friends about how the fire fighters built their own bunk room, and how they disposed of the asbestos and mold. The construction worker relayed the story to his boss who promptly filed an unfair labor practice grievance against the city.

When the city's administration discovered details of the incident, Chief Smith was given the choice of quietly retiring or being publicly removed from office.

- Were Chief Smith's actions ethical?
- Did his intentions justify his actions?
- What, if any, responsibilities did Smith fail to honor?

Introduction

Imagine that a particular building product was injuring several dozen fire fighters per week from around the country. Most of the injuries are relatively minor, but at least a few times a month there are reports of serious injuries that are ending careers. Your department has suffered through at least three similar incidents in the past 10 years. Each occurrence not only injured a fire fighter but also had a significant economic impact.

Would your department ignore this threat to its fire fighters? Do you believe that the fire service, as an industry, would overlook a problem with such nationwide impact? Unfortunately, the fire service is ignoring a widespread threat to its fire fighters. Every day fire fighters are being disciplined for ethics breaches. On a regular basis, inappropriate behavior is ending careers. Bad behavior by fire fighters not only has dire consequences for them, but it can also destroy your department's reputation and cost thousands of dollars in legal fees. Still, fire departments, and the fire service collectively, continue to treat ethics problems as isolated local issues.

This response is similar to the fire service's attitude toward personal injury just a generation ago. For too long, the fire service treated fire fighter injury as an inevitable consequence of the job. Each department treated fire fighter safety as a local issue. For over a century there was no cohesive national response, nor was there guidance for fire departments in the protection of their fire fighters.

Fortunately, dedicated and determined individuals eventually educated and motivated the fire service to become serious about fire fighter safety. A concentrated and collective effort was launched to educate fire departments on life safety initiatives; legislation was passed; and, most important, attitudes toward safety were eventually changed. Safety became a priority in the fire service. Fire fighter safety is now included in most training sessions, and it is a primary concern in the design and purchase of equipment. Fire fighter safety now permeates all levels of the fire service. Every fire fighter in every fire station hears the message, and every national organization rallies resources to ensure a safer profession.

In short, we are building a safety-oriented culture within the fire service. It is time for a similar response to fire service ethics. We are losing good people to bad choices. Ethics needs to become a priority at every level of the fire service. Ethics training needs to be included in candidate testing, promotional testing, training, and education.

Most important, individual departments need to be conscious of reputation management and the necessity of articulating and enforcing fire service values and behavior. As with safety, the fire service must change fundamental attitudes toward fire service ethics to the creation of ethical cultures.

What Is an Ethical Culture?

The concept of an ethical work culture can be very hard to describe, yet it is easy to recognize. It is even easier to recognize an unethical culture. Culture permeates entire organizations. It reflects how people feel about what they do. Culture is a dynamic entity; it shapes actions and simultaneously is shaped by action. It is at once a reflection of values, and it is a powerful force in shaping those same values.

An ethical work culture goes beyond the flowery language of a value statement, the often vague language of an ethics policy, or the didactic language of a rule book. Policies and documents express how the fire chief wishes a department to behave. In reality, the culture of the department is a description of "how things really work." An accurate illustration of a department's ethics culture describes the department's personality. The ethical culture of the fire department reflects its actual values and priorities. It is the product of the combination of departmental policy and traditions, as well as its members' personalities.

The most straightforward understanding of an **ethical work culture** is that a positive ethical culture encourages correct behavior, whereas a negative ethical culture facilitates poor behavior (Meinert, 2014).

It is important to recognize that an ethical culture and an unethical culture affect behavior differently. The presence of an ethical culture has a direct influence on behavior, whereas an unethical culture has a passive effect on behavior. The difference appears subtle, but it is significant. The presence of an ethical culture can reshape individual attitudes toward correct and incorrect behavior, whereas an unethical culture facilitates behaviors that, in another context, would still be considered inappropriate (Kantein, 2012).

To better understand the effect of culture, let us substitute ethics with clothing. In this example, please keep in mind that we are substituting clothing for ethics, not discussing the ethics of wearing clothing. A modest person would likely never consider wandering around the mall in his or her underwear. Modesty restrains the behavior, much as values restrain unethical conduct. Modesty springs from cultural norms and is reinforced by culture. Likewise, a positive ethical culture reinforces restraint consistent with ethical behavior. However, place a modest person on a beach, and being scantily clad becomes normalized. In a context where modesty is not expected, two things are likely to happen. The *new beach culture* no longer reinforces modesty and so facilitates the possibility of immodest behavior by removing restrictions. Second, because of the "new" norms, the normally modest person will likely feel social pressure to emulate the others on the beach. He or she may even rationalize acting on a previously felt, but suppressed, exhibitionist desire. The beach "culture" did not change the individual's view of modesty but rather facilitated a change of context that allowed immodest behavior. The beach environment does not cause a modest person to remove his or her clothes; it facilitates the lifting of personal sanctions. The individual still likely recognizes modesty as a virtue and, in normal circumstances, an appropriate behavior. As such, reentering a modest culture will immediately reinforce traditional behavior, and so previously held views of appropriate behavior are reestablished.

Clarification

An ethical culture directly promotes ethical behavior by shaping and reinforcing values, whereas an unethical culture facilitates the undermining of values.

This example demonstrates how positive and negative behavior models affect conformity to cultural ethics standards. An ethical culture (modest) directly promotes ethical (modest) behavior by shaping and reinforcing values, whereas an unethical culture (immodest) facilitates the undermining of values. This demonstrates the power of culture in the promotion of ethics. The presence of an ethical culture creates norms that strongly influence individual restraint from engaging in inappropriate behavior. Unethical behavior tends to occur in situations where strong ethical influences are absent or where strong motivations are encouraging unethical behavior. The presence of a strong ethical culture tends to marginalize those motivations. The hallmarks of an ethical fire department culture include: priority on public welfare, valued fire fighter safety and welfare, attitude of service before self, priority of competency and dedication, high regard for virtue, and open and honest communication.

Highest Priority to Public Welfare

A concept often repeated in this text is that any understanding of fire service ethics must be rooted in the fact that we are a public service agency. We exist for the sole reason of providing for the public welfare; we are financially supported for that exact reason. We are afforded a high degree of public support because our mission literally encompasses life-and-death responsibilities. Fire service ethics and the ethical culture of any given department should be rooted in an understanding of that momentous responsibility. Our first obligation must be to the public welfare, an obligation that goes beyond all other considerations except that of fire fighter safety. Any behavior or policy that is contrary to the public welfare is a breach of public trust and inherently unethical. Constant focus on the fire department mission creates a foundation for an ethical culture by promoting an attitude of service over self.

Fire Fighter Safety and Welfare

Fire fighters routinely risk injury and even death. A fire department has an ethical obligation to take all steps possible to ensure the safety and welfare of its fire fighters. This obligation represents a solemn contract between the fire fighters and the department. Failure of the department to protect its employees from unnecessary risk creates an environment of mistrust. Trust is a hallmark of an ethical organization. In the absence of trust, unethical behavior is too easily rationalized by those acting out of self-interest. Fire service administration and rank-and-file fire fighters must understand and respect the concept that ethical behavior is a bidirectional exchange. An ethical culture will be severely challenged in an environment of distrust.

Service Before Self

The fire service has a long tradition of valuing public service and teamwork. Departments exist specifically to provide service to others, and their ability to do so is dependent on each individual fire fighter's willingness to put the mission of the department ahead of his or her self-interest. The creation and maintenance of a pervasive attitude focusing on responsibility and accountability to others is paramount in maintaining an ethical culture. The concept of ethics is rooted in three fundamental principles: responsibility, personal virtue, and seeking best consequence for others. A work culture with a pervasive attitude of self-interest first invites breaches of ethical protocol.

Competency and Dedication

A fire fighter is burdened with a tremendous responsibility for the welfare of the public and his or her fellow fire fighters. The consequences of incompetency can be catastrophic. Each fire department must strive to maintain a culture that reinforces dedication to competency as a condition of service. The obligation for competency rests on the three hallmarks of an ethical culture described earlier. Further, a recognition of competency enhances self-image. A positive self-image is an important reinforcement for positive behavior.

Virtue Is Held in High Regard

Integrity, loyalty, and respect are all virtues that build trust. Trust is essential in maintaining a culture of ethics. The presence of virtue in day-to-day activity reinforces the value of those activities. There is a saying that "ethics begets ethics." Conversely, poor behavior encourages poor behavior. Setting high standards of behavior is a powerful incentive to participating in an ethical culture. Consistent reminders that fire fighters are part of a highly respected profession reinforces the will to live up to those expectations.

Open and Honest Communication

Communication of values, coupled with trust, creates an atmosphere that encourages ethical behavior while simultaneously removing incentives for inappropriate activities. In many cases, poor ethical behavior is rooted in perceived injustices and rationalization. The human desire to maintain a positive self-image necessitates justification of poor behavior. Those justifications are diminished in a culture hallmarked by honest communication and trust.

The Importance of an Ethical Culture

As stated earlier in this text, the fire service has a significant investment in reputation management. The image of the American fire fighter is nearly iconic. That image is a major factor in the recruitment of personnel and a significant adjunct in completing our mission of "saving lives and protecting property." For many fire fighters, the pride associated with the profession and the respect it garners is an important part of their self-identity. Fire fighters are justifiably proud of what they do and what the career represents Figure 9-1. That sense of pride and respect is mirrored by the public. Public support of the fire service has been an important factor in maintaining staffing levels and appropriate benefits for fire fighters. Clearly, the fire service has a keen interest in maintaining a high standard of behavior for its fire fighters.

Ethical breaches occur far too regularly in the fire service. Some are caught and addressed, some are overlooked, and some are never noticed. In each instance, a fire department dealing with a behavior problem has a body of rules and an enforcement mechanism, yet ethics issues exist despite enforcement. Similarly, crime exists despite law enforcement and the judicial system.

Rules, and the enforcement of them, have a limited effect. As you have learned in earlier chapters, in many cases individuals performing unethical acts do not see their actions as inappropriate, or they rationalize their behavior as expedient or necessary. Rules are often limited by their scope. Try as they might, fire chiefs cannot write a rule that covers every possible scenario. It is essential that fire fighters have the ability to make informed and intelligent ethical decisions, and that decision-making process must be supported by a positive culture.

Figure 9-1 For many fire fighters, the pride associated with the profession and the respect it garners is an important part of their self-identity.

© Blend Images / Alamy Stock Photo

Figure 9-2 An ethical culture is also a high-achieving culture.

© E.D. Torial / Alamy Stock Photo

Available resources further limit rule enforcement. If we depend only on the fire officers to enforce rules, then compliance is limited only to observable behavior. We must assume that fire fighters obey rules even when officers are not there to enforce them. This requires a culture supportive of self-regulation.

We depend on voluntary compliance to the policy because we can rarely enforce policy 100 percent of the time. Because a culture is pervasive, an ethical culture can influence behavior constantly. It has an unlimited ability to influence behavior even in the absence of supervision or policy. People who habitually do the right thing act more out of internal guidance and motivation than as a result of external supervision or direction. Creating a positive ethical environment fosters ethical intelligence and reinforces the exercise of the internal moral compass.

The presence of an ethical culture has benefits beyond behavior management. An ethical culture is also a high-achieving culture. Employees who are working in an impoverished ethical culture are much less likely to be loyal to the department; they tend to lack initiative and are less liable to garner meaningful satisfaction from their job. Pride in one's work and in one's coworkers creates a positive work environment that encourages excellence in performance as well as behavior Figure 9-2.

Conversely, the absence of an ethical work culture can have detrimental effects on a department. Beyond the obvious problems associated with violation of rules, departments with a high incidence of discipline problems tend to have poor morale. Poor morale is a prime motivator for further unethical behavior. Even ethical employees working in an unethical culture suffer. The negative effects of an unethical culture causes worker dissatisfaction, a breakdown in trust, and poor performance. Organizations (including fire departments) with impoverished work cultures have higher turnover rates, higher equipment damage, and loss incidence.

Departments with high levels of inappropriate behavior and subsequent punishment tend to have poor public images. Public cynicism toward a fire department can erode funding and staffing levels.

Building an Ethical Culture

Establishing a Positive Work Environment

Establishing an ethical culture is nearly impossible if it is not supported by a positive work environment. Fire fighters who feel they are marginalized, powerless, or under siege feel no particular loyalty to the department, its managers, or its culture.

There is a field of study known as **organization development** that is dedicated to expanding the knowledge and effectiveness of people to accomplish more successful performances. An essential tenet of organization development theory is that workers within a positive work environment are more productive and more likely to be vested in the organization's best interest.

Every fire fighter has personal values and attitudes; they are primary in guiding behavior. Individuals with strong moral conviction are likely to behave appropriately even when dissatisfied or when around coworkers setting bad examples. However, the fire fighter's relationship with his or her department influences the context with which behavioral decisions are made. As we have discovered in other chapters, behavior attitudes are often about context. Behavior that is deemed inappropriate in one's personal life may be viewed as acceptable in a toxic work environment. For example, a fire fighter who is usually truthful may find falsifying reports contextually acceptable if the behavior is part of a department's normal routine.

Fire fighters in negative environments where there is a general belief that workers are underappreciated or treated unfairly are more likely to act in a self-serving manner. This attitude is supportive of unethical behavior. Conversely, fire fighters who are self-actualized (see Maslow's hierarchy of needs in Chapter 3, *Influencing Behavior*) are more inclined to concentrate on conceptual principles such as self-sacrifice, teamwork, empathy, loyalty, and personal responsibility. As noted earlier in this chapter these qualities are all essential to an ethical culture.

Clarification

A positive work culture features strong leadership that places as much emphasis on how things are done as is placed on results.

Texture: Eky Studio/ShutterStock, Inc.; Steel: © Sharpshot/Dreamstime.com

A positive work culture is leadership-centric and, by extension, focused on meeting organizational obligations by the empowerment of those tasked with meeting them. A positive work culture features strong leadership that places as much emphasis on how things are done as it does on results. A strong ethical culture approaches organizational operations with a focus on the integrity of process. Being part of an organization that consistently meets high ethical standards instills a sense of pride.

Fire fighters who feel a sense of pride in being a fire fighter are more likely to engage in appropriate behavior than those who do not. By extension, fire fighters who are proud of their fellow fire fighters' competency and values are much more inclined to feel a sense of pride in their connection to the department.

The relative importance of effective leadership cannot be overstated in developing a positive work culture. In turn, a positive work environment supports a positive ethical culture. Positive work cultures instill a sense of optimism in their employees. There is trust that leadership is competent and well intentioned. Fire fighters in positive work cultures feel there is a sense of justice. They tend to believe that qualified people are promoted, exceptional effort is recognized, and high performance translates into success. Positive work cultures also reinforce a sense of connection to the department. Fire fighters believe they are part of something bigger than themselves and, as such, they are willing to displace their short-term personal interests for the good of the department and its mission.

The creation of a positive work environment is relatively simple; however, it can be very demanding of departmental leadership. Effective leadership rarely relies on authority alone to achieve results. Overpowering employees with rank is usually counterproductive to effectiveness, and nearly always detrimental to workplace culture. Still, when a person has the authority to tell people what to do, it is an attractive alternative to the effort necessary to motivate, teach, and mentor employees into doing the right thing. Unfortunately, some fire officers mistakenly believe that empowering employees somehow diminishes personal power. Fire officers who do not make an effort to lead and who are threatened by the success of subordinates create environments supportive of inappropriate behavior. The creation of a productive workplace, supportive of an ethical culture, involves the following:

- **Focus on mission.** Constant attention to the fire department organizational mission maintains a sense of purpose. Fire fighters who understand why they are doing what they are doing are much more likely to accept responsibility. The consistent articulation of mission also diverts attention away from insignificant distractions and problems.
- **Allow for organizational voice.** Fire fighters who believe that they have input regarding their work conditions and who believe that their ideas are given consideration are more likely to remain committed to departmental success. They are also less focused on personal grievances. The scalar paramilitary organization of fire departments is highly effective for emergency scene operations. However, it has significant limitations in employee engagement that must be recognized and overcome.
- **Hold leaders to high standards.** Rarely does fire fighter performance, or attitude, exceed that of their officers. Fire officers, especially company officers, are critical in shaping the culture of an organization. Troubled departments often have officers who do not support department initiatives and fail to model good behavior or exhibit leadership qualities.
- **Create learning organizations.** Employees who sense that the department is improving likely feel as though they are part of a positive outcome. Further, employees who believe that they have growth possibilities, either through promotion or added responsibilities, typically focus on positive achievement rather than perceived injustices or grievances. Achievement is viewed as connected to effort rather than circumstance; this tends to inhibit the rationalization of unethical behavior as a necessary means to an end.
- **Maintain optimism.** Fire fighters who believe that they are members of a department under siege or who are underappreciated are likely to become frustrated and develop a victim mentality. This is counterproductive to employee development and job satisfaction. In general, the attitude of "us against them" creates a potential for inappropriate behavior as it readily encourages rationalization of self-interest.

Actively Promote Ethics

Having a positive work culture can positively affect employee behavior. However, it is not sufficient by itself. There is a popular notion articulated in fire service leadership courses that, in one form or another, expresses the wisdom that a fire chief "gets the department he or she builds." Certainly, in the area of ethics, this wisdom is absolute. As with all other elements in a department, that which is given priority and resources is accomplished. Considering the importance of ethics, and the potentially disastrous effects of unethical behavior, it is reasonable to assume that ethics should be a priority within a department.

Unfortunately, ethics is rarely discussed in most departments. Many departments have no ethics training whatsoever, and those that do usually employ a self-directed web-based program. If a fire fighter receives ethics training once a year and even that training is given a low priority, it seems reasonable to assume that ethics is likewise a low priority. Clearly, it is not that the department endorses unethical behavior, but rather it places little importance on the necessity of maintaining high ethical standards. Compare this to the time, effort, and resources dedicated to fire fighter safety, basic skills competencies, and specialty trainings.

An important first step to building an ethics-oriented culture is to articulate behavior expectations. We must talk about ethics, teach ethics, test for ethics, and reward ethical behavior. Fire fighters will not take ethics seriously as long as departmental leadership does not **Figure 9-3**. Issuing behavior rules and not offering training about them will likely have the same result as issuing safety policies that are not reinforced by training. Both become "background noise" (Wolde, Groennendaal, Helsloot, et al., 2014).

Figure 9-3 Fire fighters will not take ethics seriously as long as departmental leadership does not.

© Radius Images / Alamy Stock Photo

Hold People Accountable

Most fire administrators understand that fire fighters must be held accountable for policy violations. Paramilitary structure in the fire service is a long history of issuing rules and punishing infractions when they occur. Certainly, policies have little or no meaning if not enforced, and poor behavior must be addressed. It is also important to recognize that poor behavior is as much a symptom as a problem. The development and maintenance of an ethical culture requires addressing causes of unethical behavior, and there must be efforts to address failures of organizational oversight. In most cases, those failures involve supervising officers.

Consider the example of an adolescent who deliberately throws a rock through a neighbor's window. The parents are financially responsible for the damage, even though they had no part in the action. Their culpability is rooted in the assignment of responsibility. Parents are responsible for raising properly socialized children and supervising their actions. Imagine if the parent attempted to deny responsibility for the child's actions, stating that the child alone is responsible for the behavior—and as a result they should not be held accountable. Obviously, such tactics would have little legal or social support.

The same principle is at work in the firehouse. Fire officers and fire fighters do not have a parent-child relationship. However, fire officers do accept responsibility for the supervision of their fire fighters. Just as important, they are also responsible for shaping the attitudes of those fire fighters.

Officers recognize the signs of impending behavior issues and address problems before they happen. This author, like so many other fire chiefs, has been "taken to the woodshed" by city officials for the behavior of his fire fighters. While it is entirely frustrating to be rebuked for the behavior of others, it is not entirely inappropriate. Chief fire officers are entrusted by government officials to maintain order and discipline. It is a responsibility inherent in the position. Sadly, too many fire officers fail to appreciate the importance of that responsibility and, as a result, preventable incidences occur. Building an ethical culture requires diligence. Departments must reinforce the importance and obligation of doing so.

Ethics Must Pervade Throughout the Department

Similar to safety concerns, ethics must be consistently addressed within all aspects of departmental operations. A major factor in building an ethical culture is hiring and promoting ethical people. Ethical training, testing, and policy must be specific to activity and rank. The ethical implications of every positional responsibility must be clearly defined and included in job descriptions. The ethics of responsibility should be included in training sessions. Ethics testing should be included in the selection and promotional processes. Ethics training should be provided to incoming fire fighters as part of orientation. It should further be included in preparatory classes for new fire officers at every rank.

A department that is dedicated to maintaining an ethical culture recognizes that it is not enough to state rules and punish infractions. We must empower employees with knowledge and ethical wisdom. We must, as best as possible, prevent breaches of behavior before they occur. This can only be accomplished if ethics awareness becomes a constant companion in everyday operations. It is not enough to hang value statements in dayrooms and place rules in operations manuals. Ethics must be ingrained in process. Ethics must be a fact of life.

Make Sure Policy Does Not Interfere with Values

Several years ago, a fire department began to become heavily invested in the concept of customer service. The overriding intent was that fire fighters embrace the idea of servant leadership and in being a positive force in the community. The chief was convinced that community engagement would have a positive impact on fire fighter attitudes toward their work and on job satisfaction.

Not long into the program, a company officer picked up a pedestrian whose car had broken down and as a result was walking in the rain to work. The officer and his crew drove the individual to work on the fire apparatus. In general, the officer and crew felt good about their efforts. The chief, however, had concerns about liability and consulted with the city's legal staff. As was expected, the city's lawyers expressed concern over the liability of having civilians riding on fire apparatus. As a result, the chief issued a policy restricting private citizens from riding on fire apparatus, including putting an end to the very popular "ride along" program for Explorer Scouts.

The policy had the unintended consequence of undermining the public service initiative and also inhibiting the department's involvement with its own Explorer post. The attorneys were correct in their assessment of liability; however, they were not asked to assess values. Their concern was strictly limited to assessing risk; it is not their responsibility to assess benefit. The pursuit of eliminating risk was at the expense of the desired value.

Very often policies seek to eliminate risk or discourage a particular behavior. Similar to the previous example, policy is sometimes incongruent with values. In that scenario, the fire officer was acting compassionately. The chief's resulting modification of the policy not only sent a contrary message regarding community engagement but likely made the fire officer and his crew feel as though they had done something wrong. The likely results are disengagement and distrust of the chief's commitment to the type of culture to which he verbally subscribed.

In reviewing this example, there seems to be no direct correlation to community service initiative and ethics. However, ethics is rooted in attitudes toward the department and its leadership. Just as important, the department's culture shapes individuals' attitudes toward their self-image. In making policy, department leaders are encouraged to assess direct outcomes while at the same time considering the consequence to culture. In order to attain political cover, a fire chief may institute a drug-testing program assuming all his fire fighters are clean. That same fire chief may be surprised by the amount of animosity generated by the policy because it is seen as a breach of trust. We tend to write policies to solve problems, and too often we fail to recognize their unintended effect on culture.

Use of Rewards and Discipline

The creation of an ethical culture promotes appropriate behavior, and so appropriate behavior should be incentivized by reward and recognition. For example, many departments issue annual awards for achievement in training, public service, and valor. Rarely do departments have an awards program recognizing high standards of ethical behavior or ethics leadership. It is assumed that ethics is a condition of employment or expected behavior. Arguably, the same can be said for competence, public service, and valor. That we recognize high standards of these assumed behaviors—and not ethics—speaks to a general lack of emphasis. It is a well-known adage in management that we reward behavior we wish to encourage, and we discipline behavior we want to discourage. For this reason, it is highly recommended that a reward system be implemented for ethical conduct.

Conversely, the department has an obligation to maintain discipline when inappropriate behavior occurs. Disciplinary processes are well documented in fire department management texts. However, it is advisable that readers remember that the primary goal of discipline is instruction. Inappropriately lenient discipline, draconian discipline, and the inconsistent application of discipline are all counterproductive to the purpose of ethical instruction. Discipline is a reinforcement of positive cultural values. Just as reward demonstrates the department's desire for an ethics culture, the department's application of discipline demonstrates its commitment to ethics (Kaptein, 2012).

Approaches to Culture Building

Chances are you have either participated in or watched a pickup basketball game. The game may have had anywhere from three to five players per team and might have been played half-court or full-court. What each game has in common is that there is no referee. Even so, the game runs relatively smoothly. The players obey the rules, they do not double dribble, they refrain from traveling for the most part, and they avoid fouling each other. Moreover, if there is an infraction, the players involved generally consent to other players' judgment. In many cases players will call infractions on themselves. The integrity of the game requires voluntary compliance with rules. Each player knows that if the rules are not followed, the game descends into chaos, and there is no real point in playing it.

An odd thing happens, however, when the referee is injected into the game. The players will adopt the attitude that infractions are okay if not caught. No one would ever call a foul themselves if the referee did not blow the whistle. Players will deliberately test referees to see how closely they are calling the game, willfully stretching the limits of rules. In short, their attitude toward following rules changes from voluntary compliance to compliance based on enforcement.

This is an example of two approaches to ethics: compliance-based and integrity-based ethics. In **integrity-based ethics** the individual self-regulates based on a shared understanding of behavior standards and with the intention of contributing to the common good. **Compliance-based ethics** utilizes specific policies and the enforcement of imposed standards.

Nearly every fire department utilizes both approaches to ethics. Compliance-based ethics is most easily recognized. However, every fire department assumes, and even depends on, voluntary compliance with behavior standards even absent supervision. A department cannot function without rules and policies, nor can it completely enforce those policies without the fire fighters' cooperation.

Compliance-Based Ethics

Compliance-based ethics systems operate on a clearly defined set of rules that are enforced by oversight and the potential for discipline. Within the fire service, and most other organizations as well, compliance-based ethics is commonplace. Every department has a rule book, standard operating procedures, and assorted operational policies. Departments also have a process for investigating policy infractions, and a disciplinary procedure intended to enforce policy compliance. Compliance-based ethical approaches are the primary methodology by which public organizations maintain behavior standards.

Organizations (including fire departments) tend to rely on compliance-based ethics for several reasons.

Compliance Ethics Is Straightforward in Its Approach

It is easier to identify behavior that is unacceptable than it is to describe generalized concepts associated with virtuous conduct. Further, it is easier for management to punish bad behavior than to educate and motivate positive behavior. Comparatively speaking, it is simple to identify behavior, make a rule against it, and then punish those who break the rule. As an analogy, when teaching children the fundamental principles of appropriate behavior, we begin with simple, straightforward restrictions. It is easier for a child to understand the admonishment against lying than to explain the virtue of truthfulness. Hence, we forbid lying and punish the behavior when it is discovered.

Contrary to popular belief, bureaucratic institutions similar to fire departments thrive on simplicity. Fire chiefs, union leaders, and legal staff place great value on **bright lines** when it comes to behavior. Compliance systems are more easily interpreted, and easier to enforce, litigate, and defend. By virtue of simplicity, protracted grievance procedures and legal proceedings can be more easily avoided by use of compliance-based ethics systems.

Compliance-Based Ethics Gives the Organization Cover

Having a rule forbidding sexual harassment is an important first step in protecting the organization from legal culpability, regardless of the effectiveness of the rule's existence in dissuading the behavior. These types of policies are sometimes described as defensive policies. Although well intentioned, their existence is more to demonstrate the department's intent rather than to modify behavior.

Further, the existence of rules and policies are a prerequisite for disciplinary action. The most common defenses provided by fire fighter representatives in defending an offense is that there was no specific rule forbidding it. The existence of policies allows for enforcement and disciplinary action (Roberts, 2009).

Compliance-Based Programs Significantly Reduce Pressure to Implement Integrity-Based Programs

Culture management, coaching, and mentoring require a significant investment of time and effort. In contrast, compliance-based systems are relatively easy to manage (Roberts, 2009). The creation of a robust departmental code of conduct is typically seen as the first step in managing behavior, and for many departments, it is also the final step. Departments with impoverished leadership place responsibility for rule compliance entirely on fire fighters, excusing themselves from the obligations of culture management and value shaping.

Elements of a Compliance-Based Ethics Program

A successful compliance-based program is dependent on the promulgation of a behavior policy that is thorough, easily understood, and enforceable. Fire fighters must understand behavior standards, the importance of following policy, and the consequences of prohibited behavior. Fire officers must understand their obligation to support and enforce the policy. The department must ensure that policies are legal, justified, and applied fairly. Discipline must be evenhanded, appropriate, and consistently applied. The following are the essential elements of a compliance-based ethics system (Meinert, 2014; Taylor, 2016).

- A comprehensive policy guide covering all anticipated behavior issues is the first step in the creation of a compliance-based system. Effective policy guides (rules) are specific, resistant to subjective interpretation, and thoroughly vetted for conformance to labor contracts and labor law.

- Competent supervision is necessary if a compliance-based system is to be effective. Supervising officers must understand the intent and scope of policy, as well as the capability to exercise judgment in its enforcement. While judgment and discretion are necessary, a fundamental requirement of a compliance-based ethics system is the willingness of officers to enforce policy consistently. This requires training of officers and disciplinary process.
- Compliance-based systems necessitate sufficient incentives for compliance if they are to be effective. Incentives may be either positive by rewarding ethical behavior or negative by punishing noncompliance.
- An institutionalized reporting and investigation system must be in place to facilitate the documentation of policy infractions. The system must be transparent and impartial. It must also be consistent with due process obligations imposed by collective-bargaining agreements and state law. Finally, reporting systems must include protections for whistle-blowers.
- Effective compliance-based systems must have an adjudication process. Evaluation of an appropriate response must be dispassionate, fair, and respectful of due process.

Disadvantages of Compliance-Based Systems

The common theory behind compliance-based ethics is that of deterrence. The threat of punishment is intended to dissuade inappropriate behavior. The entire disciplinary system of the fire service is based on this assumption. To be effective, compliance-based ethics systems must have organization-wide participation. Every fire officer and fire fighter must recognize and accept an obligation to adhere to policy. A culture of compliance not only depends on the enforcement of policy by fire officers but also by the fire fighters. Absent voluntary compliance, fire officers and fire fighters are placed in adversarial positions that are completely antagonistic to productive work environments.

Compliance-based systems are limited by both the fire officer's ability to supervise behavior and the officer's willingness to enforce policy. In discussing discipline, a common concern expressed by senior fire officers is that company officers have close relationships with their fire fighters. Those relationships can build competing loyalties concerning policy enforcement. Loyalty to the crew is sometimes of higher value to the company officer than adherence to a policy (Taylor, 2016).

Even with cooperation by frontline officers, compliance-based systems often achieve poor results. Read any daily newspaper, and you will see evidence of that fact. By itself, the existence of a law does not necessarily ensure compliance. Within the fire service, the prolific number of disciplinary actions taken each year reinforces this notion. Rules and the threat of punishment routinely fail in deterring inappropriate behavior. Most compliance systems rely on the voluntary cooperation of those affected by the rules.

In the end, compliance-based systems depend on the individual character and ethical judgment of employees. Unfortunately, compliance-based systems do not educate in moral decision making or shape values. As witnessed by the basketball example at the beginning of this section, compliance-based systems can actually normalize inappropriate action. Once an outside authority accepts responsibility for controlling behavior, it creates a **contextual standard of behavior**. As a consequence, behavior that is otherwise deemed inappropriate is recontextualized in conformance with the boundaries of the imposed rules.

Contextual standards of behavior are **ethically bounded**. Bounded ethics describes a condition in which right and wrong are defined strictly through conformance to policy. The concept is related to the bright lines referenced earlier in this section. The presence of detailed rules has the effect of creating "boundary lines" around behavior. Behavior that is within the rules (boundaries) will often be considered inherently ethical regardless of consequence or compliance with departmental values. Further, and most important, behaviors not covered by rules will also soon be considered "in bounds." If there is no rule against it, then it must be okay.

It is nearly impossible to anticipate every potential circumstance surrounding inappropriate behavior. Fire chiefs have sought to address this limitation with the inclusion of generalized rules of conduct. These clauses often have titles such as "conduct unbecoming" or "moral turpitude." The idea of these clauses is to act as a catchall for inappropriate behavior not otherwise covered by policy. These generalized policies have been the source of numerous disciplinary hearings, court cases, and grievances. They are problematic because they attempt to apply the concept of compliance ethics to nonspecific criteria under the assumption that the offending fire fighters "should know better."

As explained previously, a problem arises when the department relies on the individual ethical judgment of its fire fighters, absent training or culture reinforcement. By solely relying on the creation of behavior boundaries, absent the provision of positive behavior guidance, the department has simultaneously defined limited behavior standards while negating personal responsibility and judgment.

Integrity-Based Ethics

Compliance-based ethics systems can be described as external—rules are created and imposed on the individual. It is the fire fighter's responsibility to comply with the rules or face some sort of negative consequence. Integrity-based ethics can best be described as internally oriented. In this case, the fire fighter chooses a path of ethical behavior based on his or her moral compass relative to the fire fighter's understanding of departmental expectations.

Clarification

Integrity-based ethics systems reinforce ethical behavior through the creation of a positive ethical culture and the development of ethical intelligence among department members.

In an integrity-based ethics system, the department endeavors to reinforce ethical behavior through the creation of a positive ethical culture and the development of ethical intelligence among department members. The focus of an integrity-based ethics system is the empowerment of the individual's values rather than a focus on the restriction of the individual's behavior choices. It is imperative that any department clearly articulate its behavior expectations through the use of policy. The difference between compliance-based and integrity-based ethics is how those expectations are expressed and the approach the department uses in achieving them.

Clarification

Compliance-based systems tell fire fighters what they cannot do; integrity-based systems focus on doing the right thing.

In an integrity-based system, integrity is the governing ethic. Doing the right thing is not only expected, it is placed front and center in day-to-day operations. Virtues and values are embedded in operational policy, training, personnel evaluations, promotional testing, and new employee selection.

Typically, departments that utilize integrity-based ethics have relatively small departmental rule books. In place of restrictions on specific behavior, integrity-based systems tend to describe expected behavior through value statements and intended outcomes. As an example, in a compliance-based program the department policy manual may require departmental personnel to refrain from identifying themselves as members of the department when utilizing social media. It may also have specific policies regarding local media and public speeches. An integrity-based program would likely state that fire fighters "shall at all times conduct themselves professionally, responsibly, and in a manner consistent with departmental values."

The intent of both of the previously mentioned policies is the same: to avoid departmental embarrassment by limiting controversial or inflammatory statements. However, their approach is remarkably different. The compliance policy tells a fire fighter what he or she cannot do, whereas the integrity policy describes the desired outcome from behavior. However, integrity-based strategies are broader and deeper than simply requiring compliance. Integrity systems seek to shape and maintain attitudes toward behavior and to create an environment conducive to ethical conduct.

Benefits of an Integrity-Based Ethics System

Integrity-based ethics are **unbounded**. Unlike compliance-based ethics, which prohibit specific actions and thus tacitly allow others, integrity-based systems define proper approaches to ethical situations. As such, they are all-encompassing. By describing values and intended outcomes, guidance is provided to both anticipated and unanticipated circumstances. The principles associated with integrity ethics can be applied to any situation and offer guidance that is essentially universal.

Integrity-based ethics places the burden of behavior restraint on the employee rather than his or her officer. As a result, the officer becomes a coach rather than a referee. This change in dynamic is more consistent with a positive work culture. High standards of behavior become endemic to the identity of the department. As such, self-image and group identification become tied to those standards of behavior. The removal of adversarial roles subsequently removes grounds for rationalization of inappropriate conduct.

Challenges Associated with Integrity-Based Systems

Integrity-based ethics is demanding in its implementation. It requires effective leadership and engagement at all levels of the fire department. Fire officers must model proper behavior and insist on consistent, appropriate conduct. They must not only correct improper behavior but also identify and address the root cause. This requires engagement far beyond mere supervision. Integrity-based systems require education and mentoring in addition to enforcement and presenting positive role models (Paine, 1994).

Company officers are enormously important to the success of integrity-based ethics. As we stated before, the culture of an organization is shaped daily on apparatus floors and in dayrooms Figure 9-4. This places responsibility on company officers to prioritize departmental values. However, the company officer's effectiveness in promoting departmental values is dependent on the support provided by senior administration. Fire officers must be trained as leaders and educated in both leadership and ethics. It is also important that they must be empowered to act. In dealing with the daily instruction of department values, fire officers must be empowered to correct inappropriate attitudes without interference from senior staff. Far too many departments focus solely on the development of technical skills for their officers and too little on human resources management. To paraphrase Mark Twain, "Everybody talks about leadership, but no one ever does anything about it."

Figure 9-4 More often than not, an ethical culture is built on the apparatus floor—not in the classroom.

Table 9-1 Comparison of Integrity-Based Systems and Compliance-Based Systems

	Compliance-Based Systems	Integrity-Based Systems
Character	Conformity with externally imposed rules	Creation of culture encouraging appropriate behavior
Objectives	Prevent misconduct; negative reinforcement	Enable responsible behavior
Leadership	Supervisor oversight	Driven by fire officers with the aid of training division and policy
Methods	Education, reduced discretion, behavior controls, and penalties	Education, leadership, accountability, organization development, internal auditing, and enforcement
Assumptions	Employees guided by self-interest	Self-interest is time to focus on values, ideals, and peers

In addition to officer development, ethical integrity systems depend on departmental policy supportive of an ethical culture. Policy must reflect the values the department espouses as guiding expected behavior. Fire officers cannot be expected to effectively build and maintain an ethical culture if there is incongruence between stated priorities and administrative policy. Similar to company officers, integrity-based ethics demands much from fire administrators. Commitment to ethics training and development of ethical intelligence among fire fighters is paramount.

In comparing compliance-based ethics and integrity-based ethics programs, it is important to understand that they are not mutually exclusive. Implementing the integrity-based system does not preclude the use of compliance-based strategy, and vice versa. Either strategy is inadequate by itself. While integrity-based systems are more inclusive in dealing with ethics problems, they can lack specificity. Conversely, compliance-based systems are adept at defining specifically forbidden behaviors but can be of limited effectiveness in areas not covered by a particular policy.

A Comparison of Integrity-Based and Compliance-Based Systems

A department will benefit from utilizing both strategies, with its primary focus on building an ethical culture supported by the promulgation of rules and regulations as needed Table 9-1.

Barriers to Building an Ethical Workplace Culture

Culture management is as much an art as a science. Creating and maintaining an ethical workplace requires consistent effort and cooperation across all ranks and positions within the department. Even with diligence and best intentions, impediments to an ethical culture can present themselves. The following are some of the common challenges faced by fire officers in maintaining an ethics-first environment (Ragazt, 2012).

Lack of Support from Senior Leadership

Chief officers must demonstrate a commitment to ethics. Senior staff should be present during ethics training and actively engage in ethics-related workshops and discussions. Too often ethics training is treated as a necessary distraction in order to comply with governmental mandates. Occasional training sessions by experts are certainly useful.

However, these sessions cannot replace the benefits achieved by having senior staff directly involved in ethics training. Fire fighters need to see senior staff engaged and committed to ethics. Senior officers should develop and present departmental training. It sends the wrong message when ethics training is mandated by an impersonal email with instructions to log onto an even more impersonal website that provides little context and no feedback or discussion.

A Belief That Ethics Cannot Be Taught

There is no shortage of individuals who believe that ethics is an inherent trait or a condition of personality. There is the assumption that ethics is, therefore, unteachable. Typically, these individuals have little exposure to formal ethics training. To be clear, ethical behavior, similar to any other behavior, is learned and habituated. Ethical decision making requires a frame of reference and understanding of context. It is also inaccurate that ethics is entirely ingrained during childhood. Individuals develop moral attitudes and values throughout their lifetime and are subject to change and growth (Ragatz, 2012).

A Belief That Ethics Training Is Not Needed

It is only natural that we believe the best about our friends and coworkers. However, every department who has ever suffered through scandal or has been shocked by the behavior of a respected member likely believed that unethical behavior happens at other departments. In many cases, fire chiefs are shocked when they learn of subordinates' behavior even though, in many cases, peers saw it coming. Similar to accidents, ethics violations often catch us by surprise, but they shouldn't. The mantra of accident prevention applies to ethics: "That which can be predicted can be prevented." It is naïve of a fire chief to believe that his or her department is immune to inappropriate behavior or that breaches of ethics can't be prevented. All departments can benefit from ethics training.

Compliance Creep

Compliance creep is a term used to describe the gradual overdependence on rules to define ethical behavior. Compliance-based ethics has a certain seductiveness for senior fire administrators. The creation of new policies to address behavior issues nearly always seems logical and responsible in the moment. Almost every department has had a fire fighter observe, "When one person screws up we all get punished." However, rarely do rules address the cause of inappropriate behavior, and the prolific use of compliance-based strategies can cause **ethical fading**, which refers to a condition in which individual moral judgment and responsibility are supplanted and replaced by simple adherence to policy. As a result, instead of fire fighters thinking about doing the right thing, they focus on calculating the risk and benefit of noncompliance (Ragatz, 2012).

Closing Thoughts

Both the topic of ethics and culture building can be difficult subjects to discuss and embrace. There are few hard and fast rules in either ethics or culture management, and effectively dealing with the subjects requires as much art as science. Undoubtedly, the building and maintenance of an ethical culture can be challenging, yet it is an important element in protecting the reputation of a fire department and its fire fighters. The effective development of an ethics culture requires efforts in the development of productive policy, the motivation of fire fighters to comply with that policy, and a system to address policy infractions. However, an effective, ethical culture requires more than policy enforcement. An ethical culture requires the voluntary cooperation and a commitment by fire fighters to doing the right thing. That commitment requires education and the positive reinforcement provided by a productive work environment Figure 9-5.

An important takeaway from this unit should be that fire administrators must be proactive in ensuring an ethical workplace. Fire administrators can do this by exercising leadership in ethics and demonstrating a commitment to both policy enforcement and embedding ethics into daily work routines. The development of an ethical work culture requires a priority focus on ethics—it must be recognized as an important element in the strategic management of the department.

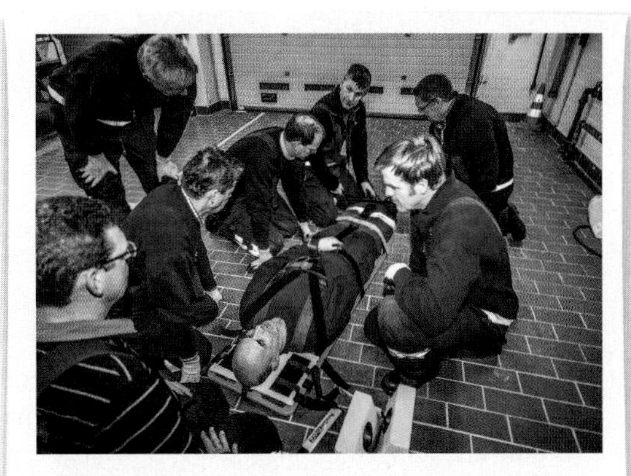

Figure 9-5 Education and positive reinforcement builds ethical cultures.

WRAP-UP

Chapter Summary

- The chapter opens with an exploration of what is meant by an ethical culture. The term "ethical culture" describes the prevailing attitude toward behavior and an expression of departmental priorities related to ethics.
- An ethical culture was found to be a condition where doing the right thing is normal and encouraged. Conversely, an ethical culture makes doing the wrong thing unusual, difficult, and unrewarded.
- The chapter summarized the differences between an ethical culture and an unethical culture and their relative effects on behavior. Ethical cultures directly influence behavior by creating social norms and peer pressure to behave appropriately. Unethical cultures were seen as indirect influences that facilitated unethical behavior rather than cause it.
- While there may be some peer pressure to conform to inappropriate behavior standards, typically these do not have the same degree of influence as does the influence of a positive ethics culture. As a result, the text urges that departments invest in the time and effort necessary to build and maintain an ethical culture.
- The hallmarks of an ethical culture were discussed. These include a commitment to the public welfare and fire fighter safety as paramount concerns. Other hallmarks include a prevailing attitude of service, a high priority on competency and dedication, and the maintenance of open and honest lines of communication within the organization.
- For an ethical culture to exist there must be a high degree of trust between employees and management, and a high level of worker satisfaction.
- The development and maintenance of a culture based on ethical principles was much more effective in the encouragement of appropriate behavior than was complete reliance on rules and regulations. It was demonstrated that rules and regulations have limitations including the ability of the department to provide adequate supervision, and supervisors' willingness to enforce policy.
- Primary among the organizational conditions necessary to support an ethical culture is that of a positive work environment. The general principles of organization development and enlightened management were explored relative to their effect on ethics.
- Self-actualized, engaged, and committed employees are more likely to act ethically than those who are disgruntled, nonchallenged, and threatened. Creating a positive work atmosphere celebrates personal growth and innovation. Employee-centered cultures are more likely to have high ethical standards than those that are based solely on competition and confrontation.
- The necessity of effective leadership was identified as important in the development and maintenance of a positive work environment and an ethical culture. The maintenance of an ethical culture requires committed leadership willing to mentor and shape values by articulating departmental priorities.
- Company officers must be examples of ethical behavior as well as teachers in ethical intelligence. Also discussed was the importance of policy in shaping and reinforcing ethical cultures.
- A requisite focus must be placed on the department's mission. The creation of an optimistic and consequence-oriented work culture is vital in maintaining high ethical

standards. The department must make a priority of ethics, and ethics must permeate throughout all levels of the department. Ethics must be a guiding feature in hiring and promotional practices, training, and policy decisions.

- Ethics training must be treated as important, an investment in proper ethics training should be made, and attendance should be required for all personnel. Training should be of a quality that is not generic—it should be job-related and involve a frank discussion of ethical principles among employees.
- Compliance ethics was introduced as a system in which rules are promulgated, enforcement is achieved by strict supervision, and negative consequences are imposed for noncompliance. The advantages of compliance-based ethics were presented as ease of implementation, its relative low demand on departmental leadership, and its compliance with legal standards.
- Compliance-based systems are usually mandated by governmental oversight agencies, and so some form of compliance ethics is almost always necessary within governmental agencies such as fire departments.
- Compliance-based ethics is one of the most common forms of ethics systems because it is readily understood by senior fire department staff who may be poorly educated in other ethical options.
- The disadvantages of compliance-based ethics were also presented. The biggest disadvantage of a compliance-based program is that it is a bounded system. Bounded systems are those that tend to draw specific lines around acceptable versus unacceptable behavior.
- Bounded systems focus on acceptable behavior as opposed to correct behavior. The goal of compliance in a bounded system is to avoid getting in trouble as opposed to seeking to do the right thing. Bounded systems also create limits on ethics.
- It is easy for fire fighters to assume the attitude "that which is not against the rules must be inherently acceptable behavior." That attitude puts immense responsibility on departmental administration to write an ever-increasing number of policies to cover all unacceptable behavior possibilities.
- The more rules that are promulgated the more likely it is that individuals will consider them the extent of moral and ethical responsibility. This is known as compliance creep.
- Extensive rules tend to disengage moral judgment by individual employees. Another shortcoming of compliance ethics is that ethical behavior is limited to those actions that can be supervised. Behavior decisions by employees tend to be modified from questions of right and wrong to assessments of risk versus benefit.
- The opposite of compliance ethics is integrity-based ethical systems. Integrity-based ethics is grounded in the belief that if properly educated in ethics, workers are more likely to behave ethically within an ethical culture absent of motivation for inappropriate behavior. Integrity-based ethics empowers employees with ethical knowledge while simultaneously removing environmental conditions that may facilitate unethical behavior.
- The disadvantage of integrity-based systems is that they put a tremendous amount of responsibility on departmental leadership. Integrity-based ethics requires culture management, mentoring, and education.

WRAP-UP

- In contrast to compliance-based ethics, which only require the promulgation of rules, integrity-based systems require significant engagement by departmental leaders at all levels of the organization.
- The primary advantage of integrity-based ethics systems is that they facilitate an ethical culture; as such, ethics is pervasive within the organization. Employees who possess a high degree of ethical intelligence and who are operating in an ethical culture are much more likely to behave appropriately even absent supervision compared to those who work in an exclusively compliance-based environment.
- The chapter closed with a discussion of barriers that might inhibit the development of an ethical workplace culture. These included: a lack of support from senior leadership, a belief that ethics cannot be taught, and the belief that ethics training is not needed within an individual organization.
- These three conditions are common in fire departments and lead to a false sense of security regarding employee behavior. Departments involved in public scandal are consistently surprised by inappropriate behavior because they assume that their employees' personal values are enough to prevent misbehavior. Evidence has demonstrated, however, that several elements including ethical culture can subvert personal values and normalize inappropriate behavior.

Key Terms

bright lines clearly identifiable boundaries of behavior.

compliance-based ethics an ethics system utilizing specific policies and enforcement of imposed standards.

compliance creep the gradual overdependence on rules to define ethical behavior.

contextual standard of behavior a situational condition that suspends normal behavior standards, replacing them with new situationally relative standards.

ethical fading the displacement of personal values and judgment as a result of overdependence on rules.

ethically bounded a condition in which right and wrong are defined strictly through conformance to policy.

ethical work culture a culture where doing the right thing is normal and easy, and doing the wrong thing is unusual, difficult, and unrewarded.

integrity-based ethics the individual self-regulates based on a shared understanding of behavior standards, and with the intention of contributing to the common good.

organization development a field of study dedicated to expanding the knowledge and effectiveness of people to accomplish more successful performances.

unbounded having no limit; all encompassing.

Challenging Questions

To check your understanding of this chapter's material, answer the following questions. It is highly recommended that you discuss your viewpoints with fellow students, peers, coworkers, and friends to discover their opinions as well.

- In your opinion, what are some of the challenges associated with building and maintaining an ethical culture?
- Most departments who have weathered a scandal as a result of inappropriate behavior

by a fire fighter argue that they never saw any hint of oncoming trouble. Keeping in mind our discussion of culture, what sort of signs are likely to indicate a potential for unethical behavior?
- What is a fire chief's role in building and maintaining an ethical culture?
- Is your department primarily driven by integrity-based ethics or compliance-based ethics?
- Based on the principles discussed in this chapter, develop an ethics training program for your department. Include a schedule and topics to be covered.
- Discuss the likely problems with overdependence on rules and regulations in developing an ethical work culture.

Case Study Conclusion

Revisit the case study at the beginning of the chapter. Spend a few minutes considering the questions posed at the end of the case study. In light of the information shared in this chapter, have any of your original observations changed? Chief Smith is not the first fire chief to feel frustration over perceived cumbersome regulation. Sadly, this case study reflects behavior that has been utilized in several departments, and with similar results.

Were Chief Smith's actions ethical? In a word, no. While Chief Smith's intentions may have been honest, he ignored the responsibilities of his office. As a fire chief, he voluntarily undertook the obligation to enforce municipal policy and to act as a faithful agent for city administration. Smith, like so many fire chiefs, confused loyalty to the department with loyalty to the position. The position of fire chief is an administrative one—his obligation is first to provide the highest level of fire protection possible, and second to act in the public interest, including the use of public funds. The important caveat is that these actions are to be carried out within the confines of relevant law and city policy. If Smith found the regulations overly cumbersome, he could seek the city council's permission to revisit the regulations. He did not have the right to ignore them, any more than his fire fighters would have the right to ignore one of his policies that they thought was counterproductive. This case study is an example of an instance in which having good intentions does not justify betraying trust.

Chapter Review Questions

1. What is compliance creep?
2. Define the term "ethical culture."
3. List three barriers to building an ethical culture.
4. What is the difference between an integrity-based and compliance-based ethical system?
5. Explain the concept of ethical fading as it relates to compliance-based ethics.

References

Kaptein, M. 2012. "Why Do Good People Sometimes Do Bad Things." Accessed June 12, 2018. https://papers.ssrn.com/sol3/papers.cfm?abstract_id=2117396.

Meinert, D. 2014. "Creating an Ethical Workplace." *HR Today*. Accessed June 12, 2018. https://www.shrm.org/hr-today/news/hr-magazine/Pages/0414-ethical-workplace-culture.aspx.

Paine, L. S. 1994. "Managing for Organizational Integrity." *Harvard Rusiness Review*.

Ragatz, J. 2012. "Building an Ethical Culture." Accessed June 12, 2018. https://ethics.theamericancollege.edu/library/news/building-ethical-culture.

Roberts, R. 2009. "The Rise of Compliance-Based Ethics Management." *Journal of Public Integrity* 11, no. 3, 261–278.

Taylor, A. C. 2016. "Ethics and Compliance-Based Leadership Models." Accessed June 12, 2018. http://files.acams.org/pdfs/2016/Ethics-and-Compliance-Based-Leadership-Models.pdf.

Wolde, A., J. Groennendaal, I. Helsloot, et al. 2014. "An Exploration Study of the Connection Between Ethical Leadership, Prototypically and Organizational Misbehavior in a Dutch Fire Service." *International Journal of Leadership Studies* 8, no. 2, 18–43.

CHAPTER 10

Ethics Responsibilities in Fire Administration

"With great power comes great responsibility"

—Uncle Ben, *Spider-Man*

OBJECTIVES

After studying this chapter, you should be able to:

- Assess the role of administrative responsibility within the fire department.
- Outline the concept of obligation as it relates to responsibility.
- Expound on the concept of accountability as it relates to obligation.
- Distinguish between internal versus external management of public accountability.
- Differentiate between subjective and objective responsibilities.
- Understand the elements of ethical dilemmas.
- Interpret ethical conflict.
- Describe the functional imperatives within the ALIR model.

Case Study

Fire Chief Bob Jones is the fire chief in Anytown USA, and he has a dilemma.

For over 10 years a beautiful 100-year-old, three-story building has sat empty and deteriorating on the old square in downtown Anytown. Through significant effort by the Anytown economic development team (of which Chief Jones is a member), the old downtown square area has seen significant redevelopment as an entertainment center. The building in question is an eyesore and an increasing fire hazard.

The good news is that a developer has purchased the building and plans to put an upscale restaurant on the main floor and luxury apartments on the upper two floors. The project will have a tremendous positive impact on the city's tax base and act as a catalyst for future development.

Unfortunately, the developer is highly resistant to the requirement for sprinklers to be retrofitted in the building. He argues that the required modifications will force the removal of many of the architectural elements that make the building historically significant. Most important, he contends the sprinkler requirement makes the renovation economically unfeasible. As a result, the developer is asking for a waiver of the sprinkler requirement, without which he insists he cannot move forward with the renovation plan. The Anytown Mayor, City Council, Historical Society, Chamber of Commerce, and the building inspector all support the developer's request for a waiver.

As a fire professional, Chief Jones is supportive of fire sprinkler regulations because he knows they make buildings much safer. He is also aware that developers are almost always opposed to sprinklers and that if he allows this waiver it will set a precedent for future developments.

On the other hand, Chief Jones is a member of the economic development team and knows how important the restaurant is to the community. He is also aware that the current condition of the building presents a real danger to its neighboring buildings, not to mention the potential occupants of the proposed restaurant and apartments. The renovation, in the absence of sprinklers, is not ideal but does represent a step forward in making the entire area marginally safer.

- What are the conflicting ethical responsibilities facing Chief Jones?
- How should Chief Jones go about prioritizing his responsibilities?
- If you were Chief Jones, what would you do?

Introduction

In Chapter 3, *Influencing Behavior*, you were introduced to the concept of responsibility. Again, in Chapter 6, *Fire Fighter Responsibilities*, responsibility was discussed as it pertained to the specific obligations imposed on fire fighters and fire officers. In this chapter we will delve deeply into the concept of responsibility, with a particular focus on how it applies to administrative personnel within the fire service.

You cannot understand professional ethics, particularly the ethics associated with public administration, without understanding the relationship between ethics and responsibility. Responsibility is a fundamental element in the consideration of ethical attachment. Consider this example: A small fire breaks out in a neighbor's shed. Some would argue that you have a moral responsibility to assist your neighbor. That moral responsibility exists as a fundamental condition of humanity to help those in need. Ethically, whether you put the fire out or not is irrelevant because you have no ethical obligation to do so. However, if you are either paid as a fire fighter or voluntarily accept the responsibility of a fire fighter for your local volunteer department, you now have accepted ethical responsibilities to assist others as a condition of your role as a fire fighter.

The ethical difference is rooted in responsibility. As a bystander, you may have a moral obligation to help, but you have no ethical responsibility to act. As a member of a fire department, you have voluntarily accepted responsibilities that create ethical obligations. Morality exists as a condition of humanity or with adherence to a moral or religious code. Professional ethics exists as a condition of professional responsibility.

Most fire officers believe they understand their responsibilities. Yet responsibilities can be complex. They can be interwoven and subtle. Responsibilities can compete with each other and conflict with each other.

Not all responsibilities are equal; some are imposed on us by others. Some responsibilities spring from a sense of obligation inside us. Some responsibilities are associated with segregated portions of our lives called "roles." Roles can make responsibility demands that align or confound professional obligations.

It is certain that to be an effective fire chief or administrative officer within a fire department an individual must understand the various responsibilities and how they relate to each other, in addition to grasping the responsibilities of others.

Ethics and Administrative Responsibility

A senior fire administrator is burdened with significant and varied responsibility. Those responsibilities define his or her role within the department and the municipality. The roles of an individual fire chief vary from department to department but will typically include the following: budget director, human resources manager, fire code expert, leader, mentor, and fire suppression expert.

The chief's roles do not stop at the station door. He or she is also a member of the municipality's senior leadership team. Fire chiefs are often involved with zoning boards, traffic commissions, economic development teams, and technical commissions such as water and electric.

In common understanding, each of the previously mentioned roles consists of responsibilities. In reality, they are a collection of tasks for which the chief is responsible. The distinction may seem subtle, but it is important. Tasks have no ethical consequence. Rather, it is the underlying responsibility for those tasks that imparts ethical relevance. For instance, brushing your teeth is a task essential to dental health, but it lacks ethical implication. Unlike brushing your teeth, the fire chief's task of judiciously reviewing and enforcing fire safety codes has ethical relevance because it is rooted in responsibility for the welfare of others. To clarify, tasks may be undertaken for any number of reasons, but their ethical relevance is determined by whether or not there is a specific responsibility attached.

Clarification

Tasks may be undertaken for any number of reasons, but their ethical relevance is determined by whether or not there is a specific responsibility attached.

The ethical attachment of responsibility is defined by a connection to an obligation to a specific duty, or as a condition of accountability to someone (Cooper, 2012). Brushing our teeth is important but it has no ethical attachment because there is no obligation or duty, nor is there accountability to anyone but ourselves for the action. Conversely, a parent's responsibility to teach a child good dental hygiene is linked to an obligation to a child's welfare and development. The child has a dependency on a parent, and parents shoulder that responsibility as a condition of parenthood. Failure to meet that obligation is a breach of responsibility and so is inherently unethical. Similarly, the fire chief has an obligation to public safety, and the public trusts he or she will act consistently with that obligation. Failure to undertake tasks related to that responsibility is a breach of that responsibility and therefore is unethical.

Obligation and Accountability

The concepts of obligation and accountability are fundamental to understanding responsibility, and they are the catalysts that impart ethical implications to a particular task role. The weight of responsibility is ultimately dependent on the conditions of obligation and accountability. Where responsibility is lacking, ethical implication does not exist. Herein lies one of the distinguishing features of ethics versus morality. Moral implications may still exist even in the absence of ethical responsibility. As you may remember from earlier chapters, morality is a concept of behavior expectation, which exerts that some actions are universally right or wrong regardless of intent or consequence and are defined by a "higher" authority. Conversely, ethics are rooted in human relationships, specifically our obligations to the tasks associated with our roles as human beings, and the various attachments that form accountability to others.

Obligation

Obligation describes a duty or commitment to a task. An obligation may be imposed externally by a code of ethics or as a fundamental element within an accepted role. For instance, a painter who is hired to paint a house has an ethical obligation to paint the house once he has accepted employment and been paid. Whether based on trust, or an actual contract, a covenant is formed that imparts responsibility. With that responsibility comes ethical attachment. The house painter's contract is external; it is a condition of the relationship.

Additionally, a conscientious painter may feel an internal obligation to complete the job and do it well as a matter of personal integrity. As we will discuss later, this sense of responsibility comes from an internal source within the painter—one's personal values. As you can see, the sense of obligation can be imposed externally or arise internally. The source of obligation defines the nature of a responsibility as either objective or subjective, as we first introduced in Chapter 3, *Influencing Behavior*.

Accountability

Accountability is an element of responsibility associated directly with another individual or group. The foundations of accountability are trust and dependency. A fire chief is tasked with several responsibilities that have a direct bearing on the welfare of others. The citizens living within his or her jurisdiction depend on the chief to perform those responsibilities, and so there is accountability as shown in Figure 10-1. The

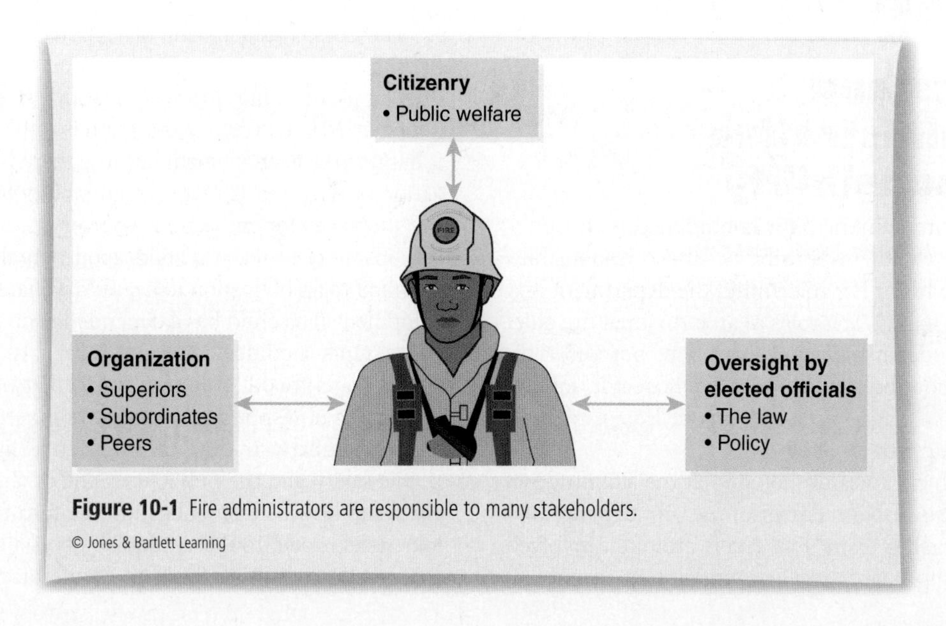

Figure 10-1 Fire administrators are responsible to many stakeholders.

mayor and city council who appoint a fire chief trust that the individual will act in a way consistent with his or her responsibilities; again, this imposes a condition of accountability on the fire chief. In this case, accountability is supervisory and may result in negative consequences if responsibilities are not met.

Accountability is rooted in obligation, specifically, with obligations associated with personal relationships. In its most common usage, the term "accountability" often has a connotation of supervision and punishment arising from nonconformance. While this is often the case, accountability is sometimes associated with self-imposed tasks as well. A fire chief may likely feel accountable to fire fighters for the quality of their living conditions within the station. There is no specific obligation imposed on the fire chief to provide comfortable station furniture, yet he or she may feel accountable to fire fighters based on an empathetic connection or because the chief wants subordinates to view him or her positively.

A challenge faced by fire chiefs is that they are accountable to many stakeholders. Most chiefs are contractually accountable to their superiors in local government for the efficient operation of the department. They are also answerable to the citizenry for the provision of public safety, and to their fire fighters for safe and suitable working conditions. Further, they are accountable to their family and, most important, they are accountable to themselves and their sense of integrity and values.

The complex web of obligation and accountability defines the ethical responsibilities within which a fire chief must operate. Obligation and associated accountability issues are often intimately connected to predefined roles. These are **objective responsibilities**. Feelings of obligation and accountability can also arise from personal values and the virtues of loyalty and integrity; these types of obligations are associated with personal or **subjective responsibilities**.

> **Clarification**
>
> Objective responsibilities are rooted in obligation to others, and are typically easily identified and described. Subjective responsibilities are self-imposed and arise from personal values. Due to their nature, subjective responsibilities are sometimes difficult to define and articulate.

Objective Responsibilities

Objective responsibilities are imposed from external sources. They tend to be clearly defined and not subject to interpretation. And so how the individual feels about a responsibility tends to be secondary to the explicit demands of the obligation. In the case of fire administrators, there are specific and unique objective responsibilities that define their role as an administrator. Those objective responsibilities define the parameters of a fire chief's job, and they represent the measure by which external stakeholders judge performance. They also contain an **ethical attachment**. The typical objective responsibilities of the fire chief are listed next. You will notice that the responsibilities are organized into three general categories. A collection of related responsibilities is often described as a role. As we will explore later, roles define the scope and breadth of professional and personal expectations as well as their attendant ethical obligations.

Acting in the Public Interest

Acting in the public interest is a fundamental responsibility of any fire official and, for that matter, any individual fire fighter. In a genuine sense, the fire service exists to provide public service. Acting in the public interest is a fundamental condition of employment and is a duty imposed by a social contract. The public supports the fire department explicitly to provide a service. That service is vital to the public welfare, and fire fighters voluntarily accept the responsibility for the public welfare. To act against the public interest is to be negligent of the job's fundamental responsibilities.

- Because emergency responders are responsible for public safety, the fire administrator's most elemental responsibility is the *assurance of the best possible response to emergency scenes as resources allow.* Public safety is essential to the general welfare of the community. In almost all instances the local fire department is the sole provider of the services it renders. As such, there is an element of dependency that makes it an ethical imperative.

- Hand in hand with the obligation to respond effectively to emergencies is the fire administrator's responsibility for the *promotion of public safety via risk-reduction strategies.* Prevention of potential risk to life, safety, and property, coupled with a response to emerging threats, is at the core of the fire department's mission and its responsibilities.
- In order to provide for life and property safety, fire departments are entrusted with access to public money. The word "trust" is key to the previous sentence. Fire administrators are entrusted with other people's (taxpayers') money. As such, *the responsible management of public resources* represents an ethical responsibility fundamental to acting in the public interest.

Protect the Health and Safety of the Department's Personnel

Many people may assume this to be a subjective responsibility because a good fire officer feels an obligation to protect those for which he or she is responsible. While the previous sentiment has merit, it is also true that there is an objective responsibility imposed on administrative personnel for fire fighter safety. The obligation is a condition of position, with accountability enforced by several governing bodies including the Occupational Safety and Health Administration (OSHA), civil courts, and individual state regulatory agencies. The relationship between fire fighters and administrators is not entirely different from that of the general population. In both cases, there is an expectancy of duty. Similar to the general public, fire fighters have a right to expect their department leadership to act in their interest regarding safety.

Within the role of assuring health and safety are some of the following obligations:

- Core competency is critical to fire fighter safety. A chief officer of a fire department *has the responsibility to provide adequate training to personnel.*
- Firefighting is a physically demanding profession. As such, there is an ethical responsibility to *provide adequate medical screening and medical care* to ensure that fire fighters are healthy enough to fulfill duties safely.
- The fire service has a significant body of documented best practices ranging from fire fighter training to standards of personal protective equipment and vehicle maintenance. *Compliance with industry standards and best practices* is an ethical responsibility in that it is necessary to both the health and welfare of fire fighters, as well as the fire department's ability to serve the public interest.
- *The promulgation of policies that protect fire fighter safety and health* is an ethical responsibility for fire administrators. Compliance with rules and regulations is an ethical responsibility of rising officers.

Professionally Manage Department Operations on Behalf of the Supervising Authority

The chief administrative officer for a fire department has an obligation to *loyally represent the interests of the jurisdiction's governing authority (mayor, council, city manager).* Elected officials are likely to regard this responsibility as the specific reason why a fire chief is hired. By assuming the position of fire chief, there is an inherent responsibility to meet the expectations of the mayor and city council. Again, as with the responsibility of providing for public safety, this is a condition of employment.

- A fire chief also has the responsibility to ensure compliance with federal, state, and local laws and policies.

- Related to his or her role managing the department, as well as serving public interest, it is incumbent on the fire chief to fairly and honestly communicate needs and plans.
- The responsibility to acquire, inventory, and maintain public assets is inherent in effectively managing the department; it is also consistent with the responsibility to act in the public interest. A fire chief is entrusted with public resources. That placement of trust imparts ethical obligation.

The hallmark of an objective responsibility is that it has a direct link to accountability. Because objective responsibilities are externally imposed, there is a necessary element of accountability to task. This assumes external oversight to ensure compliance with ethical obligations. This viewpoint of administrative ethics was championed by **Herman Finer** (Al-Habil, 2011). Finer asserted that ethical standards of behavior among government officials could only be assured by a clearly defined policy body, accompanied by significant oversight. It was Finer's contention that, absent clear direction and external oversight, administrators would invariably pursue self-interest over public interest.

Finer's approach to the assurance of ethics among administrators is often called **external controls**. External controls usually take the form of either legislation or codes of ethics. In looking at governmental ethics, it is evident that Finer's approach is almost universally adopted. Every level of government has codes of ethics, laws, or policies that clearly define and govern the behavior of government officials.

The application of external controls is nearly universal for several reasons. First, it is a measurable response that is attainable by a government. The alternative to external controls is more indirect controls that utilize in-depth education, culture management, and screening. Implementation of internal controls is challenging and a resource-intense endeavor. It is easier to forbid behavior and punish infraction than to prevent improper behavior through education (Cooper, 2012).

Second, the nature of administrative work necessarily involves ministerial discretion. The implementation of laws and codes of ethics clearly defines boundaries (Cooper, 2012).

The imposition of fines, suspension of duties, and possible imprisonment create a significant disincentive for inappropriate behavior. Each administrative officer, when faced with potential conflicts of interest, is compelled to perform a risk-benefit assessment that theoretically makes unethical behavior counterproductive to self-interest (Cooper, 2012; Waldo, 1984/2006).

Finally, the imposition of codes of ethics, ethics laws, and policies seems to positively affect individual awareness of ethics expectations. The imposition of ethical boundaries seems to support an ethical culture and bolster internal controls (Cooper, 2012).

Clearly, in Finer's view, accountability supersedes the individual administrator's sense of obligation in ensuring ethics. The earliest philosophical approaches to public administration are rooted in the assumption that with authority comes the likelihood of corruption (Waldo, 1984/2006). This view of ethics is rooted in a distrust of governmental authority. The American system of checks and balances within representative government is essentially based on that distrust. The logical conclusion from this position is that elected officials are better equipped to determine public interest than bureaucratic chiefs. Presumably, that advantage extends from the fact that they represent the public's wishes.

What Do You Think?

Consider the issue of residential sprinklers. They are universally endorsed by the fire service yet remain unpopular with the public. Under what circumstances is it prudent for a fire chief to dismiss public opinion and replace it with his or her view of the greater good?

That belief might be legitimate if a fire chief's ethical responsibilities were confined solely to the public wishes and the assumption that the public is the best judge of what is in its best interest. Among Finer's advocates, the extreme position is that public administrators should be automatons carrying out specific tasks absent moral or ethical judgment. This extreme position suggests that a fire chief should run his or her department strictly according to the wishes of elected officials because those officials represent the wishes of people who pay for the fire department. However, there is an opposing view based on a more activist role of administration.

Consider again the issue of residential sprinklers. There is a widespread belief in the fire service that beyond any other technology or intervention, residential sprinklers most effectively promote public fire safety. Yet there is significant resistance among elected officials and even members of the public regarding their implementation. The consensus opinion of people well educated in the issue is that public resistance springs from a misunderstanding of sprinkler technology and its cost-effectiveness. There is a belief within professional circles that the public and their elected officials are not the best judge of what is actually in their own best interest when it comes to fire safety.

Many fire officials believe that elected officials are not well-versed in the technical issues associated with public service provision. Individuals elected by the public are not necessarily experts on the technical aspects of fire protection, law enforcement, sanitation, and all the other technical services provided by the municipality. As such, they are no better qualified than the general public to provide direct oversight of subject matter experts in the fire service. Thus, this perspective places heavy emphasis on the expert judgment of fire officials to work alongside elected politicians, instead of simply doing the bidding of elected officials. Finally, there is also a common opinion that elected officials are not necessarily motivated by public welfare. Dependency on reelection, and most important on funding for campaigns, often creates inherent conflicts of interest between public interest and the special interests of those investing in political campaigns.

As a result, many believe that people in bureaucratic positions such as fire chief are much better educated and motivated to make ethically related decisions regarding public policy (Cooper, 2012; Al-Habil, 2011). This position is closely associated with **Carl Friedrich**, who asserted that ethical responsibilities within public agencies could best be assured by **internal controls**. Those internal controls consisted of education in ethical decision making, the building and maintenance of ethical cultures, and removal of disincentives for unethical action.

Herman Finer and Carl Friedrich were contemporaries whose opposing views were greatly debated throughout the 1960s and 1970s. Contemporary thought seems to support Friedrich's philosophy of internal controls based on several elements rooted in the modern interpretation of representative government.

Limitations of Accountability

Herman Finer's reliance on external controls and accountability has significant limits. Adherence to ethics policy is often dependent on the proximity of supervision, the likelihood of being caught, and the significance of punishment if issues are violated. The maintenance of oversight is difficult and time-consuming, and is often ineffective.

Conversely, Friedrich's view tends to promote ethical control within the organization, focusing on the reinforcement of ethical decision-making skills at that point in time where ethical dilemmas may arise. Friedrich believed that unlike supervision, conscience-driven ethics policies have the advantage of being consistently present in the decision-making process.

A Lack of Consensus Regarding the Public Interest

The last 20 years in American politics has seen increasing divisiveness as to what constitutes the public interest. A major political divide between conservatives and progressives is the role of government in assuring justice, equity, and public interest. The conservative viewpoint is that of minimalist government, assuming that the public must share a significant burden for its welfare and well-being. As an example, the individual must make a choice whether or not to invest in the safety provided by a residential sprinkler system.

Conversely, progressives tend to advocate a much more proactive role for government. In their viewpoint elected officials have an obligation to guide policy that directly affects and, in some cases, limits individual choice with the intention of creating a greater good.

Regardless of your view about the role of government, the amount of accountability and the effectiveness of government oversight are negatively affected by inconsistency within the government as to what actually constitutes the public interest.

A Perceived Erosion of Moral Authority

Another significant development in government is a perceived erosion of moral authority among elected officials. While the average citizen still assumes that fire chiefs, police chiefs, plumbing inspectors, and other government officials are committed to the duties of their respective fields of expertise, there is a growing belief that elected representatives are somewhat less committed to duty as they are to political affiliation. The increasing hyperpartisanship within politics has eroded public confidence in elected officials as ethical agents and so calls into question their ability to direct and oversee independent policy within technical professions such as the fire service.

As public mistrust in elected officials to act exclusively in the public interest grows, there tends to be increasing pressure on the leadership of bureaucratic institutions (fire departments) to exercise ethical judgment regarding their numerous ethical responsibilities. This judgment is almost certainly based on their interpretation of greater good and on the values that they feel have priority within their realm of public service.

A Lack of Adaptability in External Controls

A major criticism of external control methods is that laws and codes of ethics establish only a moral minimum. By doing so they create a bright line of acceptable behavior based on minimal compliance, with little or no incentive for higher degrees of ethics principles. For instance, if ethical law limits the accepting of gifts costing more than $50, then gifts costing less than $50 intended to influence decisions become "ethical," at least under the law. Ethics laws and policies tend to identify negative behavior as a basis for ethics expectations rather than the promotion of positive behavior. If there is no rule against an action, then it is by default acceptable. Any ethics policy or law is only as effective as its scope of applicability.

Generally, it is understood that no policy law can cover every possible eventuality, as demonstrated in Figure 10-2. It is a common belief that ethical behavior can only be guaranteed by the good intentions of the public administrator. Supporters of internal control suggest that it is the individual conscience

Figure 10-2 Ethics policies tend to focus on defining bad behavior, rather than promoting good behavior.

that drives ethical decision making, not a policy that ensures ethical behavior, and so there must be a subjective analysis of the demands of responsibility.

Subjective Responsibilities

Subjective responsibilities focus on fire administrators as human beings, and not just administrators of policy. Certainly, there are clearly defined expectations of obligation associated with being a senior officer within a fire department. Just as certainly, there are those *feelings* of obligation and accountability that arise within each of us based on our personal beliefs, attitudes, and values.

For instance, consider training. Objectively there are certain training requirements imposed by various external governing bodies. These may include the NFPA, OSHA, and the Insurance Service office (ISO). These impart objective responsibilities. But, in addition to these objective responsibilities, a fire administrator may act on his or her subjective sense of what is appropriate training. For instance, a fire chief may feel strongly regarding succession planning and may require mandatory training in fire administrative areas. Another chief may feel that it is the responsibility of fire fighters to prepare themselves for advancement through enrollment in college programs. A sense of obligation to training and accountability for fire fighter development springs from a chief's personal values regarding education and so are subjective responsibilities.

Given that subjective responsibilities come from within and are often not enforced by external reward or punishment should not imply that they are less important, nor less powerful, as motivators. As we learned in earlier chapters, individuals have a tremendous need to align their personal values and their perceived responsibilities. To act against your character causes emotional dissonance, which must be resolved.

The typical perceived subjective responsibilities of a fire chief include:

- An obligation to act in the public's interest as interpreted by the organization's professionally accepted best practices and within the scope of authority. The primary responsibility is that of ensuring public safety.
- An obligation for the personal safety, health, welfare, and professional development of department members.
- An obligation to sustain the department's traditions and future viability.
- A duty to professional standards, values, and practices as articulated by professional organizations and advocacy groups.

Because subjective responsibility springs from our intrinsic values and priorities, theoretically they must be mediated by our internal moral compass. The exercise of conscience, the application of virtues, and subjective moral decision making are most closely associated with the internal control systems described by Carl Friedrich. Friedrich's approach to public ethics also has its problems and limitations.

Shortcomings of Internal Controls

As mentioned earlier in this chapter, there are many who struggle with an exact definition of the public interest. Is a fire chief to act as a patriarch or matriarch who, based on expertise, makes value judgments as to what is best for the citizens, or is he or she bound to comply with the consensus opinion of the public? Is a fire chief a leader in public policy or an agent of the people's will? A conservative view supports acting according to the public's wishes. A progressive view supports an assumed best outcome for the public.

Complicating the issue is that there is rarely consensus on exactly what the public wants or what part of the public's interests are to be served. Different constituencies within the public have different needs and expectations.

Further, the concept of a "best outcome" is problematic in that it is open to subjective interpretation. Given carte blanche, there is no doubt that some fire chiefs would pursue aggressively outlawing all forms of open flame including candles and fireplaces. Others would advocate for fire stations on every corner staffed with 10 fire fighters per unit. Herein lies the difficulty with subjective responsibilities—they empower the public administrator to make decisions about others' welfare based on his or her interpretation of what the public "good" actually is. Values clearly complicate subjective responsibility. Unfortunately, one person's idea regarding subjective responsibility may not be consistent with another's (Kernaghan, 1990).

Recent trends in the ethics of public administration have become more individualistic, and relative to context, intent, and outcome. However, many people suggest there is an issue with permitting subjective value judgment in solely guiding ethical decision making because the concepts of context, intent, and outcome are moving targets. While the administrator may have what he or she considers the best of intentions, those affected by the actions may have entirely different views as to the worthiness of intention or the value of the outcome, as seen in Figure 10-3. What the fire chief considers good and what a local business person considers good may be two entirely different things, and at least at some level the fire chief is accountable to the public he or she serves. As a result, several communities have seen significant pushback against what are often considered governmental intrusions into private affairs, and overregulation.

Figure 10-3 In the last 20 years, we have seen a significant increase in public resistance to activist government, including fire codes and fire safety regulations.

Internal controls are not necessarily reliable in assuring conformance to responsibility. Absent any oversight there is no guarantee that the fire chief will not intentionally or unintentionally act in his or her self-interest, or in the interest of someone to whom he or she feels accountable. If internal controls are the primary assurance of ethical compliance, then ethical intelligence, the creation of an ethical culture, and the reinforcement of ethical behavior become critical. Unfortunately, ethics education, culture creation, and motivational techniques are difficult to implement and harder to measure.

As we will discuss later in this chapter, another problem with subjective responsibility is that even the well-intentioned fire chief is likely to have competing values shaping the decision-making process. For instance, a fire chief who places a great deal of value on loyalty may feel a heightened sense of obligation to those to whom he or she reports (mayor or city administrator). At the same time, that chief will likely be motivated by a deep sense of responsibility for the welfare of his or her individual fire fighters. In dealing with budget cuts that may lead to staffing reductions, these competing values may cause considerable internal conflict.

Implications for Fire Administrators

Much has been written about differentiating between objective and subjective responsibility (Cooper, 2012; Waldo, 1984/2006; Kernaghan, 1990). Although comparisons and contrasts are inevitable, it would be an error to assume that they are not mutually exclusive. A fire administrator cannot rely entirely on his or her subjective evaluation of responsibilities to run a department. It is evident that there are objective obligations imposed on any public official. Attention to duty is often demanded regardless of personal opinion. Still, it is important to remember that in many ways a fire administrator's subjective responsibilities are important and valid. Subjective interpretation of responsibility is often supportive of objective obligations. Objective responsibilities can often conflict with each other, and, as a result, decision makers require benchmarks to decide "best" choices. Subjective responsibilities often guide administrators regarding how best to serve the public interest and which public interests are given priority (Cooper, 2012).

Administrative ethics is nearly impossible without the inclusion of conscience. It is our subjective analysis of responsibilities that brings values, experience, and internal virtues to bear. When responsibilities, loyalties, and interests conflict, it is a holistic best approach to sorting out ethical dilemmas that appears to work best. The combined benefits derived from the dispassionate instruction provided by objective responsibilities coupled with subjective analysis derived from conscience tends to bring balance to solving dilemmas.

Conflicts of Responsibility

An **ethical dilemma** is a condition where an individual must make a choice between two competing values or between two or more actions, each of which have negative ethical consequences. The most common context in which fire chiefs face ethical dilemmas is in conflicts of responsibility ("Ethical Dilemma," 2015). The obligations inherent in the functions of a fire chief are as varied as they are numerous. Whether subjectively or objectively, a fire chief feels an obligation to the law and to public safety. He or she likely also feels a subjective responsibility to the organizational status quo, to the profession and professionalism, and to fire fighter safety. There are also personal responsibilities such as to family and friends and, of course, self-interest. It is apparent that conflicts of obligations within responsibility, or conflicts of the responsibilities, are not only possible but likely.

Conflicts of Obligation

Conflicts of obligation require a choice between two legitimate concerns or duties. Typical elements involved in obligation conflict include objective and subjective values, purposes or intentions, public goods, self-interests, organizational interests, and personal relationships. The following are some common conflicts of obligation.

- **A conflict between obligation and values.** For example, the obligation to enforce discipline may conflict with personal values of empathy and sympathy.
- **Conflicts between obligation and purpose.** It is not uncommon for the obligations inherent in being a fire chief to be at odds with well-meaning intention. For instance, the obligation to utilize required preapproved vendor lists may conflict with the intention of saving money by utilizing nonapproved purchasing sources.
- **Conflicts of obligation with other obligations.** For instance, a fire chief may be instructed to freeze spending, which may conflict with the provision of mandatory training for a hazmat team.
- **Conflicts of obligation and self-interest.** As a cost savings measure, a fire chief may be obligated to reduce his or her travel budget halfway through the completion of the Executive Fire Officer Program.
- **Conflicts of obligation and personal relationships.** Many fire chiefs have been forced to choose between attending an important meeting or attending their child's championship soccer game.

Conflicts Between Responsibilities

As explained earlier in this chapter, obligations are tasks that support responsibilities. Responsibilities are by their nature multifaceted and complex. Conflicts between responsibilities can also be especially difficult. Adding to the difficulties is that, even in the case of objective responsibilities, conflicts have personal connotations. Responsibility conflicts are not only *felt* with regard to their obligations but also within the underlying values associated with those responsibilities (Cooper, 2012). For instance, a fire chief is compelled to provide for the public safety through fire prevention activities as well as fire suppression activities. The inability to fund both not only presents a logistical problem to be solved but also induces emotional stresses springing from personal feelings of accountability and professional values.

Because responsibilities can be objective (external) or subjective (internal), a sense of dissonance can occur within the fire chief extending from both the expectations imposed on him or her by the benefit of the position and his or her internal sense of justice, equity, and personal beliefs about "greater goods."

Internal and external responsibilities are not always mutually exclusive. Often a particular responsibility is externally imposed and internally embraced. Conflicts among such responsibilities can be particularly vexing.

Because of two decades of economic pressure on municipalities, staffing reductions have created an all too common conflict between the objective and subjective responsibility to provide for public safety, and the objective and subjective responsibilities of maintaining fire crews' security. These are two competing responsibilities of significant weight, and they carry significant consequence to the chief as a decision maker.

Fire chiefs also must often weigh responsibilities inherent with their position as the municipality's managerial representative, versus stewardship of the department's mission and capabilities. It is exasperating when city managers and elected officials mandate policies that are contrary to public safety. Certainly, public safety is a prime responsibility. However, as fire chief, there is a responsibility to faithfully carry out the will of those who have legitimate authority to regulate the department's activities.

Not all responsibilities are connected to professional conduct. Being a fire chief can be an all-consuming occupation. There's a constant tension between balancing work life and home life. Placing personal interests above professional responsibility is usually presented negatively. However, taking care of yourself and your family is a legitimate form of responsibility. Protecting your economic interests is not necessarily self-centered. For many fire chiefs, there are others who depend on them and their economic welfare. Unfortunately, many fire chiefs treat themselves much more unfairly than they treat their subordinates. It is not unusual for fire chiefs to spend weekends at the office, work overtime without compensation, and travel without reimbursement. Those sacrifices usually are a result of a heightened sense of responsibility to the department's needs and may represent a compromise of their personal responsibilities to family.

Balancing the legitimate responsibilities among public service, ministerial responsibilities, and the welfare of the department and its members is a daily exercise. Fortunately, that balancing act can usually be accomplished within the confines of ethical behavior. Unfortunately, there are exceptions.

Conflicts of Loyalty and Accountability

As described earlier, responsibilities are defined by obligation and accountability. Obligation is relative to a specific duty, whereas accountability is relative to a specific relationship. Senior fire administrators are accountable to numerous legitimate stakeholders. There are, of course, fire fighters, elected officials, the public, and professional affiliations. Complicating these relationships is that, even within these groups, there are factions that may have competing needs, viewpoints, and agendas. Loyalty and a sense of accountability may spring from a sense of obligation, genuine affection, or a self-serving need of advancement or self-preservation. Regardless of its source, loyalty can complicate responsibility.

- **Conflicts between accountability and purpose.** Many fire chiefs have faced difficult decisions regarding the maintenance of relationships with elected officials and the need to "fight" for needs within their department.
- **Conflicts of competing internal loyalties.** Various stakeholders have various needs and expectations. As an example, a fire chief may feel an equal obligation to support both the training chief and the fire marshal as they compete for

department resources such as money and the increasingly precious commodity of time.

- **Conflicts of external loyalty and accountability.** Consider the example of a fire chief who is determining the site of a new fire station. He or she likely feels an obligation to place the station in the location most conducive to efficient fire operations. However, residents in the area are expressing legitimate concerns regarding the negative impact on their quality of life by having a station close by. For a fire chief, those concerns should not be ignored. The good public servant represents all stakeholders and their interests, including those contrary to task.
- **External conflicts of loyalty.** Related to the previous discussion, there are times when public stakeholders are in conflict with each other in determining public interest. Spending hawks may object to the expansion of the department to accommodate newly annexed neighborhoods. The residents in the newly annexed area may likely support the required expansion as it has direct economic benefits in the form of reduced insurance rates. This places the fire chief between two opposing viewpoints held by two legitimate stakeholders.
- **Conflicts of external accountability.** Perhaps no ethical dilemma is more frustrating than inconsistency of purpose among those to whom a fire chief is tangibly accountable. Almost every fire chief has had the experience of elements within the city council having opposing views regarding public safety needs and resulting expectations. These debates among legitimate authority can leave a fire chief squarely in the middle of a high-stakes political feud. Unfortunately, an inconsistency of purpose is common among elected officials.

Conflicts of Interest

Chapter 11, *Ethics and the Law,* discusses conflict of interest from a legal perspective. Every state and most local jurisdictions have some form of policy and regulation regarding outside interests and the serving of public interest. Unfortunately, the law only addresses minimum expected ethical behavior, and it does not define the ethical intelligence required to ascertain what exactly is meant by serving the public interest. Not all **conflicts of interest** are necessarily a choice between serving the public good and promotion of self-interest. There are instances when legitimate competing interests exist.

- **Competing public interests.** The previous example regarding the expansion of fire service into a newly annexed area also illuminates competing public interests. A fire chief has a compelling objective responsibility to protect taxpayers' economic interests. There is also an inherent responsibility to provide for public safety. Fire chiefs must constantly weigh the pros and cons of spending funds relative to financial accountability and public interest. For instance, is it in the public interest for a local fire department to provide emergency medical services if that same service can be accomplished more economically by a private contractor? On the one hand there is a measurable cost savings to the taxpayer. On the other hand there are intangible benefits to the public provision of emergency medical services. Those may include assurance of quality, greater accountability to the citizenry, and future cost control.
- **Competition between public interest and department interests.** At first glance it may seem the public interest must absolutely supersede department interests. Yet, as any fire chief knows, dismissing the interests of the department and the people who work in it is not so easily done. Again, let us visit the issue of the provision of emergency medical services. The public interest may indicate that contracting a private EMS service best represents the taxpayers' economic interests. However, that decision may likely have a negative impact on the department's personnel,

their employment status, and morale. A good fire chief feels some responsibility for the welfare of those individuals who faithfully served the department in exchange for an expectation that the chief represents their interests. Fire fighters' property interest in employment, the economic welfare of their families, and the commitment they made to the public welfare cannot be easily dismissed within a cost-benefit analysis. Fire fighters have a legitimate expectation of a fire chief's loyalty. A further argument can be made that what is good for the department generally serves the public interest in the better performance of public safety initiatives.

- **Professional and personal conflicts of interest.** There is a fine line between seeking to profit from the position of fire chief and protecting self-interest. Within the confines of loyalty, it is not entirely reasonable to assume that an individual will commit professional suicide in support of a higher ideal. Assume for a second that you are a fire chief in a city where the elected officials just passed a budget that will require the downsizing of your department by half. You are convinced this will have a catastrophic impact on public safety. You're also convinced that it places an unjust burden on the hardworking and dedicated fire fighters you lead. As a matter of principle, should a fire chief be expected to resign rather than implement the draconian cuts? At the very least, is a fire chief obligated to vigorously object even if it may jeopardize the relationship with his or her employer? By leaving in protest, or even by publicly criticizing superiors, the chief places his or her career in jeopardy, which, in turn, places the chief's family's economic welfare in jeopardy. By staying, the chief is ignoring his or her personal beliefs and values. He or she is also placing self-interest above that of the public's welfare. At what point does self-preservation cross over into self-interest at the expense of public interest?

Dealing with Ethical Dilemmas

In sorting out competing values and loyalties, there are some guides that may be useful, as shown in Figure 10-4.

- **Morality.** Many principles stand at face value. Ethics can often be entangled by a sense of obligation and conflicting loyalties. Disentanglement can often be achieved by focusing on overriding principles of right and wrong independent of professional relationships. Consider the chief who is struggling with the decision of whether to misrepresent spending needs to protect the department's discretionary spending. Loyalty to his or her responsibilities as an administrator and to the department may present a false dilemma. While loyalty to the department and its fire fighters may offer a sense of legitimacy to the decision, the overriding moral implications of the misuse of funds and deceiving superiors are clear.

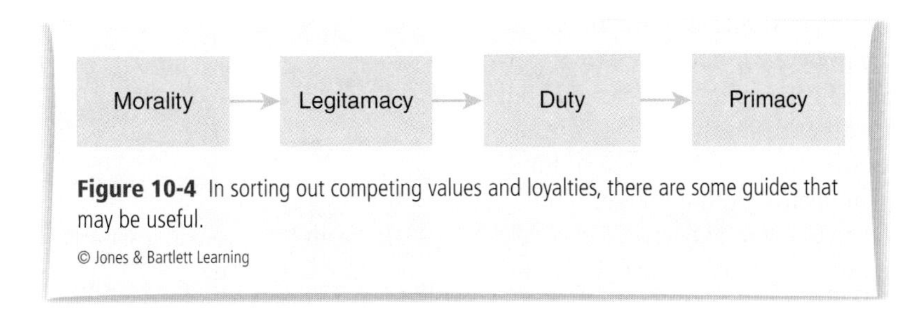

Figure 10-4 In sorting out competing values and loyalties, there are some guides that may be useful.

- **The legitimacy of stakeholders**. When conflicts of internal values and loyalties present themselves, asking who has a legitimate interest in the outcome can be instructive. Who has a legitimate claim on loyalty to obligation versus those *perceived* obligations based on internal values that serve self-interest? As an example, a fire chief has a legitimate right to expect loyalty to policy from a company officer compared to the company officer's desire to keep fire fighters happy by "relaxing" the rules. The fire fighters have no inherent claim on policy loyalty. As a result, the company officer's perceived ethical dilemma is illegitimate, and his or her concern over fire fighters' happiness is actually based on the self-serving desire for popularity.
- **Duty.** It is normal, well, and good that we seek the best outcome for others when making decisions with ethical implications. In your personal life this is a legitimate ethical concern. However, as a public official, a fire chief has a higher responsibility to duty and process. It is inherently unethical to substitute your personal assessment of "best outcome" in place of the clearly defined duties. As an elected public official accountable to the citizens of your community, process and law must take precedence.
- **Primacy.** Fire administrators should have a clear understanding of those responsibilities that are inherent within their position versus those that serve lesser needs. For example, a fire chief should feel a subjective responsibility to the health and safety of his or her fire fighters. The value of human life argues that this is a prime responsibility. Conversely, the subjective responsibility to develop career growth opportunities for fire fighters may be substantial but subordinate to health and safety. As a result, the fire chief who must choose between funding personal protective clothing versus sending aspiring officers to command school may face a dilemma, but the choice is clear given the transcending value on human life.

Imperatives of Ethical Administration

A different approach to the analysis of administrative ethics issues can be accomplished by the deconstruction of an ethical dilemma and evaluating its components against the universal concepts of responsibility (aka *functional imperatives*). It is believed by many that ethical decision making is enhanced when specific criteria are applied. A particularly useful set of **functional imperatives** is contained within the **ALIR model** (Makrydemetres, 2002). The ALIR model, displayed in Figure 10-5, incorporates accountability, the rule of law, integrity, and responsiveness in ethical decision-making criteria.

(A) Accountability

The first functional imperative is *accountability* to democratic legitimacy and processes. The concept is similar to Herman Finer's assertion of external control. Bureaucratic leaders such as fire chiefs are granted legitimacy to act in the public interest by elected officials, who represent the wishes and best interests of the public. Any action that subverts first, the public interest, and second, the public wishes, is therefore ethically suspect. Values and personal interpretation of responsibility can guide ethical

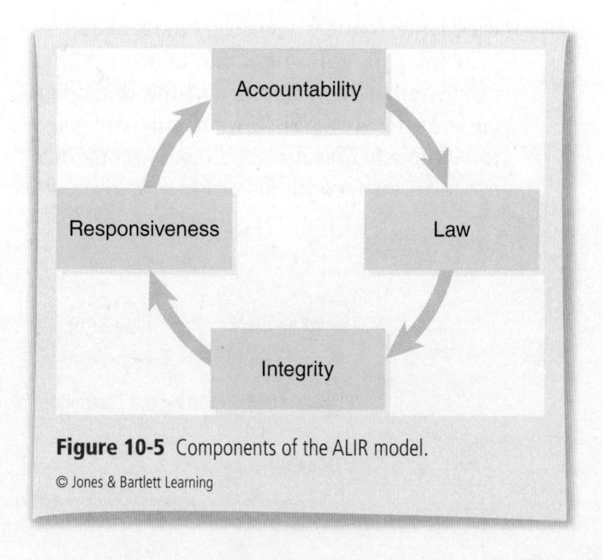

Figure 10-5 Components of the ALIR model.

reasoning but cannot and should not displace those responsibilities imposed by objective administrative responsibility. This element is particularly relevant to the case of the Kentucky county clerk who refused to issue same-sex marriage licenses on the grounds that she personally felt that same-sex marriages were immoral. Public administrators must subordinate personal ideology to legitimate authority.

(L) The Rule of Law

The rule of *law* only imposes a minimum ethical standard. However, that standard is absolute. Conformance to legal principles manifests a sense of constitutionalism and therefore is essential to legitimacy. The rule of law is theoretically based on social equity, fairness, and process. That public administrators, including fire chiefs, are bound by law rather than personal whim is a cornerstone of the democratic process. It is for this reason that most unethical behavior (relative to the public administration) is in fact illegal. Most philosophers would agree that law does not necessarily supersede morality. In other words, a law may be immoral even if the public administrator is ethically bound to adhere to it. Consider slavery. Lincoln, as president of the United States, was ethically bound to follow the precepts of the Constitution. As such, he could not ethically emancipate slaves in states remaining loyal to the U.S. Constitution; he could do so only in those states in rebellion. Because Lincoln considered slavery immoral, his ethical obligations and moral obligations were in conflict. He was then left with only two ethical choices, resign office or uphold his constitutional obligations.

(I) Integrity

The third functional imperative is that of *integrity*, or more generally the application of virtues, values, and principles. Effective public administration cannot act effectively in a moral vacuum. The individual administrator is often faced with ethical dilemmas that require personal character. The responsibilities of a fire chief are diverse and often discretionary. While accountability to the public and law creates boundaries within administrative duties, value judgments within those boundaries are necessary. The fire service has an absolute need for professionalism, and the values and virtues associated with it.

(R) Responsiveness

Accountability to supervising authority is not the sole extent of responsibility for a fire chief. He or she has a mandate to be *responsive* to public need. As community leaders, there exist requisite responsibilities associated with leadership. The fire service exists as a public service agency; it is funded entirely for the purpose of meeting community needs. It is as much a social covenant as a contract for services. While community expectations of the local water department likely do not exceed the reliable delivery of clean water, expectations of the fire service are much broader and have greater social significance. There is an expectation of civic virtue and community engagement. Any action compromising that social covenant is likely unethical.

WRAP-UP

Chapter Summary

- The responsibilities of a fire chief and the fire department senior administrative staff are significant. This chapter introduces three primary objective responsibilities associated with fire administration: the provision of public safety, the responsibility to protect the lives and well-being of the fire fighters who serve their community, and the the efficient and responsible management of the department and its assets.
- We introduced the concept of responsibilities imposed by outside authorities, otherwise known as objective responsibilities. In addition, some responsibilities spring from personal values and beliefs. These are known as subjective responsibilities.
- The individual fire chief's relationship with both objective and subjective responsibilities ultimately determines the ethical competency with which she or he performs her or his duties.
- With these responsibilities comes a myriad of obligations to individual tasks and priorities. Those obligations carry with them ethical requirements inherent within the position.
- The fire chief's obligations are numerous and complex, and they often compete for attention and resources. This can be the source of ethical dilemmas. The chapter reviewed the fire chief's core responsibilities and associated obligations.
- Responsibilities are not only guided by obligation; there are also personal and professional relationships for which a fire chief is accountable.
- Accountability to various constituencies can lead to competing demands and creates ethical dilemmas. The chapter reviews the typical relationships to which a fire chief is accountable.
- Inherent within the idea of accountability is the notion of oversight and control. Fire departments are public agencies and must answer to the public.
- The methodology by which fire department agencies are supervised, and so held answerable to the public interest, relies on two strategic approaches: external controls and internal controls.
- External control is a form of accountability dependent on supervision of the fire department by elected officials. The negatives of external accountability were explored and described as rooted in the competency of elected officials to understand, and therefore judge, technical decisions made by public administrators, including fire chiefs.
- Also discussed was the lack of assurance that elected representatives are necessarily invested in the public good as opposed to self-interest. The advantages of external control are that it increases transparency, is easier to implement, and is consistent with democratic principles.
- Internal control methodologies tend to focus on the selection and promotion of ethical individuals, education in ethical decision making, and the removal of disincentives for ethical behavior.
- The primary advantage of this type of decision making was found to be its immediacy to the decision-making process. Critique of internal control was offered as the unreliability of individuals to resist acting in self-interest.
- Ethical dilemmas include conflicts of responsibility, conflicts of obligation, conflicts of loyalty or authority, and conflicts of interest.
- The chapter closed with an introduction to two ethical evaluation methods. The first method introduced was the use of functional imperatives including morality, legitimacy,

duty, and primacy. The second methodology introduced the ALIR model. This methodology utilizes the concepts of accountability, law, integrity, and responsiveness.

Key Terms

accountability an element of responsibility associated directly with another individual or group.

ALIR model a guideline for assessing ethical dilemmas rooted in the application of fundamental public responsibilities to public accountability, respect for law, personal integrity, and responsiveness to public need.

Carl Friedrich a theorist who asserted that ethical responsibilities within public agencies could best be assured by internal controls.

conflicts of interest ethical dilemmas springing from conflict between responsibilities or loyalties and legitimate self-interest issues.

conflicts of loyalty ethical dilemma springing from inconsistency of needs and expectations of stakeholders with legitimate expectations of loyalty or authority.

conflicts of obligation ethical dilemma springing from competing obligations.

ethical attachment the conditional assignment of ethical responsibility.

ethical dilemma a condition where an individual must make a choice between two competing values or between two or more actions, each of which have negative ethical consequences.

external controls a form of accountability relying on the supervision of public agencies by elected officials, policy, and law.

functional imperatives guidelines for assessing ethical dilemmas rooted in the concepts of morality, stakeholder legitimacy, duty, and primacy of responsibilities.

Herman Finer a social scientist who asserted that ethical standards of behavior among government officials could only be ensured by a clearly defined policy body, accompanied by significant oversight.

internal controls a form of accountability relying on individual ethical behavior reinforced by ethics education, the maintenance of ethical cultures, and the removal of disincentives for ethical misbehavior.

objective responsibility a responsibility rooted in obligation to task that is externally imposed.

obligation a duty or commitment to a task.

subjective responsibility a responsibility that springs from internal values, beliefs, and priorities.

Challenging Questions

To check your understanding of this chapter's material, answer the following questions. It is highly recommended that you discuss your viewpoints with fellow students, peers, coworkers, and friends to discover their opinions as well.

- The text discusses subjective and objective responsibility. Review the subjective and objective responsibilities that affect you.
- The chapter introduced the concepts of internal and external control. Which methodology seems to be most prominent within your local department?
- A fire chief has clearly defined professional responsibilities. As a human being, a fire chief has a right to pursue his or her success and personal happiness. Sometimes professional

interests can compete with personal interests. At what point does the pursuit of personal interests become unethical?

- The quote that opens this chapter implies that with knowledge and authority comes responsibility. Critically analyze the attachment of responsibility and explain why one person is responsible for the welfare of others.
- Subjective responsibility plays a prominent role within this chapter. The word "subjective" is noteworthy and means "subject to interpretation." Give an example where a public official was misguided in his or her ethical judgment based on a subjective analysis of his or her responsibilities. Explain why the public official's thinking was wrong and how it misguided his or her understanding of responsibilities.

Case Study Conclusion

Revisit the case study at the beginning of the chapter. Spend a few minutes considering the questions posed at the end of the case study. In light of the information shared in this chapter, have any of your original observations changed?

The dilemma facing Chief Jones is an excellent example of a conflict of responsibilities. Jones has an objective responsibility to enforce fire codes and most important to assure that buildings are fire safe. As a member of the fire service profession, he feels a subjective responsibility to support fire sprinkler initiatives and would find it embarrassing to contradict his public support of them.

Conversely, as a member of the city's senior leadership, Chief Jones has a responsibility for the city's general welfare. There is no doubt in his mind that the development of the building represents a significant step forward in the city's efforts to revitalize its downtown core. Chief Jones is also aware that having the building occupied and renovated presents some safety gains for the area, although they may be at the expense of individual tenants within the building.

Subjectively, Chief Jones also feels an obligation to be a "good team member" with regard to economic development. He feels a duty to work cooperatively with the mayor and the other city agencies involved in the project.

The ethical requirements of responsibility faced by Chief Jones are as follows:

- As a fire chief, he is obligated to point out the fire dangers of a residential mixed-use occupancy, especially one above a restaurant, that does not have sprinklers. The fire safety of the building and its residents is entirely his responsibility. As such, he cannot overlook fire code requirements unless he has realistically satisfied himself that the building is reasonably safe without the sprinkler system.
- As a member of the senior management team, he is obligated to work with city officials to determine the greater good and to take what steps he can to reasonably assure public safety as best as possible.
- As a member of the economic development board, Chief Jones has an obligation to assist in the facilitation of the restaurant, assuming that it does not pose an undue life safety hazard. As such, he can ethically explore alternate solutions that protect life safety.

In the end, there is no actual way to align all responsibilities bestowed on Jones. His only ethical choice is to first prioritize life safety. In order to meet his secondary responsibilities, he may, if possible, seek alternative solutions that could reasonably assure life safety other than strict adherence to code.

Chapter Review Questions

1. What is meant by the term "accountability"?
2. How do objective and subjective responsibilities differ?
3. Give an example of an external control method.
4. Give an example of an internal control method.
5. Name two of the four functional imperatives described in this chapter.
6. What are the component parts of the ALIR method?
7. Herman Finer is associated with what type of control method?
8. Carl Friedrich is associated with what type of control method?
9. What are the two parameters of responsibility?
10. List the three fundamental objective responsibilities associated with fire administration.

References

Al-Habil, W. I. 2011. "The Administrative Ethics Between Professionalism and Individual." *Business and Management Review* 1, no. 10, 5.

Cooper, T. 2012. *The Responsible Administrator: An Approach to Ethics for the Administrative Role*. San Francisco, CA: Jossey-Bass.

"Ethical Dilemma." 2015. *Merriam Webster's Dictionary*. Accessed June 14, 2018. http://www.merriam-webster.com/dictionary/ethic.

Kernaghan, K. 1990. *The Responsible Public Servant*. Quebec, Canada: Institute for Research on Public Policy.

Makrydemetres, A. 2002. "Dealing with Ethical Dilemmas in Public Administration: The 'ALIR' Imperatives of Ethical Reasoning." *International Review of Administrative Sciences* 68, no. 2, 251–266.

Waldo, D. 1984/2006. *The Administrative State: A Study of Political Theory of American Public Administration*. London: Routledge.

CHAPTER 11

Ethics and the Law

"For what will it profit a man if he gains the whole world and forfeits his soul?"

—Matthew 16:26

OBJECTIVES

After studying this chapter, you should be able to:

- Summarize the fiduciary responsibilities in the fire service.
- Understand the complexities of conflicts of interest.
- Discern government-imposed ethics restrictions on:
  - Reporting requirements
  - Gifts and payments
  - Outside employment
  - Post-government employment
  - Political activities
- Differentiate ethics practices in labor relations:
  - Contract negotiations and conformance
  - Grievance procedures
- Examine ethics related to discipline:
  - Fire fighter bill of rights
  - Due process
  - Privacy rights
  - Whistle-blower protection
- Assess the ethics of government transparency.

Case Study

Assistant Chief Jones oversees the procurement and maintenance of his department's equipment and supplies. As part of his duties, Jones writes a budget, seeks bids, and submits purchase orders to the administrative office. By local ordinance, his department is required to receive three vendor quotes for purchases of more than $1,000, and sealed bids for purchases over $3,000.

Jones's department needs six new sets of bunker gear. He contacts several vendors and finds that sets run from $1,500 to $2,100 depending on the supplier. Bob Smith runs a local company called Acme Fire Equipment. He is a friend of Chief Jones, and they had done business together on numerous occasions.

Jones likes doing business with Bob Smith. He sells good equipment and has excellent customer service. Plus, Jones just feels more comfortable doing business with people he knows. Unfortunately, although competitive in pricing, Acme was not the lowest bid for the fire equipment. Jones is relatively confident that he will not be able to justify bypassing the low bid price for the gear based solely on his preference to deal with local vendors.

As a result, Jones decides to buy the sets of gear "à la cart." Instead of making one large purchase of $9,000 to $12,000, he instead decides to bypass the bidding requirement by purchasing coats and pants separately spread out over twelve purchases. By doing so, he stays under the $1,000 bid requirement. Although the department will be paying more for the equipment via this arrangement, Jones justifies his actions with the following rationale: first, he believes that there is real value in dealing with local companies. Second, he has not asked for, nor will receive, any personal benefit from the purchase.

- Are Jones's actions ethically questionable? Why or why not?
- Are Jones's actions legal?
- Does the fact that Jones does not receive any personal benefit from his actions mitigate his ethical responsibilities?
- How should Jones have best approached this issue?

Introduction

Throughout most of this text, the issue of ethics has been approached as a personal and internalized phenomenon. The correctness of actions has been assessed based on one or more philosophical approaches to ethics. However, ethics standards can also be imposed from external sources through policy. In these cases, the correctness of an action is based solely on its compliance with law or policy. This is especially relavent for administrators working in the public sector where nonconformance with ethics laws can result in criminal charges. As a result, in addition to making decisions based on an internal moral compass, fire chiefs are also required to comply with numerous ethics regulations.

This chapter will explore many of the most common and important ethics regulations with which fire chiefs must comply. The discussion will focus on ethics regulations in a general sense, and you are reminded that ethics regulations vary in each state and municipality. It is imperative that fire officers familiarize themselves with the ethical obligations imposed by their jurisdiction's policies and regulations.

As you move through this chapter, it is important to understand two basic principles associated with the relationships among ethics, morality, and law.

First, remember that the law represents a mandated ethical minimum. Ethics rules and regulations must be adhered to at risk of significant penalty. The individual's personal beliefs and values cannot be substituted in place of legal requirements absent personal jeopardy. For example, a public servant may come to believe that violating an ethics law may serve some greater good. In such cases, the individual may argue some moral authority, which he or she feels outweighs ethics law. In these cases, the individual's interpretation may be consistent with personal moral standards, but nonconformance remains unethical and subject to penalty. In other words, morality may not equate to ethics.

A second principle regarding ethics and law must also be understood: While the law may define an ethical minimum, it is still only a minimum. The fact that an action may be within the boundaries of law does not necessarily mean that an action is in itself ethical or moral. Laws tend to be static; ethics tend to be fluid. As an example, ethics law in a state may require that gifts with a value of more than $100 must be reported. However, it remains unethical for a fire fighter to accept a $99 gift intended to influence a fire inspection report. The intent of the rule is to avoid undue influence, and the $100 minimum represents an arbitrary threshold. Legally, $100 is a "bright line," which delineates an action that is punishable by law. If one accepts gifts of less than $100 as a condition of quid pro quo, the action remains unethical—albeit legal. To place the issue in the common vernacular, "Just because it's legal doesn't mean it's right."

Fiduciary Responsibilities

Fiduciary responsibilities are rooted in the concept of trust. A **fiduciary** is a person who holds an imperative ethical or legal responsibility for the welfare of another person (the **beneficiary [aka principal]**) or for a person's assets (Temchenko, 2016). A fire chief, or any other member of the fire department who is responsible for the management of public funds has a fiduciary responsibility (Rabin, 2003). A fiduciary duty can be defined as an imposed mandate for the highest standard of care under the law. A fiduciary is expected to act solely for the benefit of the beneficiary, and there must be no conflict of duty or interest between the fiduciary and the beneficiary. Further, a fiduciary must not profit from any activity made upon the beneficiary's behalf, without express permission from the principal (U.S. Legal, 2016).

A guiding principle in public administration ethics is the necessity for transparency. The public has a right to expect that any elected or appointed official responsible for the use of public money is acting in the best interests of the public. As a result, every elected and appointed government official has an ethical obligation to report financial interests. As you will learn later in this chapter, a fiduciary must be free of conflict of interest, and so it is important that public officials thoroughly report their business and personal financial obligations and incomes.

Clarification

Fiduciary Responsibility

- Defined as a mandate for the highest standard of care under the law.
- A fiduciary is expected to act solely for the benefit of the beneficiary.
- There must be no conflict of duty or interest between the fiduciary and the principal.
- A fiduciary must not profit from any activity made on the principal's behalf, absent the principal's specific permission.

In the case of the fire service, the fiduciary is the department's administration, while the beneficiaries are the taxpayers and the taxpayers' elected representatives. The fiduciary responsibilities placed on fire department administrators are many, but in general, they can be divided into two groups: acting in the public's best interest regarding public safety, and managing public funds responsibly. As public safety providers, a fire department and its administration have a fiduciary responsibility to act in the public interest regarding public safety. The ethical elements of that responsibility have been discussed at length in numerous chapters of this text. As public administrators, there is also a fiduciary duty for the responsible management of public funds.

In general, there are three categories of financial transactions routinely undertaken by fire administration. These include the acquisition and accounting of funds through budget management, purchasing procedures, and the dispersion of obsolete or excess assets.

Ethics and Budgeting Practices

For many new fire chiefs, one of the most challenging aspects of department administration is the process of budget management. The size and complexity of departmental budgets vary with department size. However, even relatively small departments have budgets in the millions of dollars. Funds are likely distributed throughout several accounts, and each of those accounts is likely to have numerous line item classifications. The rules regarding the acquisition and dispersion of funding can be incredibly complicated Figure 11-1. So are the associated ethical responsibilities. As stated earlier, it is imperative that fire department personnel tasked with financial responsibilities become familiar with the laws, rules, and policies specific to their jurisdiction. However, there are some general ethical principles that apply to almost all jurisdictions.

Writing budgets is a painstaking and often aggravating process. For most departments, there are more needs than funds and fire departments must compete with other city agencies to acquire needed resources. As a result, there can be tremendous temptation to play some of the more common "budget games." A typical strategy is inflating budget requirements under the assumption that elected officials will automatically cut certain percentages. Another common practice is the placement of additional funds in accounts that receive less scrutiny, so they can later be moved to accounts that typically are not supported by local government. For instance, a fire chief may put a few extra thousand dollars in the insurance budget with the intent of later moving that money to the office supply account to purchase a new copy machine.

Some fire chiefs will aggressively pursue discretionary spending early in the budget cycle to spend money in anticipation of the dreaded ten-month budget cuts. Other chiefs have gone on "spending sprees" to spend down accounts at the end of the year under the impression that "if you don't use it, you'll lose it." Both of these practices can lead to unnecessary or even frivolous purchases.

Motivation for such tactics is usually not self-serving; rather, chief officers may often come to feel that such tactics are necessary to protect department interests and so are just "part of the game."

Figure 11-1 The rules regarding the acquisition and dispersion of funding can be incredibly complicated. So are the ethical responsibilities.

Department heads may feel that other departments are doing it, and elected officials almost expect such behavior. While such practices may be useful in acquiring or protecting assets, they are deceptive. Further, they represent a common trap in public administration ethics: a loss of focus on fiduciary responsibility. It is quite easy for a fire chief to become so committed to serving the best interests of his or her department that he or she forgets that their first and primary obligation is the honest and responsible use of public money. A fire department exists for the good of the people, not the fire fighters, and so a fire chief's first ethical responsibility is to the taxpayers. Second, as a member of the municipal management

Clarification

It is quite easy for a fire chief to become so committed to serving the best interests of his or her department that the chief forgets his or her greater obligation is the honest and responsible use of public money.

team and as a subordinate to city administration, the fire chief has a responsibility to faithfully and honestly work with his or her superiors.

In general, fire chiefs have a fiduciary responsibility to budget money based on accurate anticipation of needs, spend money as it was intended, and support local government officials in their efforts to effectively manage the overall city budget. In a very real sense, fire chiefs are spending other people's money and so have an ethical burden to act accordingly.

The following are some general guidelines regarding the budgeting process:

- Budgets must accurately reflect needs.
- Line item descriptions should match intended usage.
- The creation of an "unofficial" and unreported discretionary fund, also known as "slush" fund, is inherently unethical.
- The approval of a budget does not equate to permission to spend.
- Financial agents have an obligation to be judicious in managing public funds and remain aware that financial emergencies can and do happen.
- Cash donations and gifts in kind must be promptly reported according to state and local policy.

Purchasing

The general concept of ethics described in writing budgets also applies to purchasing goods and services. Unforeseen circumstances may require budget adjustments throughout the year, but in general fire departments have an obligation to use funds as they are intended per the budget.

The golden rule of treating fire department funds as other people's money should guide purchases Figure 11-2. While the author is not suggesting that public agencies should operate in abject poverty, there is an ethical obligation to use public funds for only needed items. It is not uncommon for fire officials to confuse priorities by coming to believe that their first loyalty should be to the welfare of the department, as opposed to the welfare of the public that the department serves. Unfortunately, it is common for fire fighters to judge a chief's effectiveness by his or her ability to acquire the newest and best equipment. As a result, chiefs often fall into the trap of measuring their own success via purchasing ability. It is imperative that fire chiefs remember their ethical responsibility to be guardians of taxpayers' money. They should look for the best value in making purchases, and they must limit purchases to only those items that are necessary or have been justified within the budget process.

Fire chiefs need to be aware that how they spend public funds is as important as on what they spend those funds. There are numerous ethical and legal

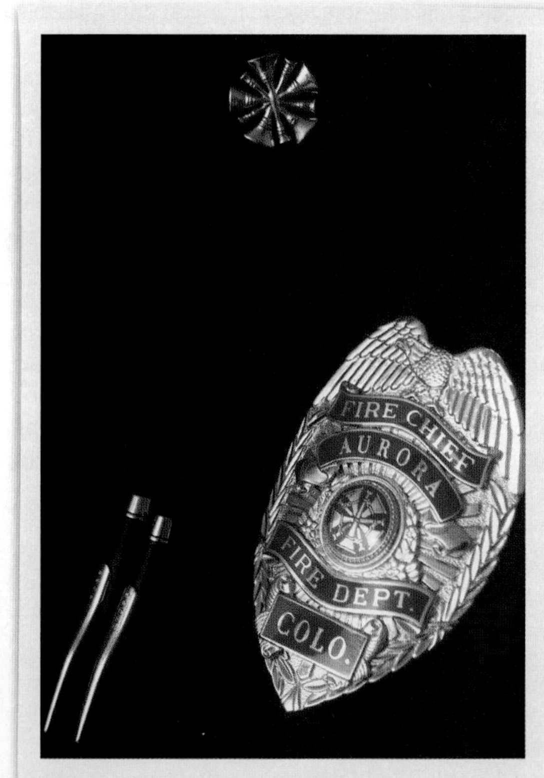

Figure 11-2 A golden rule of ethical financial management: Always remember that you are asking for and using other people's money!

implications associated with the purchasing process. Limits may vary from state to state or from city to city, but all jurisdictions have some form of guidelines regarding purchasing requirements. In many cases, purchases of less than $1,000 can be bought through a discretionary process, while purchases over $1,000 require some form of approval from a city manager or council. Most jurisdictions also have a bidding requirement. In most cases, local policy requires getting three vendors to quote prices for substantial purchases, while capital purchases may require a formal sealed bidding process. The intent of purchasing rules are first, to assure oversight of purchasing procedures. Second, bidding rules ensure equitable vendor access to government contracts. Purchasing laws and policies reflect ethical intent, but they only represent a minimum required action. Even for small purchases not covered by purchasing rules, fire departments have an ethical obligation to meet the intent of those standards. All purchases must be in the public interest, best values should always be sought, and there should be no favoritism in selecting vendors.

The following are some general guidelines for purchasing:

- Purchases should reflect respect for department priorities.
- Where deviance from budget must occur, such deviations must be justifiable and approved by the proper authority.
- Practical efforts should be made to obtain "best values" when making purchases.
- State and local guidelines for purchasing procedures including bidding requirements must be followed to the letter of the law and within the spirit of their intent.
- The purchasing process must be transparent—all records must be kept according to local or state policy.

Asset Maintenance and Disposal

A third area of fire service fiduciary responsibility rests in the maintenance, inventory, and disposal of departmental assets. There are few, if any, ethics policies or regulations regarding the maintenance of property or equipment, but there are certainly ethical obligations attached to these responsibilities. Just as public funds should be considered other people's money, so should fire department assets be considered other people's property. Although not often considered, you must remember that taxpayers bought and own the buildings, the apparatus, and every other piece of equipment owned by the fire department. There is an ethical obligation to inventory and maintain fire department assets Figure 11-3.

Fire administrators must be responsible caretakers of public property. Likewise, they must impress upon fire fighters the responsibility of inventorying, maintaining, cleaning, and protecting department assets. As a former fire chief, the author is familiar with the frustration of replacing lost or unnecessarily damaged equipment.

Even the best-maintained equipment eventually needs to be replaced. Fire departments often have a surplus of equipment that has become obsolete or unserviceable. Because such equipment is technically the property of the taxpayers, there are ethical and legal guidelines for its disposal. Local jurisdictions have specific regulations regarding

Figure 11-3 There is an ethical obligation to inventory and maintain fire department assets. In a real sense, they are other people's property.

legal disposal of surplus public property, and there must be adherence to those policies. However, there are some ethical issues outside the parameters of policy requirement.

Surplus equipment should not be "given" to fire department members. Doing so has the appearance of impropriety and can encourage inappropriate behavior. The dispersal of surplus equipment should be done so without favoring any individual or group.

The fire department has an ethical responsibility to assure that equipment dispersal does not adversely affect the safety of those acquiring it. Departments often get requests from other departments for access to surplus gear. A guiding principle should be: *If the equipment is not safe for your fire fighters, it is likely not safe for anyone else's either.* As a result, damaged or worn out personal protective equipment, apparatus, and suppression equipment should be destroyed and properly disposed of.

Clarification

There is an ethical obligation to ensure that surplus equipment is disposed of legally and that unserviceable or dangerous equipment is not distributed in a method that poses a hazard to others.

Texture: Eky Studio/ShutterStock, Inc.; Steel: © Sharpshot/Dreamstime.com

Many departments dispose of equipment by donating it to training organizations, fire explorer posts, and fire-related educational programs. This practice is acceptable and encouraged, assuming that it is used for training and demonstration purposes only and is clearly marked as such.

Surplus equipment that is still usable should be sold, auctioned, or donated according to state and municipal regulations. In all cases, disposal must be approved by the proper authority. Further, official permission, dispersion details, and inventory adjustments must be thoroughly documented. Any income from the sale of public property must be properly reported and deposited.

A Final Note About the Handling of Public Funds

Ethics violations regarding financial practices are one of the most common ethical crises faced by fire chiefs. Few fire chiefs steal money outright or set out to personally benefit from unethical financial practices. The far more common scenario is that of a fire chief who, either out of frustration from lack of funding or what he or she felt were overly restrictive regulations, decided to cut corners. An Internet search using the term "fire chief investigated" will yield an extensive list of tragically poor decision making (Figure 11-4). Some have subverted the bid process to expedite a purchase, and others have used inappropriate accounting methodology to liberate funds for an unapproved purchase. In many cases, their intent was defensible; in all cases, their actions were not.

The greatest threat a fire chief may face is a false belief that outcomes are more important than procedures. When fire chiefs get caught bending the rules, the headline will not read "Well-Intentioned

Figure 11-4 Ethical mistakes often arise from self-serving motivations. Sometimes they are caused by lack of reflection.

Courtesy of Scott Walker.

Fire Chief Bends the Rules to Further Department." Rather, it will likely read "Fire Chief Is Indicted for Misuse of Funds." Fire chiefs must continually remember that when it comes to public finance, ends do not justify means, and procedure is more important than outcome.

Listen Up

When it comes to public finance, ends do not justify the means. The procedure is more important than the outcome.

Poor ethical decision making is often prompted by a confusion of responsibility and priorities. It is not unusual for fire personnel to eventually regard the department as an entity unto itself and to become emotionally invested in the welfare of the department as an institution—independent of its purpose. Ethically speaking, it is helpful to think of a fire department as an end to a means. Fire departments exist to provide public safety services. A fire administrator's first loyalty should always be to the fire department's mission and the people it serves, not the organization. This may seem like a fine distinction, but it is important nonetheless. Spending priorities must focus on the public's welfare, not necessarily on the department's welfare in and of itself.

Conflict of Interest

"The notion of public interest is for the public servant what justice and liberty are for the legal profession, or what healing and mercy are for the medical profession" (Kernaghan and Langford, 2014). As public officials, fire administrators, as well as fire fighters, must avoid conflicts of interest. A **conflict of interest** can be defined as a situation in which a person is in a position to derive personal benefit from actions or decisions made in his or her official capacity.

Fire departments are public agencies funded by taxpayer money for the sole purpose of maintaining public welfare. As such, fire departments are ethically bound to act in the public interest first, and no other endeavor must conflict with that responsibility. Acting in the public interest is inherent within the ethics of public institutions including the fire service. Beyond ethical responsibility, federal, state, and municipal governments impose ethics law to ensure that the public interest is served. These ethics laws tend to revolve around the general concepts of reporting requirements for the sake of interest transparency, gifts and payments, abuse of position and resources, outside employment, post-government employment, and political restrictions.

Reporting Requirements

A fundamental prerequisite for the assurance that public agencies are acting in the public interest is transparency. The public clearly has a right to know if there are any financial or personal interests that may conflict with the public servant's primary duties. As a result, most state and local jurisdictions have a requirement that civil servants (including fire department personnel) report financial and business interests that may intersect with the responsibilities associated with departmental obligations. Financial interest statements are necessary to maintain transparency and the effective facilitation of government oversight.

Reporting requirements may vary from jurisdiction to jurisdiction, and it is imperative that fire department personnel comply with specific reporting requirements Figure 11-5. Beyond legal responsibility, there is an ethical obligation for all personnel to report any potential conflicts that may exist even if they are not technically required under the jurisdiction's policies. For instance, a member of the department's *cause and origin team* may also "moonlight" as a contracted investigator for an insurance company. A conflict may exist when investigating a fire for her or his department, where the insurance company is the underwriting agency. Although not likely required under most jurisdictional interest

Figure 11-5 Transparency is a fundamental requirement for the assurance that public agencies are acting in the public interest.

reporting requirements, the potential for conflict would ethically require the investigator to report the possible conflict to his or her superior officer. In dealing with ethics law, the spirit and intent of policy outweigh the actual wording.

Gifts and Payments

The ethical issues surrounding gifts and payments are varied in circumstance and scale. They may range from a restaurant's simple offer of a discount to fire fighters in uniform to a blatant bribe made to an inspector to overlook a code violation.

The Latin term **quid pro quo** translates to "something for something," and in the public administration ethics forum it generally describes the act of bribery. A fire fighter's acceptance of money, gifts, or services in exchange for some perceived benefit related to the official capacity of his or her position is unethical. Some examples are obvious, such as the aforementioned acceptance of money as an incentive to overlook a fire code violation. Some instances of quid pro quo may seem more innocuous, such as a fire chief being offered an exceptionally good deal on a new pickup truck as thanks for the department's recent purchase of three new utility vehicles.

Consideration of gifts and payments are not limited to instances of quid pro quo. Companies, vendors, and stakeholders who routinely interact with government agencies have frequently offered "gifts" to fire officials without asking for anything in return. Their intention is usually to create a favorable impression or to gain influence. A public administrator must remember that the appearance of impropriety can be as damaging to the public interest as an actual impropriety. A government's ability to function effectively is dependent on the trust of the public. As a result, ethics laws and policies restrict gifts and payments to assure the absence of the appearance of impropriety. As a result, the acceptance of gifts is usually unethical even in situations where no explicit mutual consideration exists. Typical restrictions on gifts are as follows (Illinois General Assembly, 2003b; State of Massachusetts, 2016):

- A government official (including fire fighters) may not accept gifts, discounts, services, or gratuities not readily available to the general public.
- An employee may not give (or solicit contributions for) a gift to an official superior or accept a gift from another employee who receives less pay, subject to certain exceptions.
- Municipal employees may not accept gifts and gratuities valued at $50 or more that may influence their official actions because of their official position.
- Accepting a gift intended to reward past official action or to prompt future official action is illegal, as is giving such gifts. Accepting a gift given to you because of the municipal position you hold is also illegal.
- Meals, entertainment event tickets, golf, gift baskets, and payment of travel expenses can all be illegal gifts if given in connection with official action or position, as can anything worth $50 or more.
- A number of smaller gifts together worth $50 or more may also violate these sections. Note: The $50 limit expressed above may vary in some jurisdictions.
- Exemptions to these rules may include gifts from family or friends given for reasons not related to official duties. Payment or reimbursement for travel, meals and related expenses to and from events in furtherance of the public interest, or for educational purposes, are also exempt.
- Acceptance of gifts: Certain gifts may be accepted so long as no quid pro quo is implied and gifts are properly reported to the agency having jurisdiction. It is usually required that either a cash donation or donation in kind be made of equal value to an appropriate charity.

As with other ethics laws, it is important for fire officials to become intimately familiar with the reporting requirements and restrictions of their particular jurisdiction. It is also important to understand the spirit and intent of the laws as they apply to ethics.

Fire department employees must always be aware of how an action looks, and the potential consequences associated with accepting a gift, even a legal one. For instance, let us assume that Restaurant A sponsors an annual Fire Fighter's Day for all local fire fighters. They are offered a free meal and a 50 percent discount on all purchases made that day. The owner extends the offer solely out of appreciation for public service and does not expect, nor seek, any consideration in return. Even in this innocent gesture, there is a significant potential for public misperception.

Assume that fire department personnel are inspecting Restaurant B and they find many violations that require immediate remediation. Out of frustration and anger, the owner of Restaurant B wonders aloud whether or not he would be in this position if he also had offered discounts to fire fighters. The charge is false, but local participation in Fire Fighter Day at Restaurant A has potentially undermined the department's credibility with the owner of Restaurant B. That single incident of mistrust can easily cascade into a full-blown scandal.

Clarification

It is important that fire departments and fire department personnel never place themselves in a position where appearances may indicate indebtedness or favoritism.

Abuse of Position and Resources

Another variant commonly covered under state and municipal ethics laws is the prohibition of positional abuse of position. Typically, **positional abuse** can be defined as using official status to acquire benefits for yourself or your immediate family. Most ethics policies specifically limit the following (Illinois General Assembly, 2003b; State of Massachusetts, 2016):

- **Seeking favors.** The use of official status to acquire special consideration or benefits is unethical and likely a violation of jurisdictional ethics policy. Examples may include asking for discounts or being excused from speeding tickets.
- **Misuse of subordinates**. It is generally unethical and illegal to order or permit subordinates to do personal bidding. This may include having fire fighters wash the chief's *personally owned* car or requiring staff to do personal errands. It should be noted that even if subordinates are willing to perform personal errands or tasks, allowing them to do so on company time is inherently unethical.
- **Promoting personal interests or profit.** Using influence, access to information, or direct positional power to promote personal interests or those of friends and family is contrary to public interest and therefore unethical. For example, a fire fighter who runs a contracting business while off duty would be ethically prohibited from handing out business cards to fire victims. Many years ago, the author was familiar with a deputy chief who required that fire fighters purchase a particular brand of uniform sold exclusively by a local business owned by his wife. His imposing that requirement was an abuse of position.

  In situations where personal interest overlaps with public duty, officials must recuse themselves from the decision-making process. Additionally, fire fighters should divest themselves of business interests likely to produce an interest conflict and avoid entering into business agreements where a conflict of interest is likely.
- **Misuse of departmental resources.** One of the more commonly reported conflicts of interest is abuse of departmental resources. Incidents may include misuse of department vehicles, departmental computers, Internet access, and tools. Certainly, restrictions vary from jurisdiction to jurisdiction, but in general, the use of publicly owned equipment for personal use should be prohibited.

One of the most overlooked abuses is that of time. Fire fighters and staff are paid to perform duties and have an obligation to be relatively productive during the workday. In particular, staff positions working eight-hour schedules have an obligation for productivity. The public does not pay employees to waste time on Facebook or conduct personal business **Figure 11-6**. Again, appearances are important. Many fire chiefs come and go as they wish and have the privilege of taking off when they want to. Still, when the public sees the chief at the golf course every morning there is an image of abuse of position.

The author recognizes that "down time" is built into a fire fighter's schedule. While it is unreasonable for the expectation of 24 hours of uninterrupted productivity, personal projects should never interfere with duties. Every firehouse seems to have one or more individuals who are more interested in television than productivity. Contrary to the belief of some members of the public, fire fighters are not paid to sit and wait for an alarm to sound.

Figure 11-6 Fire fighters are paid to be productive and have an obligation to be productive.

Studying for promotional exams, impromptu training, equipment inventory, and maintenance should all take precedence over recreation and personal projects. As has repeatedly been stated throughout this text, fire fighters have an ethical obligation for competency and to act in the public interest. In other words, "Eat when allowed, rest when you can, and work when you should."

Outside Employment

Fire fighters typically work under a variant of the 24-hour schedule. As a result, secondary employment is convenient and common. While there is nothing inherently unethical about outside employment activities, there are some restrictions usually expressed in ethics policies, which may include (Illinois General Assembly, 2003b; State of Massachusetts, 2016):

- **Secondary employment must not create a conflict of interest.** There can be no overlap between personal financial interests and primary employment duties. For example, it would be unethical for a code enforcement official to inspect a restaurant in which he or she holds a financial stake.
- **Inside track restrictions are typically prohibited.** An example might be a fire fighter who owns and operates a fire equipment company. It would be unethical for that fire fighter to do business with his or her local department.
- **Secondary employment must not create a conflict of loyalty.** For instance, a fire inspector may not accept employment as a code consultant with an architectural firm that does business within his or her jurisdiction.
- **Secondary employment must not interfere with the performance of primary duties.** Interference may include, but is not limited to: scheduling conflicts, physically exhaustive activities, and high-risk activities likely to produce lost time.
- **A conflict of priority is unethical.** This is an area of some contention. The author asserts, however, that the work of fire fighters has special responsibilities and obligations that make it a priority over other forms of employment. To be a fire fighter is to accept those responsibilities.
- **Outside employment should never interfere with a fire fighter's commitment to the community and to his or her fellow fire fighters.** Fire department employment must come first, including training exercises, emergency calls, and even nonemergency callback. *To be blunt, if you are not a fire fighter first, you are not much of a fire fighter.*

Post-Government Employment

Most ethics regulations include a **revolving door** clause. These provisions typically restrict government employees (including fire fighters) from accepting employment, for a given time period, with agencies that deal directly with their former employer.

Further, ethics policies usually include a provision permanently restricting former government employees from working on projects or contracts that were under their purview during government employment. As an example, a retired fire chief would likely be prohibited from taking employment with a construction company to which he awarded a contract to build a new fire station.

Typically, ethics laws restrict employment with companies doing general business related to municipal employment for one year. Further, most ethics policies impose lifetime restrictions on post-municipal employment dealing specifically with projects and contracts under the previous employee's direct supervision.

Political Restrictions

The **Hatch Act** was a federal law passed in 1939 that limits certain political activities by governmental officials. The intention of the law is to separate political influence and campaign activity from public service and duty (U.S. Office of Special Counsel, 2015). As federal legislation the Hatch Act represents a minimum requirement for state and local employees. State and local governments can and have imposed additional requirements. It is the responsibility of fire department employees to make themselves aware of the political restrictions in effect within their local jurisdiction. Common restrictions include (Illinois General Assembly, 2003b; U.S. Office of Special Counsel, 2015):

- Use of official authority or influence to interfere with, or affect the results of, an election or nomination.
- Directly or indirectly coerce, attempt to coerce, command, or advise a state or local employee to pay, lend, or contribute anything of value to a party, committee, organization, agency, or person for political purposes.
- Usage of publicly owned equipment, including telephones, computers, Internet connections, and office supplies, for political or campaign purposes.
- Participation in political events or campaigning while on duty.
- As of 2013, government employees (including fire fighters) may seek political office. However, fire fighters are responsible for maintaining separation between elected obligations and professional responsibilities. Conflicts of interest are likely and must be dealt with according to state and local regulations.

Fire fighters are first and foremost citizens and so have the right to free speech and participation in the electoral process. However, while such activities are inherently ethical, discretion is advised. As with conflicts of interest, appearances are important. The electoral process is adversarial and can become personal. Hyperpartisanship by municipal employees can give the impression of bias to the general public. Further, it can lead to unfair, but likely, retaliation by political adversaries. Unfortunately, others often suffer from the recklessness of a few.

Labor Issues

Perhaps one of the most complicated relationships a fire administrator must manage is with his or her department's union. Since the development of the **National Labor Relations Act** in 1935, public policy has guaranteed workers the right to self-organize and form unions. The relationship between employers and unions was contentious and even at times violent as a result. In 1947, the federal government passed the **Labor Management Relations Act**, which sought to normalize labor management relationships by identifying "unfair labor practices."

Over the succeeding 60 years, there has been a general attitude that the ethics of labor issues is equivalent to compliance with labor law (Adler and Bigoness, 1992). As has been reviewed earlier, the law represents a minimum compliance standard, whereas ethics refers to an idealized behavior. The law defines what one *must* do, whereas ethics defines what one *should* do. Labor relations tactics in the public sector have tended to follow the same general principles of conduct as that in the private sector. Within the business community there is a widespread attitude that deception, omission of inconvenient facts, exaggeration, and bluffing are ethically acceptable because they have been normalized as "part of the game" (Adler and Bigoness, 1992). From an ethical standpoint, this is a false assumption when applied to labor relationships. In the public sector, and especially in the fire service, labor and management must not only coexist but must also do so in a way that assures the continuation of public safety. Honesty, integrity, and good faith are requisite to the greater public interest.

At first glance, the relationship between a fire chief and the department's union likely seems adversarial. Yet to say a fire chief has an adversarial relationship with the local union is to say that he or she has an adversarial relationship with those people that the chief is responsible for leading, managing, and protecting.

Consider for a moment the ethical requirements of a fire chief in relation to his or her fire fighters. A primary obligation is to assure the health and safety of the fire fighters he or she employs. Further, if the long-term success of the department is to be achieved, succession planning is also important. Therefore, a fire chief also has an obligation to facilitate the professional development of fire fighters. Finally, if a fire department is to excel at service delivery and be a positive force within the community, then fire fighters must be engaged, inspired, and committed. The development of those qualities requires leadership. Effective leadership is not only a professional benefit, it is essential to meeting the responsibilities of the department's officers. As such, department leaders have an ethical obligation to exercise leadership as a necessary skill for the job for which they are paid. Simply put, a fire chief cannot afford to be in an adversarial relationship with those he or she is to protect, nurture, and lead **Figure 11-7**. This requires a positive relationship with her or his workforce.

Figure 11-7 A fire chief cannot afford to be in an adversarial relationship with those he or she is to protect, nurture, and lead.

In comparison, union officials are also tasked with assuring the safety and welfare of their members. They are also responsible for assuring equal access to opportunity, assuring professional development opportunities, and maintaining a positive work culture. All of these primary responsibilities are consistent with and shared by fire department leadership. When you consider the principal obligations of both union and fire administration, it becomes apparent that there should be a recognized commonality of purpose **Figure 11-8**.

However, it is naïve to believe that unions will always agree with a fire chief's chosen path of action. Fire officers are often compelled to seek efficiency, productivity, and cost savings as managerial functions. In contrast, union officials are obliged to seek benefits and wage concessions for increased productivity and longevity. The pursuits of both sides are legitimate and even ethically mandated. Fire chiefs have an ethical responsibility to local government to represent their interests and effectively manage the department. Union officials have an ethical responsibility to represent the interests of those who elected them.

An important first step in maintaining an ethical labor relationship is for both sides to recognize and respect the legitimacy of the other's purpose.

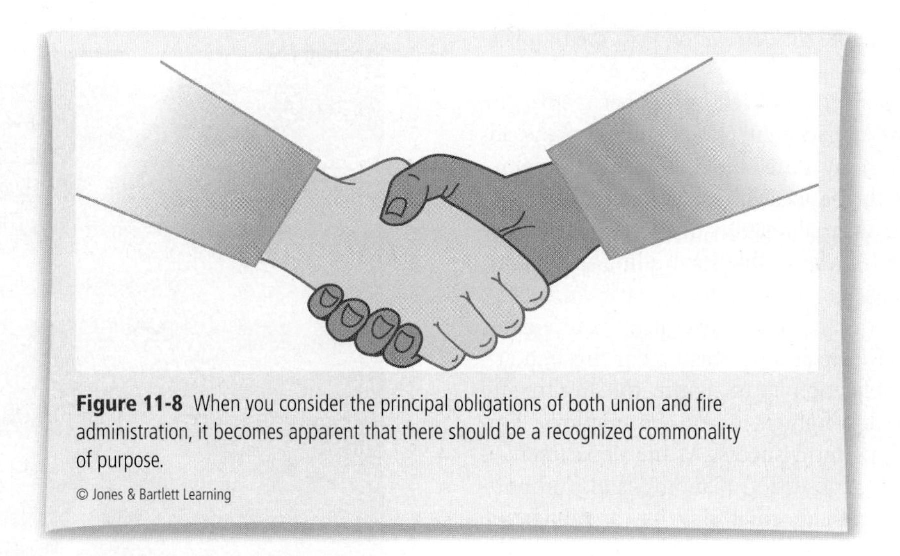

Figure 11-8 When you consider the principal obligations of both union and fire administration, it becomes apparent that there should be a recognized commonality of purpose.

Clarification

It is necessary for both sides of the labor relationship to recognize and respect the legitimacy of the other's purpose.

The failure to do so may negatively affect department performance, in addition to the public's safety as well as that of the fire fighters by promoting infighting and distrust. Assuming that both labor and management have legitimate pursuits and that security and welfare of the public are at stake, both sides have an ethical obligation to treat each other in good faith. There is data to suggest that collaborative relationships founded on ethical behavior produce value-added benefits not realized in conflict-laden interactions (Adler and Bigoness, 1992). It is therefore reasonable to assume that the public interest is served better when its fire department is operating smoothly, and time and money are not being diverted to labor disputes.

Contract Negotiations

Contract negotiations are usually stressful, and often contentious. Contentiousness can be aggravated when one or both parties adopt an attitude of "I must win, and you must lose." While both sides are ethically obligated to represent the interests of those they represent aggressively, there is also an ethical obligation to deal fairly with the opposing side. Beyond ethics, there seems to be evidence that "hard-nosed" negotiating tactics utilizing deception and bullying are ultimately counterproductive (Bowie, 1985). Such tactics may ultimately lead to organizational division and lost productivity. Within the realm of public safety these conditions are not in the public interest. There is also an equal obligation on both parties to assure the continued efficient operation of the department and support of its primary mission of public safety. Any demands and concessions by either side that inherently jeopardize the lives and property of the public, or the safety of fire fighters, is a failure to recognize those primary responsibilities and so are unethical. In participating in contract negotiations, both fire department administration and union officials are ethically and often legally bound to some common rules.

- **Both sides are ethically and legally mandated to bargain in good faith.** Deception, misrepresentation of facts, threats, and bullying are all outside the bounds of ethical negotiation practices.
- **Both sides must represent their constituents' interests.** Demands or concessions must be consistent with those responsibilities. Negotiators must not seek to personally profit from negotiations absent their constituents' express permission.
- **Fire fighters are typically legally barred from strikes and so are usually protected by the binding arbitration process.** It is unethical for fire fighters to circumvent the no-strike clause with work slowdowns or "sick outs." Conversely, it is unethical for cities to abuse the arbitration process by capriciously running up costs in hopes of "breaking the union's bank account."

Contract negotiations can be particularly problematic for fire chiefs because there is a nearly inherent conflict of responsibilities. The fire chief's primary task is to manage the fire department. As a senior member of the city's management team, a fire chief has an ethical responsibility to represent the interests of local government faithfully. Conversely, because of the nature of the work done by fire fighters, fire chiefs have an ethical obligation to protect the safety and welfare of their fire fighters. Additionally, effective leadership requires representing the fire fighters' interests. It is normal and right that a fire chief should want his or her fire fighters to be fairly paid, have the best possible equipment, and live and work within well-maintained stations.

As a result, many fire chiefs seek to be excused from the bargaining process entirely or limit their activities to that of a consultant Figure 11-9. Assuming consent from city management, there is no ethical requirement for a fire chief to participate in the negotiation process. However, should the city's administration insist on the fire chief's participation, he or she is ethically required to participate fully and effectively.

Contract Conformance

A collective bargaining agreement between the department and union is both legally and ethically binding. Because the contract is made in good faith, it must be adhered to in good faith, and so both sides are ethically bound to adhere to the spirit and intent of the agreement. For whatever motive, a fire chief who blatantly violates contract provisions and thereby forces unions to file grievances is acting unethically. Conversely, unions have an ethical obligation to honor the terms of the contract as well. Filing capricious grievances and initiating work slowdowns in response to contractually appropriate management initiatives are also unethical choices.

Grievance Procedures

Fire administrators and union officials have an ethical obligation to operate within the limits of the contract as they understand them. However, differences of opinion regarding contract wording and intent can arise. For this reason, most contracts have a formalized **grievance procedure**. Again, a collective bargaining agreement is a legal and ethically binding agreement between the fire department and the fire fighter's union. If both sides are legitimately attempting to live within the confines of the agreement, then the purpose of the grievance procedure is to clarify contract wording and intent. Grievances should not be used to force changes to the contract, nor should they be necessary as a response to breaches by management.

The guidelines for handling grievances are usually spelled out in the collective bargaining agreement and so have the force of law. Additionally, both sides have an ethical responsibility to participate in the grievance process in good faith and with the intent of resolving differences.

- Grievances should only be filed when the union believes that the contract's spirit and intent have been violated. The city administration is

Figure 11-9 Contract negotiations can be contentious and create conflicts of responsibility. Many fire chiefs choose to recuse themselves from the negotiation process.

ethically required to abide by the contract and so should not deliberately cause provocation for a grievance.

- Both sides should participate in the grievance procedure in good faith by refraining from misleading statements, distortions of truth, or tactics intended to subvert the process. This mandate applies equally to all steps of the grievance process up to and including mediation and arbitration.
- Both sides are ethically required to abide by the results of a grievance that has been properly adjudicated.
- As both parties have obligations to represent the interests of their constituency faithfully, it is inappropriate to seek retribution for unfavorable grievance results.
- Unions have the additional responsibility of representing the interests of nonmember constituents as well as members.

An Obligation for Positive Labor Relations

As stated earlier, both city administrators and union officials have responsibilities to both their constituency and to the core mission of the fire service. It is imperative that fire chiefs recognize a legally sanctioned union's right to exist and its right to represent the interests of its members. Any attempt by city officials or a fire chief to disrupt a union, seek its abolition, threaten or demoralize its members, or take retribution against its leaders is inherently unethical.

IAFF leaders have fiduciary and moral responsibilities to their membership. Due to the nature of the position, union leaders are subject to higher ethical standards than the average member or employee. These standards must be upheld in order to maintain the trust of the members and run an effective local. Failure to do so can result in loss of credibility, loss of union position, or even criminal charges. However, most IAFF leaders are also public employees. As firefighters, it is their responsibility to represent their members' interests to the extent that they do not interfere with or damage the department's ability to function properly. Unions are not tasked with management responsibilities; therefore, a union leader should not use his or her position to interfere with a fire chief's responsibilities or management prerogatives.

Unfortunately, there's a long history of fire chiefs and union leaders whose relationships became so adversarial they became toxic. Such destructive relationships have led to ethical breaches on both sides, as combatants sought to undermine the ability of the other to function. The incidence of "labor warfare" can destroy morale, interfere with efficient operations, jeopardize public safety, and ruin reputations. At times union officials and fire chiefs have become so concerned with "beating each other" they lose sight of their primary obligations.

In summary, both fire chiefs and union leaders have ethical responsibilities to maintain a productive work relationship Figure 11-10. That ethical responsibility is rooted in their obligations to their constituents and intrinsic to their position. The hallmarks of a productive labor relationship are rooted in the following precepts:

- Both union leaders and fire service officers have an ethical obligation to represent the interests of their constituencies.
- While fire chiefs have an ethical duty to represent the interests of the city, they are also ethically bound to protect the safety and welfare of their fire fighters.
- While unions have an ethical obligation to represent the interests of its members, as fire fighters they also have an ethical obligation to the department's core mission. It is important for union leaders to behave in an ethical manner so that they can best serve the needs of their members. Ethical behavior is vital for maintaining the trust of the membership and running a sound, effective organization. Leaders who follow ethical principles serve as good role models for their members, the citizens, and their departments.
- Both unions and fire chiefs share an ethical obligation to deal honestly and fairly with each other.
- Both unions and fire chiefs must recognize the other's rights to pursue their positional responsibilities. As a result, they must not interfere with the other, nor seek retribution against the other as a result of those duties.

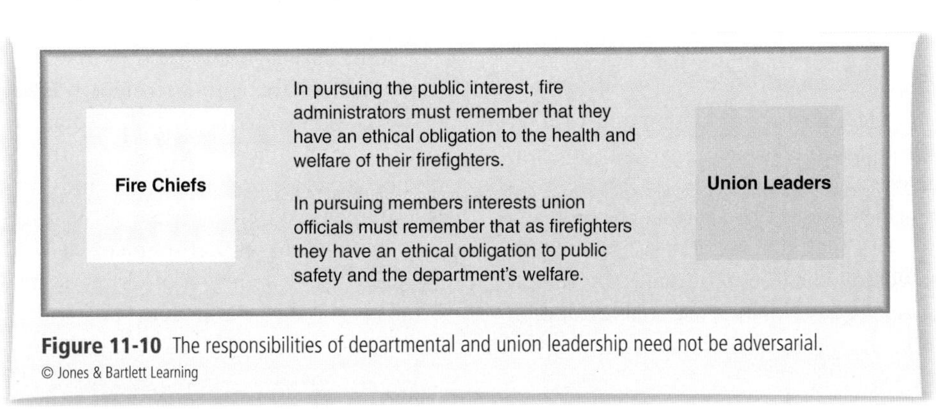

Figure 11-10 The responsibilities of departmental and union leadership need not be adversarial.

Discipline

A fire officer's ethical obligations regarding discipline are relatively straightforward. A fire officer has a duty to maintain discipline. In administering discipline he or she also has a duty to treat employees fairly and adhere to applicable law and policy. In keeping with a fire officer's obligation to the welfare and development of subordinates, disciplinary measures should be intended as corrective and instructive, not punitive or vengeful.

Fire officers have an ethical obligation to comply with relevant disciplinary laws and policies. This obligation applies to spirit and intent as well as actual wording. Under the principle of avoidance of unnecessary harm, it is unethical for a fire officer to impose punishment unfairly, unjustly, or under false circumstances. This imposes an ethical obligation even in the absence of a specific law protecting fire fighters' rights.

Fire Fighters' Rights

Jurisdictional requirements for the ethical administration of disciplinary process may vary from department to department. It is the obligation of a fire officer to understand the specific mandates within his or her jurisdiction. As public servants providing essential services, and in recognition of a professional's property interest in employment, many state and municipal governments have extended specific legal rights to fire fighters facing disciplinary charges. Additionally, collective-bargaining agreements commonly have wording regarding the disciplinary process. Many states have codified a version of a **Fire Fighters' Bill of Rights**. The following are the tenets typically included in these documents (Bennett, 2016).

- Fire fighters shall not be subjected to interrogation without receiving written notice of charges.
- Fire fighter questioning shall take place in a facility where the employee or investigator works.
- Questioning shall be performed at a reasonable time of day and for a reasonable duration.
- A fire fighter being questioned shall not be exposed to offensive language, insults, inducements, or threats.
- A complete record of all investigatory and interrogation activity shall be maintained.
- Fire fighters shall not be disciplined, demoted, denied promotion or seniority, transferred, reassigned, or otherwise disciplined for exercising their rights.
- Fire fighters shall have access to union representation or legal counsel when being questioned.
- Fire fighters are typically exempt from any requirement for self-incrimination.
- In many jurisdictions, fire fighters may not be compelled to participate in a polygraph test or chemically enhanced questioning without their express written consent.

Due Process

In addition to the "rights" listed previously, many state and local jurisdictions have **due process requirements**, which must be followed when investigating and disciplining fire fighters. **Due process** procedures serve two basic goals. One is to produce, through the use of fair procedures, more accurate results and to prevent the wrongful deprivation of interests. The other goal is to make people feel that the authority has treated them fairly by, for example, listening to their side of the story (UMKC, 2018). Due process requirements prescribe investigatory limitations, hearing rules, evidentiary procedures, and appeals processes. Whether provided by law or policy, fire officers have an obligation to abide by the letter of the due process policy as well as its intent.

Some jurisdictions, such as the state of Ohio, have statutes that protect career fire fighters from termination except in cases of serious transgression. The theory behind the limitation is that termination is not corrective or instructive and, as a result, it should be used only in a case of last resort.

There is an ethical element at work as well Figure 11-11. As a matter of fairness and justice, punishment should be commensurate with the offense.

Figure 11-11 Chief officers have tremendous authority to impose punishment; therefore, they have a commensurate responsibility to exercise judgment and restraint.

Chief officers have tremendous authority to impose punishment; therefore, they have a commensurate responsibility to exercise judgment and restraint. Reactionary and heavy-handed punishments that are disproportionate to offenses are a breach of trust.

Privacy Rights

Disciplinary issues can be emotionally charged, embarrassing, and damaging to reputations. As a result, privacy issues can be a significant complication in the disciplinary process. As a matter of principle, the privacy rights of fire fighters under investigation must be respected. However, the public may have legitimate cause to insist that details of fire service disciplinary investigations be made public. It is important that chief officers be familiar with federal, state, and local policy regarding freedom of information requirements related to employee discipline. (Further information regarding freedom of information mandates is available in the next section of this chapter.) It is essential that senior fire staff consult with legal counsel regarding the ethics requirements relative to disclosure of investigatory information. The previously mentioned fire fighters' bill of rights, due process, and privacy acts represent common elements of fire fighter protections. Jurisdictions may exclude some protections or have additional protections not listed.

Whistle-blower Protection

In 2002 Congress passed the **Sarbanes-Oxley Act**, which mandated agencies with fiduciary responsibilities to adopt a code of ethics. While much of the act is specific to financial instruments, there is wording that applies to public bodies regarding altering documents, obstructing government investigations, and taking retaliatory action against employees who publicly expose agency wrongdoing or cooperate with government investigations (Bennett, 2016). A **whistle-blower** is a person who exposes any kind of information or activity by an organization that is deemed illegal or unethical. The last element of the Sarbanes-Oxley Act is also known as **whistle-blowing protection**. It should be noted that specific whistle-blowing protections are often included in state and municipal ethics codes (White, 2014).

It should be remembered that the tenets of fire fighters' rights and due process represent a legal minimum, and their absence does not displace ethical responsibility. The concepts of integrity, justice, fairness, and the avoidance of harm all impose an ethical obligation for fire officers to handle disciplinary processes professionally, fairly, and compassionately. This obligation exists even if it is not mandated in a particular jurisdiction by law, contract, or policy.

As repeated frequently within this chapter, ethics laws can be complex, and the ethics requirements for handling disciplinary actions are no exception. Fire chiefs and senior staff are advised to familiarize themselves with local requirements and to work closely with their municipality's legal staff.

Transparency in Government Requirements

Transparency in government is a concept implying that stakeholders have the capacity to oversee government actions, which includes the ability to witness government decision making and access documentation and records (Nadler and Schulman, 2015). The impetus for government transparency is that it promotes ethics. It is generally accepted that government that is not transparent is more prone to corruption and malfeasance Figure 11-12. Transparency promotes the common good by ensuring fairness and equal access and by discouraging undue influence by special interests.

A democracy's organizing theory is that of public participation and oversight. Spending priorities, equal access to justice, and public policy are all theoretically shaped by representatives acting in accordance with the public preference. This precept requires public confidence that representatives are acting in their best interest, and this can only be assured by oversight (Nadler and Schulman, 2015). Similar to any other public official, fire department personnel are ethically mandated to act on the public behalf. As beneficiaries of public funding, they are also ethically required to provide the public access to documentation of activities and spending. The ethics concepts behind these requirements are held so important that federal, state, and local governments have imposed specific laws intended to assure transparency. These laws include: open meetings requirements, record-keeping requirements, and freedom of information request policies. Collectively these are known as **sunshine laws**.

Figure 11-12 Transparency in government is an important safeguard against ethical breaches.

Courtesy of James Madison

Open Meetings Requirements

Every local jurisdiction is subject to some form of government transparency regulations in the form of open meetings mandates. Specific requirements may vary from municipality to municipality, and fire personnel must familiarize themselves with the specific requirements of their jurisdictions. The intent of open meetings requirements is that constituencies can witness government in action Figure 11-13 . At the local level, these requirements not only pertain to city council meetings but also any governmental body conducting deliberative business meetings. These may include electrical commissions, building commissions, zoning boards, school boards, and even fire and police commission meetings. As an example, assume that a fire chief happens to bump into two of the three local fire commissioners at a social gathering. As long as no official department business is discussed, the meeting does not fall under the open meetings act. However, if the fire chief asked the commissioners to approve a request, or if the commissioners ask the fire chief about a pending purchase, then the impromptu meeting becomes an official meeting that is covered under the open meetings act. As such, all the provisions in the following list would be necessary, including a requirement of prior public notice and note keeping. It should be further pointed out that in most jurisdictions meetings are defined very liberally and typically include any gathering of a quorum of committee members, video conferences, audio conferences, and most recently chat rooms. Generally, open meetings requirements embrace the same general concepts. Examples of typical requirements include (Frosh, 2017; Illinois General Assembly, 2003a):

- With some noted exceptions, the public generally has the right to attend all official government meetings in which official action may be taken.
- Official action includes, but is not limited to, spending approvals, the creation of policy, deliberation or judgment by committees of

Figure 11-13 The intent of open meetings requirements is that constituents can witness government in action.

public requests, approval of licenses, and the adoption or modification of local ordinances.

- Jurisdictions are typically required to provide public notice of scheduled deliberative meetings.
- A meeting agenda must be publicly distributed or made available, and with some exceptions, groups may not act on any item not included on the agenda.
- Records, reports, and documentation referenced during public meetings are typically considered public record unless they fall under specific exceptions.
- Minutes must be kept. Any time a government committee or deliberative body meets, a detailed written, audio, or video record must be maintained.
- The requirement for archiving records varies from jurisdiction to jurisdiction—although one year is fairly standard. In most cases, meeting minutes are in the public domain and may be accessed by the public upon request.
- In most cases, the public has the right to attend meetings but must receive special permission to speak. Typically requests to address committees require at least 24-hour notice.

Exceptions to open meetings requirements are extended to deliberative bodies when it is contrary to the public interest that certain information is publicly disseminated. In most cases exceptions to open meetings requirements are limited to specific conditions and not subject to interpretation. Typical exemptions include (Frosh, 2017; Illinois General Assembly, 2003a; Shuette, n.d.):

- Human resource issues including compensation, discipline, performance appraisals, or dismissal of a specific employee.
- Collective bargaining matters while negotiations are in process.
- Deliberations regarding the selection of persons to fill public office.
- Discussions regarding the lease, purchase, or sale of real property.
- Discussions regarding the acquisition or disposition of securities, investments, and investment contracts.
- Security procedures, school building safety and security, and any reference to undercover law enforcement activities are generally privileged.
- Meetings pertaining to litigation concerning any suit brought by the government agency against an individual or group, or conversely suits brought against the government agency.

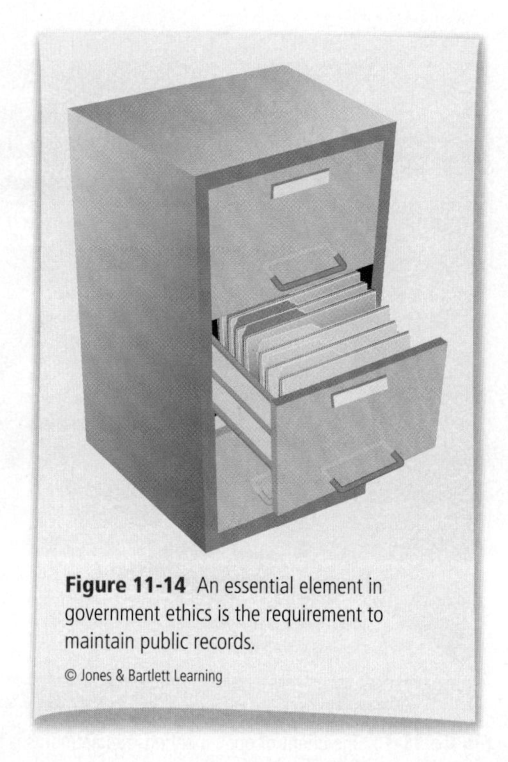

Figure 11-14 An essential element in government ethics is the requirement to maintain public records.

Record-Keeping Requirements

An essential element in government ethics is the requirement to maintain public records (Figure 11-14). Unfortunately, record-retention requirements are often poorly understood and commonly unintentionally violated. The requirements for public record maintenance can be complicated and even overwhelming. For this reason, many municipalities have staff specifically assigned to act as a resource for agency administrators.

The definition of public records is surprisingly broad and inclusive. As an example, in the state of Illinois public records are defined as "any book, paper, map, photographs, or other official documentary material regardless of physical form or characteristics; that are made produced, executed, or received by any agency or officer pursuant to law or in connection with the transaction of public business. . ." (White, n.d.). Included within the description of public records are any e-mails memos, instant messages, social media posts, videos, and faxes. In many jurisdictions, agency Facebook page content, Twitter posts, and blogs are also covered if produced by public officials and related to agency business (Illinois General Assembly, 2003a).

In general, government agencies, including fire departments, are required to keep and maintain records for specific amounts of time depending on the record type. Moreover, access to those records is granted to the public through the freedom of information process described next. As with the open meetings regulations, there are specific exceptions to public access, and they tend to follow the same general guidelines. Access to public records is described in detail in the following section.

An important aspect of record-retention policies is that, similar to other government agencies, fire departments are required to seek specific permission from a state or federal oversight agency before disposing of some records. In most cases, paper records have specific disposal methodology requirements or even specific vendors who must be utilized for the disposal process (Illinois General Assembly, 2003a; HR Hero, 2018).

Clarification

Fire departments are often required to seek specific permission from a state or federal oversight agency before disposing of some records.

Requirements for how records are kept are also common. Jurisdictions typically have specific rules regarding the security of paper documents, digitizing paper documents, and cyber security requirements for digital public records. As an example, in Illinois, some human resources and medical records must be stored in a lockable room with limited access (Illinois General Assembly, 2003a).

The greatest complication in dealing with record maintenance regulations is that each record type has different requirements for retention. For instance, transitory e-mails that deal with incidental communication regarding policy changes may only have to be kept for one year, whereas fire fighter medical records may be required to be stored for up to 20 years past retirement or separation.

As with many federal, state, and local ethics laws, it is important that fire administrators familiarize themselves with the records requirements of their local jurisdiction and regularly consult with their legal staff or other appropriate resources. In the typical routines of office management, document-retention procedures may seem unnecessarily cumbersome and unimportant. Such thoughts can be quickly dispelled when particular records are subpoenaed as a result of an investigation or litigation. Officials should be warned that penalties for noncompliance can be substantial.

Freedom of Information Requests

The federal **Freedom of Information Act (FOIA)** was enacted in 1966 and was the first law that gave Americans the right to access the records of federal agencies. Since its inception, state agencies have adopted similar legislation that applies to state, county, and municipal bodies of government (Electronic Frontier Foundation, n.d.).

FOIA legislation requires that public bodies receiving requests for records must provide those records if they are properly requested. Properly requested usually means that a specific form must be filled out, and the records requested are specific and clearly described. FOIA requests can be made by any individual regardless of citizenship status (U.S. Department of Justice, 2014). In most cases FOIA requests do not need to be justified by the individual making the request, nor is an explanation for the intended purpose of the information required.

Typically, agencies have 20 days to comply with proper information requests. The agency is also usually required to acknowledge requests within 10 working days. Exceptions to the 20-day rule may be made if records are not locally stored, or the scope of the application is beyond the administrative capabilities of staff because of volume or security clearance (U.S. Department of Justice, 2014).

While each state has its own standards about what information is considered public, information available under the FOIA commonly includes:

- Census records
- Criminal records
- Consumer protection information
- Court dockets, government spending reports, legislation
- Professional and business licenses
- Real estate appraisal records and transactions
- Voter registration
- Purchase records and bidding documents
- Official correspondence

Exclusions to records covered under FOIA are classified documents, personnel records, medical records, most documents involved in litigation, work product by contracted vendors, and documents covered by attorney-client privilege.

Of particular importance to fire departments are those documents unique to department operation considered public records. Documents subject to FOIA dissemination include: fire investigation and cause and origin reports not currently under litigation; inspection reports; material data safety sheets; budget documents; and purchasing records, including bids. Excluded from mandatory FOIA disclosure are fire fighter personnel, disciplinary, and medical records.

The determination of whether or not a request is covered under FOIA regulations should be referenced with the jurisdiction's legal team or FOIA compliance officer. Many fire departments and local government units routinely defer *all* FOIA requests to the agency's FOIA compliance officer.

Clarification

Remember, all forms of communication, including personal e-mails relative to departmental business, are considered public records.

WRAP-UP

Chapter Summary

- While adherence to ethics laws is mandatory, they represent a minimum, not a maximum, ethical obligation.
- Responsibility to act in the public interest, to act with integrity, and to respect professional responsibilities transcends the limitations of legal wording.
- Public officials are obligated to familiarize themselves with the requirements of their local jurisdiction and to work closely with their legal advisors and local ethics officers.
- The intent of the laws is to assure that officials act in the public interest, conduct their affairs with integrity, and distribute public services without prejudice.
- It is incumbent on fire administration personnel to adhere not only to the letter of the law but also to its spirit and intent.
- The ethical obligations of serving the public interest, protecting their lives and property, and assuring the health and safety of fire fighters exist independent of legal obligation and are in effect even where specific ethics regulations are lacking.
- The chapter was divided into five general areas of ethics policies. These include fiduciary responsibilities, conflicts of interest, laws requiring governmental transparency, ethics requirements and labor relations, and disciplinary procedures.
- Fiduciary responsibility centers around an obligation of trust. Fiduciary responsibility requires that a public official maintains the highest standard of care under the law and acts solely for the benefit of the public.
- Fiduciary responsibility also mandates avoidance of conflicts of duty, interest, or loyalty. Finally, fiduciary responsibility requires that public officials may not profit from fiduciary activity without the express permission of those they serve.
- Fire personnel responsible for budget management have a legal responsibility to manage funds strictly in the public interest. With that responsibility comes several ethical requirements regarding budgeting, departmental spending, and management of departmental assets.
- A common ethical error by fire personnel is a confusion of loyalty. Although a fine distinction, loyalty to the department must not be confused with loyalty to the department's mission, followed closely by a recognition of responsibility to use public funds judiciously.
- The concept of conflict of interest was introduced, with the overriding theme that public officials are required by ethics law to place public interests above their own.
- To prevent corruption, ethics laws define restrictions on receiving gifts and payments that may have undue influence on policy decision making.
- The body of ethics law also restricts using public office and resources for personal gain. Activities such as seeking favors, using public funds and resources for personal interests, and pursuing personal interests while on duty are prohibited.
- To eliminate conflicts of interest, ethics laws typically restrict any outside employment that may cause a conflict of responsibility, divide loyalties, or physically interfere with the performance of fire department responsibilities. This can also include restrictions on post-government employment.
- Political conflicts of interest were also discussed, including the Hatch Act, which specifically limits public officials from participating in political activity while at work,

and the use of any government resources for political purposes.

- The National Labor Relations Act and the Labor-Management Relations Act set ethics parameters for unions and management in dealing with each other. Specific ethics requirements were introduced regarding contract negotiations, conformance to contracts, and grievance procedures.
- Both labor and fire service management have an ethical obligation to coexist, and each has an ethical duty to serve their constituencies. However, those responsibilities are subservient to the overriding ethical responsibility to assure the health and safety of both fire fighters and the public they serve.
- The reader was introduced to the concepts typically included in states' fire fighters' bill of rights clauses, due process requirements that protect the property interests of a fire fighter's employment, and the whistle-blower protection clauses adopted by most states.
- Fire administrators are obligated to assure that breaches of policy and law are addressed in a manner that is fair, consistent, and instructive.
- Fire administrators are entrusted with significant powers to impose penalties; they are ethically required to be judicious in their use.
- The necessity for transparency was introduced as a requirement for democratic governments to function. Public participation in local government requires information and oversight.
- Transparency ethics laws are intended to eliminate corruption, undue influence, and favoritism in government policies.
- Transparency in government assures that government officials act in a manner that serves the greatest public good and requires public access to official meetings along with keeping records of meeting proceedings.
- State ethics requirements for the maintenance of public records were also introduced, as was legislation that grants the public access to government documents.
- Ethics laws can be complex and may vary from jurisdiction to jurisdiction. It is the obligation of fire officials to become familiar with the requirements within their jurisdiction. Because of the complexity of ethics law, it is also important for fire administrators to work closely with their jurisdiction's legal staff and ethics compliance officers.

Key Terms

beneficiary (aka principal) the group or individual for whom a fiduciary is responsible.

conflict of interest a situation in which a person is in a position to derive personal benefit from actions or decisions made in his or her official capacity.

due process the legal requirement that an authority must respect all legal rights that are owed to a person.

fiduciary a person who holds either an imperative ethical or legal responsibility for the welfare of another person (referred to as beneficiary or principal) or a person's assets.

fiduciary responsibility a mandate for highest standard of care under the law.

fire fighters' bill of rights legislation intended to apply commonsense principles of fairness and professionalism to the process of investigating and disciplining fire fighters.

Freedom of Information Act (FOIA) enacted in 1966, it was the first law that gave Americans the right to access the records of federal agencies.

grievance procedure a formalized process to clarify contract wording and intent.

Hatch Act a federal law passed in 1939 that limits certain political activities by governmental officials.

Labor Management Relations Act also known as the Taft-Hartley Act, it delineated unfair labor practices imposed on unions.

National Labor Relations Act also known as the Wagner Act, it granted employees the rights of self-organization and collective bargaining.

positional abuse using official status to acquire benefits to yourself or your immediate family.

quid pro quo translates to "something for something," and generally describes the conditions for bribery.

revolving door provisions typically restrict government employees from accepting employment with agencies that deal directly with their former employer for a specific period of time.

Sarbanes-Oxley Act mandates agencies with fiduciary responsibilities to adopt a code of ethics; many affect municipal agencies.

sunshine laws a body of ethical requirements intended to promote governmental transparency.

whistle-blower a person who exposes any kind of information or activity by an organization that is deemed illegal or unethical.

whistle-blower protections prohibitions against authorities taking (or threatening to take) retaliatory personnel action against any employee or applicant because of disclosure of information by that employee or applicant.

Challenging Questions

To check your understanding of this chapter's material, answer the following questions. It is highly recommended that you discuss your viewpoints with fellow students, peers, coworkers, and friends to discover their opinions as well.

- It is no secret that the political process can be full of bluffing and gamesmanship. Does this fact ethically justify employing deceptive practices during budgeting negotiations?
- Many believe that government works most efficiently when representatives can speak frankly behind closed doors. In your opinion, is efficiency or public oversight of greater importance?
- As a fire chief, which priority would seem most important to you: being actively involved in contract negotiations in order to protect your department's interests, or maintaining a positive relationship with your employees and so abstaining from the negotiation process?
- Find out the record-retention requirements for your local fire department either by an Internet search or by contacting your state's ethics compliance officer.
- Records will include but are not limited to employee medical records, purchasing records, inspections, meeting minutes, and incident reports. Survey your department's current practices to determine compliance.
- Does your state have a fire fighters' bill of rights in place? If so, highlight the general principles and mandates contained within it. If not, select a nearby state that does.

WRAP-UP

Case Study Conclusion

Revisit the case study at the beginning of the chapter. Spend a few minutes considering the questions posed at the end of the case study. In light of the information shared in this chapter, have any of your original observations changed?

The actions of Assistant Chief Jones are unethical. As an officer in charge of handling department finances and purchasing he has a fiduciary responsibility to follow the letter and intent of the law and municipal policy. While his actions may conform to the actual wording of the policy, they are clearly designed to subvert the intention. The purchasing and accounting requirements of most states and municipalities anticipate this issue and forbid it. As a result, his actions are likely illegal. Other ethical issues are also involved. As a fiduciary agent, Assistant Chief Jones has an ethical obligation to act in the best interest of the department and the taxpayers. By purchasing equipment at a higher price than necessary, he ignores those obligations. His preference for dealing with local vendors is irrelevant. As a purchasing agent, he does not have the authority to waive bid requirements, regardless of his intentions. Jones's actions are intended to deceive his superior officers. City administrators and fire chiefs rely on subordinates to follow the letter and intent of regulations. It is unethical for Jones to deceive by either directly misinforming his superiors or omitting pertinent information. By ignoring these, Jones may place the department in jeopardy for legal actions.

The proper response for Jones would be to inform his superior officer of his desire to waive the bid requirement and to fully explain the financial ramifications of that action. He would, of course, be further obligated to abide by the decision made by his superior.

Chapter Review Questions

1. What 1966 federal law first granted public access to government documents?
2. Define the term *quid pro quo*.
3. Can state law restrict conditions of employment after you retire from the fire service?
4. Define the term *fiduciary responsibility*.
5. What does the Hatch Act limit?
6. Are personal e-mails made on government equipment subject to FOIA inquiries?
7. What is the intent of due process requirements?
8. What does the Wagner Act provide for?
9. Is it legal for a fire chief to accept a $200 gift certificate from a business if no requests for consideration are made in return?
10. Does an open meetings act guarantee the public the right to speak at meetings where government business is to be conducted?

References

Adler, R. S., and W. Bigoness. 1992. "Contemporary Ethical Issues in Labor Management Relationships. *Journal of Business Ethics* 11, no. 5, 351–360.

Bennett, L. T. 2016. *Fire Service Law.* Long Grove, IL: Waveland Press.

Bowie, N. 1985. "Should Collective Bargaining and Labor Relations Be Less Adversarial?" *Journal of Business Ethics* 4, no. 4, 283–291.

Electronic Frontier Foundation. n.d. "History of Freedom of Information Act." Accessed June 21, 2018. https://www.eff.org/issues/transparency/history-of-foia.

Frosh, B. E. 2017. *Open Meetings FAQs: A Quick Guide to Maryland's Open Meetings Act.* Accessed June 21, 2018. http://www.marylandattorneygeneral.gov/OpenGov%20Documents/Openmeetings/OMA_FAQ.pdf.

HR Hero. 2018. "Document Retention and Record Retention Laws." Accessed June 21, 2018. http://topics.hrhero.com/document-retention-and-record-retention-laws-for-employers/#.

Illinois General Assembly. 2003a. General provisions (5 ILC as 430/). Ellen White Open Meetings Act. Illinois Compiled Statutes. Accessed June 21, 2018. http://www.ilga.gov/legislation/ilcs/ilcs3.asp?ActID=84&ChapterID=2.

Illinois General Assembly. 2003b. General provisions (5 ILC as 430/). State Officials and Employees Ethics Act. Accessed June 6, 2018. http://www.ilga.gov/legislation/ilcs/ilcs3.asp?ActID=2529&ChapAct=5%26nbsp%3bILCS%26nbsp%3b430/&ChapterID=2&ChapterName=GENERAL+PROVISIONS&ActName=State+Officials+and+Employees+Ethics+Act.

Kernaghan, K., and J. Langford. 2014. *The Responsible Public Servant.* Canada: Institute of Public Administration of Canada.

Nadler, J., and M. Schulman. 2015. *Open Meetings, Open Records, and Transparency in Government.* Markkula Center for Applied Ethics. Accessed June 21, 2018. https://www.scu.edu/ethics/focus-areas/government-ethics/resources/what-is-government-ethics/open-meetings-open-records-transparency-government.

Rabin, J. 2003. *Encyclopedia of Public Administration and Policy.* New York: Marcel Dekker.

Shuette, B. n.d. *Open Meetings Act Handbook.* Accessed June 21, 2018. https://www.michigan.gov/documents/ag/OMA_handbook_287134_7.pdf.

State of Massachusetts. 2016. "Summary of the Conflict of Interest Law for Municipal Employees." Accessed June 21, 2018. https://www.mass.gov/search?q=summary+of+the+conflict+of+interest+law+for+municipal+employees#gsc.tab=0&gsc.q=summary%20of%20the%20conflict%20of%20interest%20law%20for%20municipal.

Temchenko, E. 2016. "Fiduciary Duty." Accessed June 21, 2018. https://www.law.cornell.edu/wex/fiduciary_duty.

UMKC. 2018. "Exploring Constitutional Conflicts: Procedural Due Process." Accessed June 21, 2018. http://law2.umkc.edu/faculty/projects/ftrials/conlaw/proceduraldueprocess.html.

U.S. Department of Justice. 2014. *An Overview of the Freedom of Information Act: Proceedural Requirements.* Accessed June 21, 2018. https://www.justice.gov/sites/default/files/oip/legacy/2014/07/23/foia-procedures.pdf.

U.S. Legal. 2016. "Breach of Fiduciary Duty Law and Legal Definition." Accessed June 21, 2018. https://definitions.uslegal.com/b/breach-of-fiduciary-duty.

U.S. Office of Special Counsel. 2015. "The Hatch Act." Accessed June 21, 2018. https://osc.gov/pages/hatchact.aspx.

White, J. 2014. Illinois Governmental Ethics Act. Accessed June 21, 2018. https://www.cyberdriveillinois.com/publications/pdf_publications/ipub26. pdf.

White, J. n.d. "Frequently Asked Questions Regarding the Local Records Act." Accessed June 21, 2018. http://www.cyberdriveillinois.com/departments/archives/records_management/faqlocal.html.

FIRE & RESCUE
PROTECTIVE
ADKINS
0701
FAIRFAX COUNTY FIRE &
FIRE & RESCUE
DEPARTMENT

SECTION 4

Applied Ethics for the Fire Service

CHAPTER 12

Ethical Decision Making

"Imaginary evil is romantic and varied; real evil is gloomy, monotonous, barren, boring. Imaginary good is boring; real good is always new, marvelous, intoxicating."

—Simone Weil

OBJECTIVES

After studying this chapter, you should be able to:

- Recount the foundational principles of ethical decision making.
- Assess the elements for effective decision making.
- Explain the development of ethical intelligence.
- Interpret Kohlberg's theory of moral development.
- Apply the theory of locus of control to ethical decision making.
- Discuss the approaches to ethical decision making related to philosophical orientation.
- Describe the fire service decision-making model.

Case Study

Assistant Chief Jones plays in the same softball league as fire fighter Smith. After his game, Jones stays to watch Smith's team play. In the fourth inning, Smith attempts to throw out a runner from left field. After his throw, Smith seems in discomfort. Jones observes him rubbing his shoulder and shaking his hand as if trying to relieve numbness. The next inning Smith does not return to the outfield.

Two weeks later Jones is looking at an on-duty injury report filled out by fire fighter Smith. Smith claims to have injured his (throwing) shoulder while pulling a ceiling. In a supplement to the report, Smith indicates that he has seen his private physician and will require shoulder surgery with an additional six weeks of rehabilitation.

Assistant Chief Jones is confident that Smith is falsely claiming an on-duty injury and that he was having shoulder problems dating back to at least the softball game a few weeks earlier. He is offended by Smith's apparent fraudulent claim but is unsure how to proceed. In considering his options, Chief Jones believes he has two choices. First, he can confront Smith and report his beliefs to the city's human resources director. Second, he can do nothing. After careful consideration, Chief Jones decides not to report the incident.

- What would you do in Jones's place?
- Can you identify any rationale Jones may have had for remaining silent?
- Identify the ethics issues involved for Jones.

Introduction

As children, the concepts of behavior expectations are relatively straightforward. In our earliest development, we know right from wrong because we are told what is good and what is not. As we become older, our emotional development leads us to develop our own sense of morality. We form our own values, and we develop a sense of justice that is shaped by those values. An ever-growing body of experience soon instructs us that circumstances can shape perception and that very often things are not as they seem. This may cause us to be reflective and empathetic, or it can cause cynicism, which can induce judgment and moral outrage.

As our lives become more complicated so does the complexity of the ethical decisions that we sometimes must make. Decisions have consequences, and sometimes doing the right thing can have very unfortunate outcomes. Conversely, circumstances can sometimes convince us that doing bad things can achieve good outcomes. Good and bad are often blurred at the edges rather than separated by bright lines.

Being a good moral decision maker requires more than a wish to do the right thing. It requires us to be able to recognize the right thing, which is why we study ethics. It is also why individuals need a strong sense of who they are and what they believe. Self-awareness and the ability to assess honestly the implications of our actions is necessary for the development of a reliable moral compass.

Developing ethical decision-making skills is important for all people but especially so for fire fighters. Fire fighters are public servants. They are trusted agents within the community and are extended special authorities in support of meeting their important responsibilities. They are also stewards of public funds and capital resources. As repeatedly mentioned in this text, fire fighters are held to a higher standard of behavior. The decisions fire fighters make reflect on their personal character and the ethical culture of their department and the profession.

This chapter will explore dimensions in ethical and moral decision making. It will also look at the basic premises of ethical decision making and explore how individuals come to develop morally. We will also explore personality facets that can affect ethical decision making. Finally, the chapter will introduce best practices in ethical decision making.

Foundational Principles and Ethical Decision Making

You have heard it before: Life can be messy. At some point, we are all faced with difficult choices in which there appears to be no right answer. No matter what you do, someone will be harmed. Fairness seems impossible and available options all conflict with personal values. Such are the difficulties of an **ethical dilemma**.

A common complaint among people facing ethical dilemmas is a feeling of being trapped. Many believe they have no options or at least that they have no good options. There is a well-known expression of being in a catch-22—a dilemma or difficult circumstance from which there is no escape because of mutually conflicting or dependent conditions Figure 12-1. The expression epitomizes the concept of an ethical dilemma. An ethical dilemma is a complex situation where ethical principles and moral imperatives seem to conflict with personal values—or even with each other. In an ethical dilemma, correct actions may yield undesired consequences. Desired consequences may seem to necessitate questionable behavior. Many who have been embroiled in ethical controversy plead that they had no choice or that there were no good choices available. Unfortunately, this is often true. However, in many cases, what is perceived as an ethical dilemma is actually a confusion between the absence of easy choices and the presence of only bad options.

Figure 12-1 At times, fire fighters cannot escape being criticized no matter what they do, resulting in a catch-22.

Ethics professors will likely sympathize with the feelings of frustration posed by an ethical dilemma, but they would also assert that ethical choices are always present—albeit elusive. Too often, decision makers perceive ethical dilemmas where none exist. This is caused by an infusion of emotion or, in some cases, situations are clouded by the elevation of desired outcomes to parity with necessity. We study ethics and, more important, principles of ethical decision making in an attempt to untangle the web of needs, wants, responsibilities, justice, and fairness.

Entire philosophies of normative ethics have been developed in an effort to guide human behavior. Fortunately, the principles of the various ethical philosophies usually lead to similar conclusions. A fire fighter who is pondering whether to report embezzlement of department funds by the fire chief will likely arrive at the same ethical conclusions whether he or she applies the principles of virtue ethics, deontology, or consequentialism. Although the philosophic approaches to ethics can vary, the decision-making process within each is remarkably similar.

True ethical dilemmas tend to have the following elements in common: There is no clear desirable outcome, or the desired outcome requires an action inconsistent with personal values or with established ethical principles. Some foundational elements within ethical decision making can help to clarify the parameters of choice. Two of the most important principles in ethical decision making are:

1. We all have the power to decide what we do and what we say.
2. We are morally responsible for the consequences of our choices.

These two simple yet powerful statements provide context for the analysis of decision-making processes and the results of those decision-making attempts. If we as individuals are not free to decide what we do and say, how can we be held accountable for our actions? The implication is relatively straightforward: We have choices, and those choices have ethical implications. Where circumstances do not legitimately provide choices, there is an absence of ethical responsibility. Additionally, where there is an absence of accountability, ethics and morality are irrelevant (Josephson, n.d.).

Regardless of approach or ethical orientation, all philosophical views of ethics acknowledge self-determination and personal responsibility. It holds then that as we make decisions with moral implications, we must bear in mind that each person is a moral agent with an obligation to assess right from wrong. The principles of ethics place a burden on the individual and hold each responsible for the results of her or his choices, regardless of the amount of consideration that occurred before the action. A lack of forethought does not excuse unethical behavior, even in the absence of ill intent. Although we may not always consider ramifications, our actions and speech reflect choices. Those choices shape the image that we present to the world and the one we see in a mirror.

In making ethical decisions, there are several skills, mind-sets, and frames of reference that must be present. Ethical decision making does not happen by accident, nor do wise decisions always come automatically. It is incumbent on the decision maker to be aware of an issue's scope and context.

Recognizing Important Decisions

We make decisions every day, practically all day. Most of those decisions have no ethical or moral significance whatsoever. There are, however, many occasions throughout daily life that do in fact have ethical or moral ramifications. Should we tell a white lie? Should we tell the clerk that she undercharged us? Should we keep the $20 bill the the woman in front of us just dropped?

Most of the time we rely on common sense and intuition to guide our decision making. We instinctively know right from wrong and choices are self-evident and self-explanatory. We may sometimes be tempted to do a bad thing and, in fact, might choose to do so. But we still recognize the action for what it is. Unfortunately, our sense of intuition or common sense fails us sometimes when we face difficult decisions. We must dig deeper into the elements of the problem we are facing and consider our options thoroughly. We need to reflect.

The truth is, however, that deep reflection does not come naturally to most people. The routines of our day-to-day lives reinforce the habit of making instantaneous decisions. Following our gut becomes habitual. It is important in ethical decision making that individuals learn to recognize and appreciate decisions with moral consequence. There is a proverb attributed to St. Bernard Clairvaux: "The road to hell is paved with good intentions, but heaven is full of good works" Figure 12-2 (Mainser, 2007). It is likely true that those embroiled in ethical crises often fail to recognize important crossroads along the path to crisis. Scandal and moral crisis can arise from what appeared to be insignificant choices that proved to have momentous consequences. To be a good ethical decision maker, you must be constantly alert to the presence of ethical issues. Considering the consequences before an action is a sign of prudence and wisdom. As Aristotle asserted centuries ago, virtues must be habituated through awareness and diligence.

Figure 12-2 Bernard of Clairvaux, 12th-century French abbott and force in the founding of the Knights Templar.

Saint Bernard of Clairvaux. Line engraving./ Wellcome Collection

Appreciating the Consequences of Decisions

It seems a matter of prima facie principle that actions have consequences. Yet to varying extents, some people appreciate immediate and tertiary consequences more readily than others. As expressed in the previous section, most of us go through our daily lives relying on common sense and intuition to decide right from wrong and good from bad. Typically, we can almost instantaneously understand consequences. If I choose not to tell someone that she or he dropped a $20 bill, I understand that the person who dropped the bill will be denied use of that money, and I may even feel guilty about taking it. When occasions arise in which ethical problems with obscured consequences present themselves, or when desired outcomes are in competition with each other, the skill in consequence analysis becomes important.

It is not uncommon to hear a parent scolding a transgressing child with the question, "What were you thinking?" The common reply is, "I don't know." The simple exchange demonstrates a parent teaching a child the importance of evaluating consequences and to practice forethought rather than relying solely on emotion, habit, or intuition. Sadly, even adults can fail to evaluate consequences, and too often those involved in a scandal cannot articulate in hindsight how they came to choose a course of action. The reason is that they were not thinking; they were reacting. One of the key values of ethics education and training is the development of habituation in **consequence analysis** (Josephson, n.d.).

Consequence analysis can be complex and time-consuming. However, its basic elements boil down to five questions:

1. Can my decision cause harm to others?
2. Could a decision hurt my reputation or that of my department?
3. Could a decision impede the achievement of an important goal or greater good?
4. Is my decided action consistent with my responsibilities?
5. Is an intended action consistent with personal and professional values.

There is a principle in science known as the "butterfly effect," which suggests that even the smallest actions can have ramifications far beyond the understanding of the actor. Appreciating the extent of consequence can be difficult. A teacher who fails a student may set in motion a cascading effect of consequences that could have a negative impact on the student's entire life. The individual who deceives a lover may create trust issues that, years later, destroy a marriage. We cannot anticipate all the consequences of our actions, but consideration of them is important in being an ethical agent.

Seek Effective and Ethical Outcomes

A decision is usually deemed effective based on its outcome. If it achieves the desired result, it is considered effective. If the decision yields an unintended consequence, it is judged to be ineffective. The effectiveness of choices can be further tested with the simple question, Are you satisfied with the results?

To be happy with an outcome would imply that the result is "good." Obviously, defining good can be somewhat more complicated. There is a moral peril in assessing behavior from an entirely egocentric frame of reference. It was pointed out in Chapter 1, *Introduction,* that ethics is by its nature relative to how actions affect others. As a result, "good" as a concept, becomes complicated. Questions arise: Good for whom? and Good in what way?

There are many who would argue that an effective decision must be by nature an ethical one. If the desired outcome is in itself "bad" (unethical or immoral), a decision seeking it cannot be effective. Consider a fire fighter who cheats on a promotional exam. A simplistic evaluation would argue that if cheating accomplishes the desired effect of being promoted, then the decision was effective. However, if you accept that the purpose of the promotional exam is to identify qualified candidates, then the individual who cheats to gain promotion places him- or herself in a position for which he or she may not be qualified. Consequences of that decision could have disastrous implications for the fire fighter as well as the department.

From a moral perspective, the decision to cheat was ineffective even though it accomplished the desired outcome. Its ineffectiveness lies in the failed ethics of the intention, not in execution. Moral philosophy would further assert that the act of cheating diminishes the fire fighter's character and will likely cause some degree of **moral dissonance**.

Recall that virtue ethics argues that happiness can only be achieved through the development of virtuous character. By definition, an unvirtuous individual is unfulfilled and so is unhappy. A guiding principle in ethics is that bad actions do not yield good results. (Review Chapter 1, *Introduction,* for rationale on why people should be ethical.)

Ethical decisions are those that are consistent with virtues, values, and duty. Ethical decisions are supportive of social responsibility; they tend to demonstrate respect for others and for process. Someone versed in deontology would suggest that ethical decisions promote truth telling and seek fairness and justice. Behavior choices inconsistent with these principles typically have destructive consequences and cannot, by nature, be described as effective—at least not in the long term (Josephson, n.d.).

Development of Judgment and Discipline

The ability to discern best actions requires judgment acquired through knowledge, experience, and emotional intelligence. Also important is that an individual must have the discipline to pursue those actions judged as "right." Making ethical decisions requires two abilities: knowing what to do and having the discipline to do it Figure 12-3.

Judgment is necessary for ethical decision making. As you will learn later in this chapter, not all individuals have the ability to discern best ethical actions, especially in morally complex situations. As an example, there are people who believe that a lie of omission is somehow more ethical than a misrepresentation of the truth. Still others believe that a misrepresentation of a fact is ethical, even if it is presented in a way that deceives or manipulates.

As situations become emotionally charged or when consequences are perceived as personally harmful, judgment can become clouded. In the case study at the beginning of this chapter, fire fighter Smith apparently falsifies an injury report. Smith might judge his actions as ethical if he, in fact, aggravates an existing injury while on the job.

Smith may rationalize his report as *technically* accurate if, in fact, he did aggravate his hurt shoulder while pulling the ceiling. The fact that he previously injured it off duty is not reported because it is counterintuitive to his desire to receive insurance

Figure 12-3 Making ethical decisions requires two abilities: knowing what to do and having the discipline to do it.

© Kevin Dodge/Radius Images/Getty

benefits. This ethical "sleight of hand" may be motivated by the significant economic consequences of an off-duty injury compared to an on-duty injury. Assuming that Smith has a relatively conventional understanding of ethics, he may likely feel some degree of moral dissonance despite his rationalization.

Ethical judgment can be difficult in the best circumstances. Our willingness to engage in rationalization in support of desired outcomes can make judgment particularly difficult. In the example of fire fighter Smith, *virtue ethics,* and *deontology* would clearly dismiss his logic for falsifying a report.

In rationalizing his behavior, Smith is likely to invoke the consequentialist view that a greater good is served by protecting the economic welfare of his family. He may also rationalize that little harm is done to an insurance company that garners millions of dollars in profits by underwriting injury risk.

In analyzing Smith's behavior, most readers will quickly realize that his logic is seriously flawed and self-serving. An important point to take away from Smith's example is, however, that Smith and people like him do not recognize the error in their thinking. They are oblivious to it because they do not want to see it. As you may recall from Chapter 3, *Influencing Behavior*, Smith's self-imposed blindness is a form of cognitive dissonance.

Becoming Other-Centered

An essential element in making ethical decisions is the ability to project potential consequences onto affected stakeholders. **Empathy** is defined as "the action of understanding, being aware of, being sensitive to, and vicariously experiencing the feelings, thoughts, and experiences of others" ("Empathy," 2015).

To some extent, each of us is self-centered. We often focus our decision-making processes primarily on consequences as they affect us. Young children are almost entirely self-centered in their approach to decision making. It is with maturity and emotional development that they expand their thought processes to include the impact of their actions on others. The characteristics of empathy and social obligation usually are associated with adult thinking. Even so, adults have varying degrees of empathy. Moreover, even in adults who tend to be empathetic, they may become egocentric depending on context. It is fair to state that some people are oriented to the needs of others, some are not, and many may or may not be depending on the situation.

The exercising of empathy requires an orientation toward others. The empathetic person regularly facilitates the inclusion of other stakeholders' interests when making decisions. Absent this, even a well-intentioned decision maker may inadvertently fail to realize the actual impact of the actions on others. Unfortunately, ill-considered actions are often made not because of a lack of empathy but rather a lack of due consideration. Effective consequence analysis requires not only the ability to be empathetic but also the active inclusion of others' interests in the judgment process. In comparison, a driver may have good eyesight, but the driver cannot see if he or she does not look.

Similar to other factors in ethical decision making, filtering the variables of an ethical problem through a screening effect on others requires practice and habituation. Accomplished decision makers eventually arrive at a point where they instinctively consider two questions: Who will be affected by my decision, and how will they be affected?

Each individual affected by a decision has a stake in it and, by extension, a claim on the decision maker. As pointed out earlier in the chapter, our decisions have consequences. One of those consequences is that our actions have a transactional effect on stakeholders, and those transactions carry a moral/ethical burden (Josephson, n.d.). Being aware that actions have consequences for others is one element of decision making. Being empathetic toward others and thoughtful about the consequences of one's decisions is an additional element of decision making. Being in the habit of considering who may be affected by decisions is yet a third element of decision making.

Clarification

The Elements of Effective Decision Making

- Recognizing important decisions
- Appreciating consequences
- Seeking effective and ethical solutions
- Acquiring judgment and discipline
- Developing an orientation toward others

Making wise and ethical decisions is often difficult, but the fire fighter who approaches decisions rationally and with forethought is much more likely to be successful. If decision makers are careful to recognize important decisions, consider the consequences of those decisions, and seek to identify who may be affected by them, they are much more likely to demonstrate the virtues of wisdom and compassion.

Intelligent and ethical decision making has far-reaching consequences extending beyond immediate harm. Reckless decisions can cause emotional distress, ruin relationships, and create unfulfilled lives. Ethically grounded decisions lead to greater happiness, self-fulfillment, and success.

Developing Ethical Intelligence

It is early April. After a long, cold, dark winter, it is one of those unusually warm spring days that only people in northern climates can truly appreciate. It is 75°F, the sun is shining, the birds are singing, and the great outdoors is calling. It is a perfect day to go fishing, golfing, or hiking, or just being outdoors. The problem is that today is a shift day.

Looking out the window, fire fighter Smith considers calling in sick but decides against it. His rationale is that falsely using sick days is against the rules, and he may get caught. Fire fighter Jones also considers calling in sick but is worried about what his fellow fire fighters will say. He does not want to be perceived as a slacker. Fire fighter Davis also wants to take the day off but decides against it, rationalizing, "If everyone called in sick, then where would the department be?" Training Chief Washington also considers taking a sick day. At her rank, no one would question her use of a sick day, but she has a full schedule, and it would be difficult to make up the lost time. She just feels that she cannot afford the time off.

All four fire fighters face the same choice. Should they or should they not falsely use a sick day? Each seems to be approaching the decision from a different orientation, but each has come to the same conclusion. Even though the same results are reached, each orientation says something about the individual decision maker. Their rationales indicate a certain sophistication in their approach to ethics decisions. Those approaches are indicative of the likelihood of their making future unethical choices.

If fire fighter Smith's only reason for not falsely using a sick day is fear of being caught, it is quite likely that absent that fear he may, in fact, use the sick day. Jones's rationale is similar; his concern is others' view of his actions and not the underlying ethics of the situation. Similar to Smith, he is more likely to use the sick day if he is convinced that no one will find out.

Davis's approach is somewhat more ethically oriented. Although he still is not identifying the underlying ethical issues, he does seem to understand that his actions may have far-reaching effects beyond his own concerns. He is likely to obey the rules as long as he sees a negative consequence in breaking the rules. Washington feels free to disregard the rules; however, she successfully identifies why it is imperative for people to show up to work. She has connected the principles of ethical expectation with the underlying action. She is likely to do the right thing regardless of the likelihood of being caught or even if given permission to do so by a higher authority Figure 12-4.

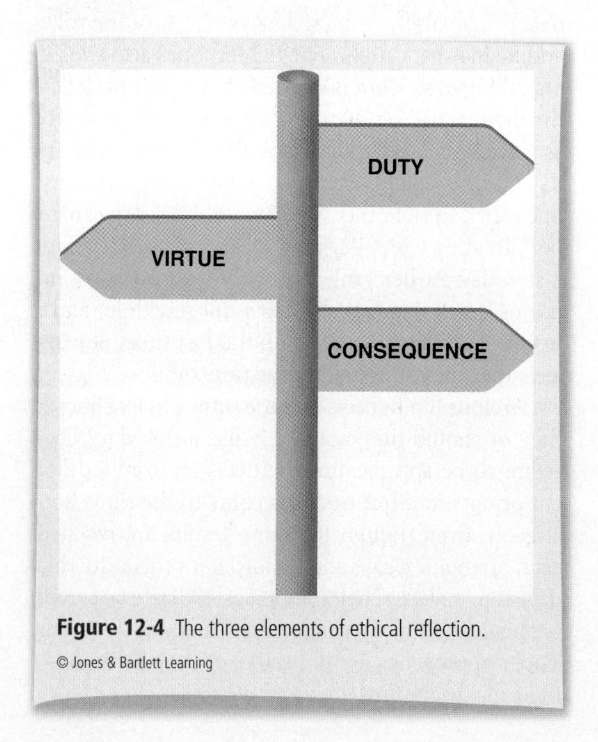

Figure 12-4 The three elements of ethical reflection.

Identifying general approaches to ethical/moral decision making is important and worthwhile, but it only goes so far. For instance, an individual may identify stakeholders and even develop an awareness of how his or her actions may affect the stakeholders, but that awareness still may not necessarily yield guidance.

There is a term called **emotional intelligence**, which refers to an individual's ability to be aware of how others feel. Emotional intelligence differs from empathy in that the empathetic person can *understand* the feelings of others *if* that person is aware of them. Emotional intelligence refers to a person's ability to read or infer the emotions of others. Being aware of consequences and anticipating the emotional response to them is still only half the story. You must determine if those consequences are relevant and if those emotional responses are valid.

Ethical intelligence uses that awareness to guide correct action. Ethical intelligence is an awareness of moral principles and their contextual application. To be ethically intelligent is to understand the general principles of correct behavior, to discern which principles are relevant, and to apply those principles in a given situation (Weinstein, 2011).

How does a person develop ethical and moral judgment? Further, having moral judgment, how does a person process decisions to arrive at proper conclusions?

Moral Development

In 1958, preeminent psychologist Lawrence Kohlberg refined a theory of moral development Figure 12-5. Similar to some of his contemporaries, Kohlberg believed that moral reasoning is the basis for ethical behavior. He further asserted that competence in moral reasoning was acquired through learned behavior, experience, and maturation. As a result, the ability or inability of an individual to act consistently ethically depends on the sophistication of that person's understanding the elements of morality and ethics (Kohlberg, 1981).

Kohlberg created a scale depicting stages of moral reasoning to illustrate how people assess and

Figure 12-5 Lawrence Kohlberg, 1927–1987, American psychologist.

justify their actions. For example, a child who resists the urge to steal candy out of fear of punishment (stage I) is acting ethically, as is someone who refrains from stealing out of appreciation for social order (stage V). Both are acting ethically, but their rationale for doing so demonstrates differing levels of ethical intelligence. The following is an explanation of each of **Kohlberg's stages of moral development** (Figure 12-6).

Preconventional Moral Reasoning

The preconventional level of moral reasoning is typically associated with children, although adults may

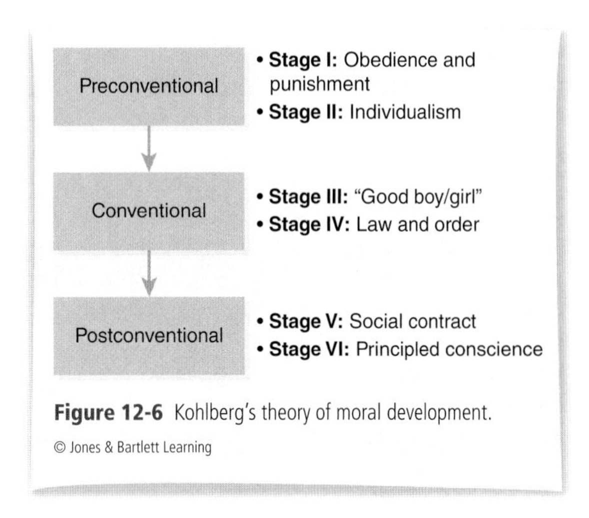

Figure 12-6 Kohlberg's theory of moral development.

also embrace the rationale associated with it. *It is important to note that, particularly in adults, decision makers may employ varying levels of moral discretion dependent on context, relevant information, or intensity of the desired outcome.* The preconventional level of development is best described as being rules-based. Rules are fixed and obeying them is important as a condition of avoiding punishment. Associated with rule-based reasoning is a sense of fairness, absent the more developed sense of social justice.

Preconventional Stage I

In the preconventional stage, the focus of moral reasoning is based on self-interest. A thing is considered ethically or morally wrong if punishment is likely. It follows then that the worse the punishment, the more "bad" the behavior is perceived. This form of moral reasoning is egocentric in that moral consequence is tied directly to self, and consideration of others' well-being is absent. Consider this example: A fire fighter may feel no moral compunction at stealing another fire fighter's leftover sandwich but would never consider stealing money from someone's locker. The actions are essentially the same, but the perceived consequences are different. Stealing a sandwich may result in criticism, whereas stealing money may result in formal charges being filed.

Preconventional Stage II

This stage of development is often described as self-interest driven. An action is perceived as good or bad based on a self-serving grading scale. A thing is good if it is good for me, whereas an action that is bad for me is inherently bad. Typically, this form of judgment is narrow and shortsighted. For instance, an action that is immediately gratifying may be perceived as good, while the long-term negative effect on reputation or personal relationships may be ignored. This level can introduce some realization of "others." However, that recognition is not rooted in empathy or social responsibility but rather in transactional usefulness. Others are recognized to have power and resources that can be obtained through either fair or unfair practices.

Conventional Moral Reasoning

Typically, individuals are first able to demonstrate conventional reasoning as adolescents. However, like preconventional reasoning, individuals may exhibit conventional moral reasoning well into and throughout adulthood. Conventional moral reasoning can be summarized as judging the morality of actions based on societal expectations. This level of reasoning requires acceptance of moral and ethical standards of community or social group. Motivation springs from a desire to be accepted as opposed to compliance to avoid punishment (level I development). Conventional moral reasoning necessitates that an individual follows society's expectations and rules even if no negative consequences are likely. The fire fighter referenced previously will not take another's leftover sandwich simply because it is a violation of cultural norms.

Conventional Stage III

Stage III development is characterized by a desire to be "good." Social acceptance motivates ethical behavior. At this level of development, individuals acquire an appreciation for respect, gratitude, and social standing. Behavior is still somewhat self-serving in that there is an expected transactional relationship between correct behavior and social acceptance. There is, however, appreciation for the necessity of being good for the sake of being good.

Conventional Stage IV

In this stage of moral development, individuals come to appreciate the need for social order and the value of ethical and moral behavior as an end unto itself. What separates stage IV from stage III is the appreciation that a thing may be right or wrong independent of other's opinions. In stage IV, the correctness of any given action is dependent on a perceived necessity to maintain social order. The decision maker may rationalize that rules exist for a reason, and they should be followed for the betterment of all. This stage of development encompasses the inclusion of a conceptual understanding of morality, but morality is still predominantly dictated by external forces. Kohlberg asserts that most adult individuals base their moral reasoning on this principle (Kohlberg, 1973).

Postconventional Moral Reasoning

Kohlberg labeled the postconventional level of moral reasoning "the principled level." It is within this level of moral development that individuals begin to assert their individual values and experiences in judging the correctness of behavior. The individual becomes a moral agent who considers the autonomy of others and the values of self-agency and justice. It is at this level that the individual may question society's laws and norms if they conflict with personal principles. Rules are not absolute dictates but rather are subject to evaluation.

Postconventional Stage V

The principles of social contracts become more important as the individual becomes aware of differing opinions, rights, and values. There is also recognition of the obligation to respect differing values. The validity of laws and social norms are judged within the context of social relationships. A rule that promotes the general welfare is deemed as "good," whereas an action that injures others is considered wrong. This stage of development is often associated with *consequentialism*, as correct actions seek to support social justice by benefiting the greatest number of people.

Postconventional Stage VI

Stage VI suggests that moral reasoning is based on an abstract understanding of moral principles. Actions are ethical only if they comply with personal values and a subjective sense of justice. There is an awareness of others' autonomy and an obligation to act in a way that respects individual rights. Stage VI development is consistent with both deontology and virtue ethics in that ethical actions are ends onto themselves and not a means to a worthy outcome.

Kohlberg's theory assumes that justice is the preeminent characteristic of moral decision making. It tends to elevate virtue and principle over intent and consequence. These assumptions are not universally acknowledged. Critics of Kohlberg's theory argue that justice, by itself, is limited in determining the ethical value of any particular action.

There is value, however, in understanding the precepts of Kohlberg's stages of development as they shed light on the elements of moral reasoning. This is particularly true if you remember that Kohlberg never intended stage VI to be considered "superior" to development stage I. Rather, the intent of the model is to articulate processes by which moral reasoning is achieved. A particular action can be judged as ethical or not regardless of the sophistication of the behavior's rationale. For instance, a child who restrains him- or herself from stealing candy out of fear of punishment is acting just as ethically as a child who abstains from theft because of an internal sense of injustice.

The consequences of both children's behavior are the same. However, the rationale for the latter's actions is more sophisticated. It can be inferred that an individual capable of postconventional moral reasoning is more likely to consistently behave ethically than a person only able to rationalize behavior at the preconventional level. The fire fighter who only obeys rules out of fear of punishment may well decide to act unethically if he or she is convinced that punishment is not likely. Conversely, the fire fighter who obeys rules out of a sense of propriety will likely feel that sense of propriety consisently and so act ethically consistently.

Loci of Control

The renowned psychologist Carl Jung is often credited with the development of a theory of behavior based on orientation to power. Jung's work was expanded by Julian Rotter in 1954 Figure 12-7. It was Rotter who coined the term "locus of control." Like Jung, Rotter believed that an individual's behavior was influenced by his or her perceived level of autonomous control.

Locus of control is defined as an articulation of the degree to which people believe they have control over the outcomes and events of their lives. Those with an **internal locus of control** tend to attribute consequences to choices. People with an **external locus of control** tend to attribute consequences to external forces such as luck, fate, or powerful individuals (Rotter, 1966). It follows then that persons with an internal locus of control tend to blame themselves for their failures, whereas individuals with an external locus of control tend to blame others Figure 12-8.

Figure 12-7 Julian Rotter, 1916–2014. Credited with the Locus of Control theory.

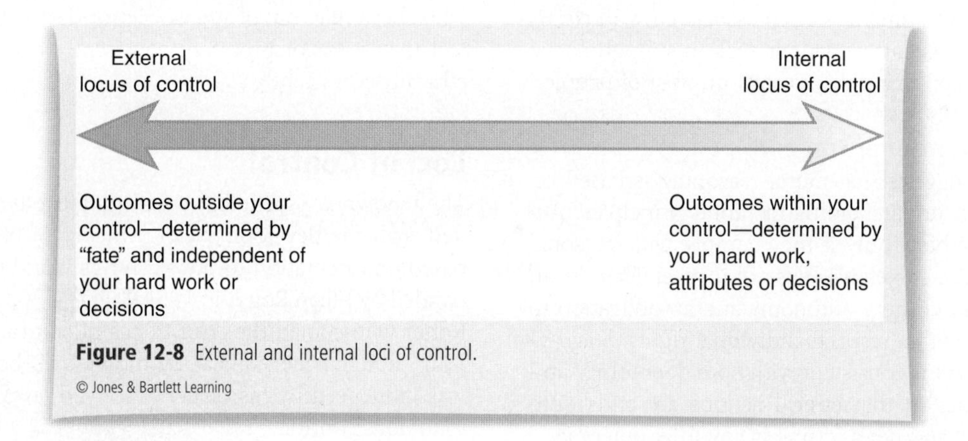

Figure 12-8 External and internal loci of control.

The concept of locus of control is firmly linked to ethics. It is believed that employees (fire fighters) with an internal locus of control tend to decide for themselves what constitutes appropriate behavior internally. They have a stronger belief in self-agency, and as a result, they tend to identify consequences with personal choices. On the other hand, people with an external locus of control tend to equate appropriate behavior with the expectations of external forces (Forte, 2005). You may note that this is consistent with Kohlberg's theory of moral development: internally focused individuals tend to reflect the tendencies of stage V and VI reasoning, whereas externally focused individuals tend to reflect stage III tendencies.

Because people with an internal locus of control see a stronger relationship between personal actions and consequences, they are more likely to assess the ethical impact of considered actions. The intimate relationship between choice and causality lends itself to a high sense of accountability and responsibility. The presence of a high sense of accountability lends itself to the referencing of personal values as well as those qualities identified within virtue ethics systems. High accountability tends to create an intimate relationship with the principles of both deontology and virtue ethics (Forte, 2005).

Conversely, those with a strong external locus of control tend to see consequences as a result of outside forces. This perception creates a tendency to evaluate actions based on circumstance (moral relativity) and necessity. Where there is a perception of limited choice, there is a diminished level of accountability. A diminished sense of personal responsibility causes a shift in orientation. Ethical decision making becomes a means to an end rather than actions being an end unto themselves. Choices then focus on the best outcome (consequentialism). The application of consequentialist ethics within the context of limited control or diminished free will increases the likelihood of **moral disengagement**. This is because the ethical decision-making process is reactive as opposed to proactive (Botti and Mcgill, 2011).

The theory of locus of control has significant relevance for the fire service. Generally, the fire fighter selection process tends to identify individuals with an elevated sense of responsibility. Additionally, the fire service's paramilitary culture, coupled with the strong emphasis on teamwork, tends to reinforce accountability. However, if a fire department has an impoverished leadership culture that demonstrates a lack of trust through micromanagement or offers few opportunities for self-actualization through professional growth, then the sense of personal accountability and responsibility can be significantly diminished.

Conversely, a department that seeks to "grow" employees through additional responsibilities, growth opportunities, and organizational voice tends to foster a higher sense of internal locus of control.

Self-actualized fire fighters (see Maslow in Chapter 3, *Influencing Behavior*) who feel that they have a high degree of autonomy tend to have a higher sense of responsibility. This creates a culture consistent with ethical behaviors.

Figure 12-9 Consequentialist thinking is often subjected to second-guessing. Those who judge our actions and those who are affected by our actions may not share the same view of what constitutes the greatest good.

Approaches to Ethical Decision Making

In facing ethical choices, there are two tasks that must be achieved. The decision maker must first define a framework for identifying the "best" ethical choices. Once that framework is identified, the decision maker must discern how that framework applies to the situation at hand.

In Chapter 4, *The Philosophy of Ethics*, we explored various theories of ethics. The purpose of ethical theories are to provide a framework for the decision-making process, and so decision making is ultimately shaped by the individual's philosophic approach to ethics.

Consequential Ethical Assessment

The premise of consequential ethics is that a "best" ethical path is the one resulting in the most "good." **Consequential ethical assessment** is familiar to fire fighters. Every incident commander has made fire ground decisions that seek to bring the best outcome in difficult circumstances. In the case of fire suppression, best outcome is defined by strategic objectives. The first priority is to save lives, the second priority is to stop the forward progress of the fire, and the third priority is to stop property loss. Within that framework, choices supporting those strategic objectives are considered the greater good. If an incident commander is presented with a problem where life safety and property loss cannot be simultaneously accomplished, the greater good is life safety (Markkula Center for Applied Ethics, 2015).

Unfortunately, most life situations do not have predefined strategic priorities. Identifying the greater good can be challenging. Our choices are driven by a complex matrix of values, anticipated outcomes, and emotional responses. As a result, the "greater good" is, in most cases, a subjective determination. It is here that the quandary lies; how exactly do you quantify the greater good?

The potential for subjectivism in consequential decision making can certainly be problematic. The value in consequential ethics decision making is that seeking worthy outcomes has universal appeal and is comfortable to our sensibilities. Seeking the greatest good or choosing what causes the least harm seems intuitively like a correct path. As such, consequential decision making is often the most easily justified rationale. However, those who judge our actions and those who are affected by our actions may not share the same view of what constitutes the greatest good. Consequentialist thinking is often subjected to second-guessing **Figure 12-9**.

Ethical Assessment Based on Fairness and Justice

Like consequentialism, fairness is an easily appreciated concept. A **justice-based ethical assessment** is defined as ethical choices made on the perceived basis of fairness or on a more advanced concept of social justice. There is a nearly universal acceptance that all should be treated equally. Even children in relatively early stages of development grasp the concept of fairness. In most cases, fairness is a reliable ethical yardstick. Fairness, however, is not universal. First, equality can be hard to measure. Modern Western philosophy embraces the idea that "all men are created equal." Unfortunately, not all people achieve equally and not all people have equal opportunity. The fact that individuals act differently and have different experiences, different gifts, and different challenges gives rise to the need for the concept of justice. Justice is a more complicated concept to assess ethically. In a larger sense, affirmative action meets a standard of justice that can argue for its fairness. However, for a candidate passed over for promotion, there is a clear and easily articulated perception of unfairness. It is understood that both fairness and justice are intrinsic to ethics. However, justice and fairness are not always in concert and, as such, can cause ethical dilemmas (Markkula Center for Applied Ethics, 2015).

Virtue-Based Ethical Assessment

Ethics based on virtue is perhaps the earliest of the philosophical systems. After nearly 2½ millennia, it remains viable today. Through the development of character, we can achieve wisdom, compassion, and empathy. These are powerful ethical tools that lend themselves to integrity, fairness, and honesty. **Virtue-based ethical assessment** is unique in that the preeminent question is not, What should I do, but rather, What kind of person will I become if I do this?

Ultimately, using virtue ethics in decision making lends itself to a high degree of subjectivity. However, its strong connection to values seems the greatest deterrent to moral disengagement and self-deception. The complexities of modern society have supported a resurgence of virtue ethics. When moral dilemmas seem to present problems that have no clear-cut right answers, reframing the question by focusing on character rather than solutions can be highly useful (Markkula Center for Applied Ethics, 2015).

Decision Making Based on the Common Good

The principles of social contracts, beneficence ethics, and care ethics can also be useful in ethical and moral decision making. Similar to consequentialism, the common good approach focuses on personal relationships. Those relationships can be with the community at large, within a social network, or one-on-one. Fundamentally it recognizes that social relationships are a value unto themselves. Just as people stranded in a lifeboat must care for each other in the pursuit of mutual survival, so too each person is responsible for the welfare of those around him or her. In this form of ethics, the greater good is somewhat less subjective in that it establishes the principle that the greater good is what serves the needs of the many.

Assessing Decisions Based on Deontology

Immanuel Kant articulated a categorical imperative or, put another way, a universal law of ethics. Simply stated, Kant held that humanity bestowed "self-agency," the ability to make free choices and shoulder the responsibility for those choices. As free agents, all people deserve respect, and it is unethical to act in any way that obstructs the natural rights of the individual. Deontology suggests that the ethical treatment of others is an obligation. Ethical decision making, then, is not based on consequence or intent but rather how well an individual's choices comply with his or her duty to respect the intrinsic humanity of others.

The **deontology-based ethical assessment** is defined as decision making that is based on compliance to Immanuel Kant's moral imperative. The deontology approach also has a significant element of practicality. It asserts that ethical truths are those that could be universally applied. In other words, if it is right for you to do, then it would be equally correct for another to do. Further, if an action is right in some circumstances, it must be right in all circumstances. Deontology leaves no room for ethical subjectivism. For instance, if lying is wrong, then it is universally wrong, and no particular circumstance can change that fact. Conversely, if lying is considered acceptable in a given situation, then lying cannot be described as "bad."

For fire fighters, deontology has several applicable advantages as a professional ethic. First, it offers specific guidance as to right and wrong. Second, its orientation in duty and obligation are consistent with fire service culture.

No particular approach to ethics has ever been deemed as the last word on the philosophy of behavior. The debate of ethical approach has been going on for centuries and will continue. Fortunately, in most cases, the theories are not mutually exclusive and tend to lead us to the same point. The consequentialist and the virtue ethicist are both likely to determine that a given action is wrong because it simultaneously does not comply with personal values, and it does not yield an appreciable good for anyone. In other words, actions that are inconsistent with personal values rarely lead to "good" outcomes. In a similar vein, actions that yield the best consequences are usually beneficial to the majority of individuals and are typically respectful of human rights.

The best approach to ethical decision making is to employ all the systems. In seeking the greatest common good, respecting rights of others, and adhering to our sense of moral virtue, we are likely to reach moral decisions via consensus.

Fire Service Decision Model

The systematic model for making ethical/moral decisions is shown in Figure 12-10. Similar to all decision models, it has limitations in that it does not cover all possible elements germane to any given issue. It is useful, however, as a template and demonstrates the thought processes required of a rational approach to solving ethical dilemmas. There is no shortage of decision-making models. Most are similar to each other in basic structure but with some unique facet. The model is an amalgamation of several such models, with some original contributions by the author.

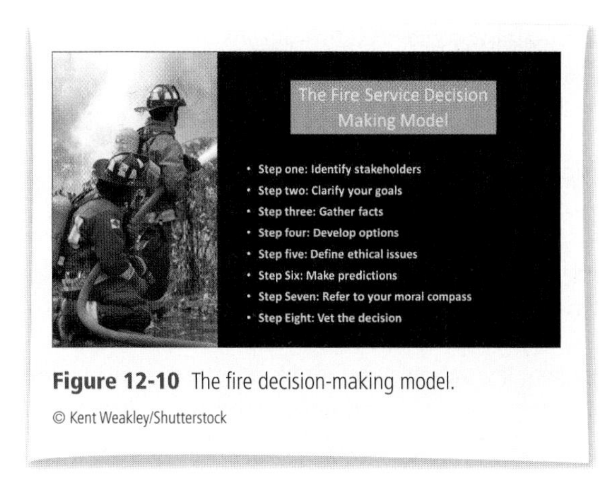

Figure 12-10 The fire decision-making model.

Step One: Identify Stakeholders

Each ethical decision is unique in circumstance and consequence. One thing all ethical decisions have in common is that your choices will affect one or more individuals. To understand the ramifications of an ethical choice, you must first understand who is affected by it.

If the decision being considered relates directly to your department, the obvious stakeholders are the public, the department as an entity, and your fellow fire fighters. Simultaneously, ethical decisions at work may have ramifications in your personal life. Due consideration should be given to the economic and emotional impact of the decision on family.

If a decision is strictly personal, stakeholders can be much more diverse in nature. Usually, family and extended social relationships are affected. Beyond that, stakeholders can involve a wide-ranging group of people affected by any particular action you may take. Due diligence must be given to identifying those individuals and their interests.

Step Two: Clarify Your Goals

Wise decisions are filtered to some extent through personal values. Additionally, decisions should be subject to identifiable goals, such as meeting obligations or attaining some need or want. As unlikely as it may seem, many people engage in decision making without a clear understanding of exactly what it is they are trying to accomplish or what is motivating the decision. Too often individuals are focused on short-term goals, even at the expense of long-term interests.

It is important to clarify what it is about a situation that is relevant, both factually and emotionally. Intentions and likely consequences are often intimately associated with responsibilities and obligations. Beyond responsibilities, our needs and wants are relevant. Once we understand what we should do and what we want to do, we can begin to assess how our personal values guide us toward a decision.

Step Three: Gather Facts

When gathering and analyzing facts related to ethical decisions, it is important to do your best to remain objective. Ethical issues are often emotionally charged, and even in the absence of emotion, objective fact finding can be difficult because we assess the relevance of facts based on our preconceived notions. It is sometimes difficult to understand our own thoughts and feelings. It can be much more difficult to understand the rationale, feelings, and priorities of others without clear understanding of their frame of reference. Clarify assumptions! Well-intended actions have often been found morally or ethically lacking because they were undertaken based on poorly understood situations.

Step Four: Develop Options

A clear understanding of the objectives and relevant facts lead to a thorough and unbiased exploration of choices. Discount no course of action until it has been thoroughly vetted. Too often, errant decisions are made before all available options are determined. A prudent decision maker determines if there is help to be had. Guidance may come from policy, a trusted mentor, or departmental precedent. The American Fire Service has been in existence for almost 300 years; there is little new under the sun. As ethics are related to the impact of decisions on others, it seems intuitive that input from others is highly useful.

Step Five: Define Ethical Issues

For each potential decision path, it is important to determine the ethical basis by which a particular course of action will be judged. Ask yourself, "Is the action being considered consistent with my values?" "Is it consistent with policy and procedure?" "Is the decision fair?" "Are stakeholders being treated equally?" "Is there a 'greater good' to be served?"

Step Six: Make Predictions

Once the decision maker becomes relatively comfortable that he or she has all the pertinent details surrounding an issue, the next step is to make a prediction. Consider all options and their likely consequences. How will stakeholders be affected, how will they react, what questions will be asked? If the decision is critiqued, will a particular option be justifiable? In addition to stakeholders, how will noninvolved and presumably dispassionate parties view an action? How will your superior officers likely react? How will local politicians and the public likely view your decision? A common defense to investigative questioning is, "You had to be there." How will a decision look to someone who was not there? People are increasingly cynical towards government, and although firefighters do enjoy public support, it can evaporate quickly when evidence of bad behavior surfaces.

Your actions may or may not be presented within context. Fire fighters have found themselves in trouble because they confused context with emotional investment. Context is a factual framework and may be a potential defense of decisions. Conversely, choices springing from emotional responses are rarely justified by a dispassionate observer, such as a board of inquiry. For instance, a board of inquiry may question a fire chief for using discretionary funds for the purchase of a generator while simultaneously bypassing the normal purchasing procedure. A contextual response may be that a power failure in the middle of the night precluded the department from being able to respond to calls. As such, the purchase was made as an emergency stop-gap procedure and thus is perhaps defensible. An emotional justification might be that the chief found the bidding process burdensome and unnecessary, and, as a result, decided the best interest of the public is served by ignoring it. This second line of defense would be discounted by a board of inquiry.

Prediction of consequences in stakeholder reactions can be difficult. It can become easier, however, with the diligent pursuit of relevant facts and dispassionate evaluation.

Step Seven: Refer to Your Moral Compass

Pay close attention to how you feel about the available ethical choices. Some people call it a moral compass, while some people call it intuition or conscience. Regardless of name, emotional response to choices can often reflect a subconscious awareness of what the rational mind has ignored. Emotions can be linked to values and self-image. While we may rationally evaluate a choice based on perceived fact and predicted consequence, feelings often indicate support or conflict with our internal moral compass. Your gut may be a good indicator of impending moral dissonance.

Consider what values are reinforced or are subverted by a decision. Is a particular course of action consistent with the person that you believe you are or wish to be? Can you live with yourself? The blush test is a simplistic yet powerful ethics tool. Would you want people you respect to know what you did or are considering doing? Regardless of your motivations or of circumstances, if you would feel compelled to hide your actions, you are likely pursuing a dangerous path. How you feel about a choice is at least as important as what you think about the choice.

Step Eight: Vet the Decision

Ask yourself the following questions:

- Is your choice consistent with your values?
- Is the choice consistent with the fire service's mission and values?
- Would people you respect do the same thing in a similar situation?
- Would I be ashamed if other people knew what I was about to do?
- Are there laws, rules, regulations, or department policies that restrict my choices? If so, am I willing to live with the consequences of breaking them?
- If a subordinate or peer acted in the same way, would I be supportive of that action?
- Are my motivations self-serving, or are they in the interest of the public, the department and fire fighters?
- Most important, can I rationally defend my decision? Internally summarize the justifications for a chosen action. If articulating that defense is difficult, there is a likelihood the action is questionable.

WRAP-UP

Chapter Summary

- This chapter focused on ethical decision making, exploring the ethical elements involved in making decisions.
- You were introduced to two cornerstones in moral decision making: people have the power to decide what they do and say, and people are ultimately responsible for the consequences of their choices.
- Ethical decision making requires certain abilities in most day-to-day decisions—to recognize decision points that have ethical consequence and reliance on common sense and intuition.
- Certain situations encompass more complex ethical issues and equally significant consequences. These situations require careful study and consideration for action.
- There are times when even the best of intentions cannot properly guide ethical action. These situations require an understanding of decision analysis and the underlying ethical principles associated with decision making.
- A fundamental skill in ethical decision making is appreciating the consequences of decisions. The most common understanding of consequence is that it is an anticipated result from an action.
- You must also consider the ramifications of consequences. An appreciation of who may be affected by your choices and how they may be affected is necessary for understanding ramifications.
- Developing the emotional and ethical intelligence to weigh options and having the discipline to pursue the best ethical choices is fundamental to decision making.
- The ability to project potential consequences onto affected stakeholders is critical in the process of making ethical decisions.
- Kohlberg's theory of moral development described three levels of moral development, encompassing six separate stages of moral and ethical orientation. Kohlberg's theory was then examined as it affects ethics in the decision-making process.
- Julian Rotter's theory of locus of control and its relevance to the ethical decision-making process was explored. Kohlberg and Rotter both have provided a framework for understanding how a person's view of the world affects his or her ability to make moral and ethical decisions.
- Ethical decision-making processes were analyzed by considering the relationship with the major philosophical approaches to ethics.
- Deontology, virtue ethics, consequentialism, justice theory, and social contract theory were all reviewed as frameworks for moral and ethical decision making.
- The eight-step fire service decision-making model was introduced. The model is an amalgamation of several popular decision-making models that were condensed and modified to better fit fire service needs.

Key Terms

consequence analysis the prediction of likely outcomes and their ramifications.

consequential ethical assessment ethical choices are made based on an anticipated result of doing the most good.

conventional moral reasoning Kohlberg's second level of moral development.

deontology-based ethical assessment decision making based on compliance to Immanuel Kant's categorical imperatives.

emotional intelligence an individual's ability to be aware of how others feel.

empathy the action of understanding, being aware of, being sensitive to, and vicariously experiencing the feelings, thoughts, and experiences of others.

ethical dilemma a complex situation where values or moral imperatives conflict with each other.

ethical intelligence an awareness of moral principles and their contextual application.

external locus of control attributing consequences to external forces such as luck, fate, or powerful individuals.

internal locus of control attributing consequences to choices.

justice-based ethical assessment ethical choices made on the perceived basis of fairness or a more advanced concept of social justice.

Kohlberg's stages of moral development a scale depicting three stages of moral reasoning in an attempt to illustrate how people assess and justify their actions: preconventional moral reasoning, conventional moral reasoning, and postconventional moral reasoning.

locus of control an articulation of the degree to which people believe they have control over the outcome and events of their lives.

moral disengagement the process of convincing the self that ethical standards do not apply to yourself in a particular context.

moral dissonance an internal conflict between an individual's actions and his or her personal values.

postconventional moral reasoning Kohlberg's third level of moral development.

preconventional moral reasoning Kohlberg's first level of moral development.

virtue-based ethical assessment basing ethical choices on perceived consistency with character.

Challenging Questions

To check your understanding of this chapter's material, answer the following questions. It is highly recommended that you discuss your viewpoints with fellow students, peers, coworkers, and friends to discover their opinions as well.

- The text reviews several methodologies for ethical decision making. If you were faced with a choice where the greater good is served by doing something inconsistent with your values, what would be your likely path?
- The text explains that duties and obligations are part of the decision-making process for fire fighters. What do you believe is a fire fighter's highest obligation?
- Identify three people that you respect. To the best of your knowledge, do they appear to have an internal or external locus of control?
- Here is a classic ethical dilemma: You are given two options. Option one: You hit a red button and a half the world dies. Option two: You do nothing and the whole world dies in five years. Using the principles within this chapter and the ethical decision-making process discussed previously, explain the rationale for your choice.

WRAP-UP

Case Study Conclusion

Revisit the case study at the beginning of the chapter. Spend a few minutes considering the questions posed at the end of the case study. In light of the information shared in this chapter, have any of your original observations changed?

At first glance it seems apparent that Assistant Chief Jones has an ethical obligation to report his suspicions to the department's HR division. Certainly, his inaction is a dereliction of duty if he supports Smith's efforts or if he simply wants to avoid confrontation. The question is, Are there other factors involved that may mitigate Jones's obligations? If, for instance, Jones believes that his suspicions by themselves may not justify potentially injuring Smith's reputation or even jeopardizing his career, then he may have ethical justification in remaining silent. If he knows Smith to be an otherwise honest individual, he may have greater justification to show restraint.

If the rationale above is Jones's justification, then Jones must also decide whether or not his reporting of his suspicions actually causes harm to Smith. By meeting his obligation to report, does Jones cause harm or simply put in motion an impartial investigation? If that investigation causes Smith harm, does Jones have an obligation to prevent it?

In this case, Assistant Chief Jones's obligation to report likely outweighs any potential harm he may cause by making the report, even if Smith is proven innocent. Further, if harm comes to Smith because of facts being placed into evidence, it is not clear that Jones is responsible for that harm. Jones's concern for harm is based on an assumption of wrongdoing. He can reasonably assume that if Smith did nothing wrong, he likely would suffer no adverse consequences by his coming forward. However, if Smith suffers negative consequences as a result of wrongdoing, Jones is not culpable and has no obligation to defend him.

Chapter Review Questions

1. What are Kohlberg's three levels of moral development?
2. Define a consequence analysis.
3. What is meant by emotional intelligence?
4. Define moral dissonance.
5. Define ethical intelligence.
6. What are the elements of a moral dilemma?
7. Describe an individual with an internal locus of control.
8. What renowned psychologist developed the concept of locus of control?
9. How do the concepts of justice and fairness contrast?
10. What is the first step in the decision-making model explained in this text?

References

Botti, S., and A. L. Mcgill. 2011. "The Locus of Choice: Personal Causality and Satisfaction with Hedonic and Utilitarian Decisions." *Journal of Consumer Research* 37, no. 6.

"Empathy." 2015. *Merriam-Webster's Dictionary.* Springfield, MA.

Forte, A. J. 2005. "Locus of Control and Moral Reasoning of Managers." *Journal of Business Ethics* 58, no. 1, 65–77.

Josephson, M. n.d. "Groundwork for Making an Effective Decision." Accessed June 22, 2018. http://josephsoninstitute.org/med-3groundwork.

Kohlberg L. 1973. "The Claim to Moral Adequacy of a Highest Stage of Moral Judgment." *Journal of Philosophy* 70, no. 18, 630–646.

Kohlberg, L. 1981. *Essays on Moral Development, Vol. I: The Philosophy of Moral Development.* San Francisco, CA: Harper & Row.

Mainser, M. M. 2007. *The Facts on File Dictionary of Proverbs.* New York: Infobase.

Markkula Center for Applied Ethics. 2015. "A Framework for Ethical Decision Making." Accessed June 22, 2018. https://www.scu.edu/ethics/ethics-resources/ethical-decision-making/a-framework-for-ethical-decision-making.

Rotter, J. 1966. "Generalized Expectancies for Internal Versus External Control of Reinforcement." *Psychological Monographs* 80, no. 1, 1–28.

Weinstein, B. 2011. *Emotional Intelligence: Five Principles for Understanding Your Toughest Problems at Work and Beyond.* Novato, CA: New World Library.

CHAPTER 13

Breaking Bad

"The road to hell is paved with good intentions."
—St. Bernard Clairvaux

OBJECTIVES

After studying this chapter, you should be able to:

- Explain the psychology of unethical behavior.
- Assess moral apathy, including contributing factors of environmental conditions, expectations, and perspective.
- Relate how the process of moral disengagement interacts with changes in values.
- Explain the cognitive disengagement process.
- Interpret moral rationalization.
- Synthesize the foundations of moral engagement.
- Appraise how enabling behavior can lead to unethical activity.

Case Study

The Anytown Fire Department responds to a Saturday night structure fire at a local bar and grill. Unfortunately, the fire destroys the building and most of its contents. Fire fighters Smith and Jones are overhauling the fire and are instructed to "clear out" the area of the storeroom to facilitate the cause and origin investigation. At the instruction of their battalion chief, they are throwing the refuse into a nearby dumpster.

While working, Smith and Jones uncover a case of vodka and a partial case of 20-year-old scotch in which the bottles are still intact. They also find an old neon beer sign that appears to be in good shape. Smith and Jones rationalize that the "debris" is going in a dumpster as spoiled consumables so it is okay for them to keep the sign and the booze. Just to be on the "safe side," they wait until no one is looking and then put the items in the crew compartment of their rig.

Lieutenant Franklin notices the cases of alcohol in the back of the apparatus and asks why they are there. Smith and Jones explained that they were saving the alcohol from the dumpster and that they were going to use the rescued liquor for a shift party. Franklin is not completely comfortable with the rationale but says, "What the hell?"

Unfortunately for Smith and Jones, a bystander takes video of them loading the alcohol into the rig and posts it on his Facebook page under the comment "Fire fighters pilfer bar." The story gets picked up by the local media, and the department's chief is greeted Monday morning by a news media request for an interview and by an angry bar owner accusing the fire department of being crooks.

- Are the actions of Smith and Jones ethical?
- Does the permission of Lieutenant Franklin validate the fire fighters' actions?
- In considering what to do about Smith, Jones, and Franklin, what elements should the fire chief consider?

Introduction

In the southeastern part of the United States, two medics are caught taking "selfies" with unconscious patients. In the Midwest, a chief is accused of illegally disposing asbestos from a fire station. In the southwest, a fire fighter stands accused of selling pictures of deceased victims to collectors. In the east, a fire chief is indicted for embezzlement.

Sadly, stories like these are all too common. Fire fighters across the country read the stories and shake their head in disgust. They ask themselves, What were they thinking? or they write the perpetrators off as bad apples. Most fire fighters are confident that such things could never happen in their department, and they certainly would never be personally involved in any actions like these. Still, the stories keep coming week after week. The reason: Most people (including fire fighters) are less ethical than they would like to believe.

One common factor in all the stories mentioned previously is that the fire fighters involved likely feel that they are moral people caught in a bad situation. In truth, nearly all individuals involved in unethical behavior think that they are relatively good people and that there were special circumstances that led to their behavior. Moreover, it is quite likely that friends and coworkers would agree that wrongdoers are "upstanding citizens" and are shocked by their behavior.

Certainly, there are bad people in the world. There are sociopaths who know that they are doing evil things and do not only feel a lack of guilt but actually enjoy doing bad things. There are the mentally ill who simply do not understand the consequences of their actions. There are also those emotionally damaged folks who, because of their upbringing, have no appreciation of moral and ethical responsibility. However, these folks are a minority. The vast majority of unethical behavior is conducted by people who consider themselves "good" and who have otherwise healthy relationships with others.

So the question is, How do good people do bad things? This chapter explores how individuals who have a basic understanding of right and wrong, and who have an intact conscience, end up making decisions that lead to unethical behavior.

When confronted with their behavior, those engaged in unethical actions are likely to plead either that they did not realize that their actions were inappropriate (ethical ignorance, see Chapter 12, *Ethical Decision Making*) or that their actions were somehow justified. This chapter will mainly focus on the latter group.

This chapter will re-introduce you to the concepts of cognitive and moral dissonance. Also discussed are the processes by which normal people manage to undermine their conscience and rationalize behavior. It is important to remember that nothing in this chapter is intended to excuse unethical actions. Rather, we provide examples and information in an effort to explain the process by which people engage in inappropriate conduct.

The Psychology of Unethical Behavior

Mental health and emotional stability are relative terms and are difficult to quantify, but there is a general understanding of what is meant by the terms. If an individual is free of mental defect and is emotionally stable, we can assume that he or she should know right from wrong. Let us refer to mentally healthy, well-adjusted people who possess a sense of right and wrong as "normal."

Normal people have a conscience and a set of internal values that guide them through their day. Theoretically, such people can be expected to act consistently with their values—they do not act out of character. Typically, normal people do not suddenly abandon rationality and morality. Rather, getting to a point where bad behavior seems appropriate is a gradual process. Self-image and internal values restrain people from acting too far out of character. When people behave badly they must neutralize those restraints.

Dissonance

We all must live inside our own head. It is fundamental to human nature that we act in accordance with what we believe to be true and right. If we act in a way that is inconsistent with our understanding of the truth, we experience **cognitive dissonance** (McLeod, 2015). Cognitive dissonance is an internal stress that presents itself when we act irrationally. Similarly, if we act in a way that is inconsistent with our personal values, or our understanding of right and wrong, we experience **moral dissonance** (Kashtan, 2013). Moral dissonance is an internal stress that presents itself when we act in a way that is inconsistent with our self-image. Moral dissonance is closely associated with having a crisis of conscience.

Logically, individuals who act either foolishly or unethically must either be unaware of the inappropriateness of the behavior or must go through some sort of mental and emotional process to justify the conduct to maintain self-image. The process is complex, but the intent is straightforward. The bad actor must align his or her actions with a sense of reality to avoid cognitive dissonance, as well as with his or her moral principles to avoid moral dissonance. As a result, we can assume that poor behavior is either accompanied by a skewed understanding of circumstance or by a rationalization of behavior. The act of avoiding moral dissonance by reshaping attitudes, or rationalizing behavior so it remains consistent with values, is known as **moral disengagement** (Spiegel, 2013).

Attribution Theory

We make sense of the world by assigning causes to events. To some degree cognitive dissonance occurs as a result of an unexplained event, and we seek explanations. Likewise, moral dissonance can arise when human behavior is not understood. It is particularly disturbing if an individual cannot find relevant meaning in the behavior. It is human nature to assign causes to actions, especially when behavior is considered inappropriate (Malle, 2011).

In most people there is a duality in the interpretation of ethical conduct. When others act inappropriately, the behavior is attributed to flaws in their character (internal attribution). When we behave unethically ourselves, we tend to attribute the behavior to external influences (external attribution). Psychologists explain this behavior as **attribution theory** (Malle, 2011).

A generous explanation of the double standard is that we tend to internally attribute behavior in others because we are less aware of precipitating circumstances. As a result, their actions are judged in terms of black and white. Even if extenuating circumstances are recognized they are more readily discarded. We are less likely to feel empathy for others' circumstances in motivating behavior than we are to feel sympathetic for circumstances driving our own behavior.

In judging self-behavior, we deal more readily with shades of gray. We tend to attribute bad behavior to external influences because we have a deeper appreciation of them. For example, if a friend is involved in an illicit relationship, we are much more likely to judge the behavior based strictly on its moral implications. The individual is breaking wedding vows, lying, and being unfaithful. An individual involved in an illicit relationship is much more likely to judge his or her behavior relative to what he or she feels are mitigating circumstances. The difference in judgment is based on the fact that the unfaithful individual feels the emotional pull of the situation in judging him- or herself, and those feelings are absent in judging others. Self-judgment is more informed.

Another explanation for the double standard in behavior judgment is that we are more tolerant of our own behavior because it is simply more comfortable. It is difficult for human beings to maintain self-image when judging inappropriate behavior absent some form of accommodation. In judging other people's behavior, we have no compelling need to excuse the behavior, and so we can place blame readily. However, it is nearly impossible for an individual to self-describe as a "bad person" and maintain a positive self-image. As a result, we create coping strategies for excusing bad behavior, thus maintaining a self-image of being a good person.

Moral Apathy

In discussing normative ethics, there is always an assumption that actions are fully considered. The assumption is that the moral compass is referenced but ignored (moral disengagement). Alternatively, actions are evaluated according to our internal sense of right and wrong and, found lacking, are pursued anyway. However, this is not always the case; some actions are not ill-considered. Rather they are not regarded at all. For some actions there is a general lack of moral engagement (not thinking about ethical consequences) or **moral apathy** (not caring). A lack of moral engagement can be a form of contextual confusion or willful ignorance (cognitive dissonance).

Some social scientists have observed that certain circumstances can debilitate an individual's propensity for evaluating consequences. Some circumstances change a person's attitudes toward ethics and pave the way for a descent down an ethical slippery slope Figure 13-1 (Kaptein, 2012). Unlike the process of moral disengagement in which known bad behavior is rationalized, these conditions tend to reshape attitudes toward behavior. Unethical actions are normalized, or an environment is created in which the ethics of the situation become irrelevant. An example is highlighted within this chapter's case study. Often, a fire fighter's attitude toward behavior reflects the culture of the fire station.

Environmental Conditions

Fire fighters who are surrounded by unethical people tend to downgrade their values or lower their expectations of ethical behavior. Reshaping ethical standards may be an adaptive behavior to maintain social acceptance. Or another's sense of values can be modified by the surrounding environment. What we normally see tends to become normal. As

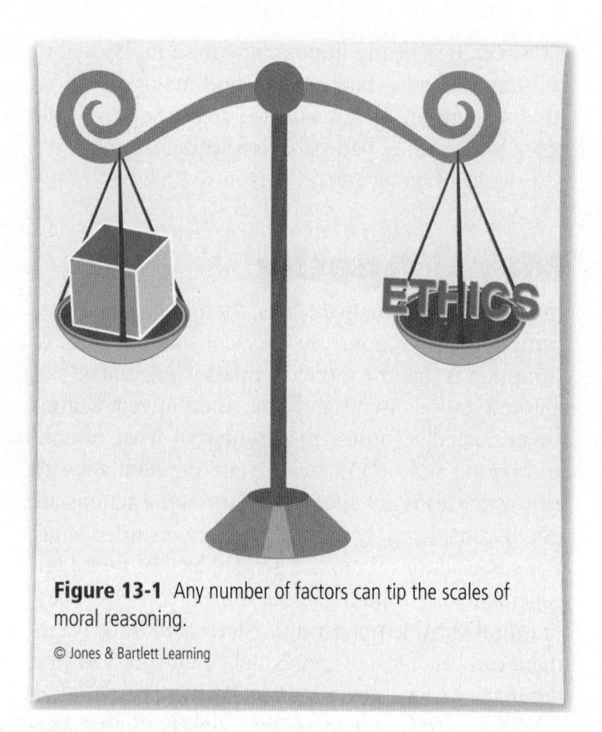

Figure 13-1 Any number of factors can tip the scales of moral reasoning.

behavior becomes normalized, personal values and attitudes can be reshaped. Our values are initially learned by observation. This process can continue into adulthood.

Also at work is an individual's wish to maintain a positive self-image through acceptance by peers. While the absence of ethical standards might encourage unethical behavior as a form of conformity, the presence of a strong ethical mentor who continuously displays moral behavior seems to have a more prolific effect on raising the behavior standards. An ethically challenged environment may passively support bad behavior. However, a work culture that is highly ethical strongly discourages inappropriate actions. (see Chapter 9, *Building an Ethical Culture*)

The impact of effective leadership on ethics behavior is significant. There is evidence that people (fire fighters) who work for unethical leaders are much more likely to behave unethically than those who are led by positive role models (Kaptein, 2012). This influence is even more profound than that of unethical peers.

Even if a fire officer is not respected or liked, he or she still represents legitimate authority within the firehouse. As a result, the fire officer's attitudes toward ethics does much to shape attitudes toward acceptable versus unacceptable behavior. The example set by officers can have effects beyond individual actions Figure 13-2. If, for instance, a fire officer falsely reports training, then fire fighters both receive the message that misleading the training division is acceptable, and that training is relatively unimportant. This general attitude can be extrapolated to other forms of reporting as well. An officer's ethics are instrumental in the modeling of an ethics culture within the fire station.

Related to leadership is a fire fighter's self-image relative to the department. Fire fighters who see themselves as valued members of their team are more likely to behave appropriately than those who feel unappreciated or unimportant. Loyalty is a two-way street. Fire fighters who believe that the department is loyal to them are more likely to behave in a way that is consistent with demonstrating loyalty to the department. That behavior is usually in the form of compliance with behavior standards.

Expectations

Workers, including fire fighters, tend to act according to expectations. Individuals who are treated as

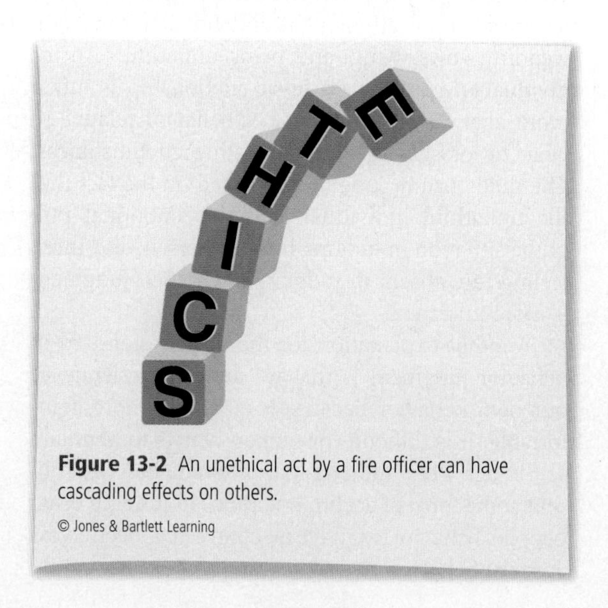

Figure 13-2 An unethical act by a fire officer can have cascading effects on others.

being dishonest or untrustworthy are more likely to be dishonest and untrustworthy. Overregulated departments tend to foster an environment in which policy enforcement and punishment exclusively control behavior. Consistent with this environment is an attitude where "that which is not caught" is considered inherently okay.

The inverse is also true. In poorly regulated environments, when small transgressions are overlooked, the questionable behavior is seen as tacitly approved. It is human nature to test limits. If breaking rule A is ignored, then rule B will likely be subsequently tested. This testing is repeated until a minimum standard of behavior is established. Contained within that standard are certain inappropriate behaviors that have been tolerated and thus normalized. As any experienced officer can attest, it's hard to "walk back" behavior accommodations once they are given. Experience has also shown that small infractions lead to bigger ones.

Perspective

There are also internal processes that can affect moral engagement. One is particularly relevant to high-ranking officers. Officers with significant authority are usually also granted a great deal of institutional freedom. They answer to fewer people, have more privileges, and are more self-directed. Some chief officers may come to believe that they are inherently different from other fire fighters and so are bound to different rules. There exists a long history of fire chiefs engaging in behavior that they would not tolerate from their fire fighters.

Examples include coming in late to work, using abrasive language, and using department resources and funds inappropriately. There have been numerous cases where questionable excesses have escalated to the point where fire chiefs, all the while, have engaged in behavior that ended their careers, assuming that they were within their rights **Figure 13-3**.

The problem may start with something as minor as including cocktails on an expense report and escalate into a criminal activity such as buying a spouse's airfare with department funds. Remember the adage, "Small indiscretions lead to big sins." Another appropriate adage may be, "Pride goeth before the fall."

Figure 13-3 People do not often turn bad overnight. They usually ease into it a decision at a time.

The phrase "tunnel vision" is a common one in the fire service. It refers to hyperawareness of a particular situation to the exclusion of the bigger picture. On the fire ground, this can lead to tragic errors with disastrous effects. In ethics, tunnel vision can have the same effect. Excessive task orientation can cause an extreme focus on outcome, which can result in ignoring the bigger ethical picture. When work cultures are highly focused on getting the job done, it becomes relatively easy to start believing that the ends justify the means. The same is true for individuals. Highly motivated, results-oriented achievers may start to believe that situations may justify, or even require, "bending the rules."

If you have heard a coworker say, "It is better to ask forgiveness than permission," then a possible impending ethics breach is in the making. When fire fighters convince themselves that playing by the rules impedes success, they are demonstrating a potentially dangerous attitude toward procedural restraint. That sort of justification can be the first step down the ethical slippery slope ending in recrimination.

Context can also affect perspective. A fire fighter who would never steal someone else's property may feel it is okay to liberate a case of alcohol from a fire scene (see the case study at the beginning of this chapter). A business owner who would never lie in his personal life may well find it perfectly acceptable to misrepresent a product to a consumer. It is not that the transgressor suppressed personal values or in some way rationalized the behavior. Rather, it's that some behaviors have become so normalized that those partaking in such behaviors have never even considered the ethical implications of the behavior (Tenbrunsel and Messick, 2004). Consider the issue of steroid use in baseball. The 2006 investigation into illicit drug use revealed some surprising results. As expected, numerous players were found to be illegally using steroids, but many outside the game were shocked by the ubiquitous use of amphetamines. Their usage was not only found in every clubhouse, it was done in the open with the full support of team management. Their use was so common that players and managers did not even think to question the ethics of the unsanctioned use of prescription drugs (Pessah, 2015).

This process described previously is known as **framing** ethical decisions (Tversky and Kahneman, 1981). In business, and in some social environments, we cognitively recognize one set of behaviors as acceptable that normally may not be consistent with individual values in other facets of our lives Figure 13-4. Priorities become somewhat different depending on context. For example, a focus on achievement in a business setting may encourage the exclusive consideration of efficiency and performance, thus leading to framing typically inappropriate behaviors for the sake of closing the sale.

In ethics, there is a different set of cognitive processes. The awareness of moral consequence and intention are heightened, and so the consequence is judged with regard to outcome *and* process. When individuals are morally engaged, positive outcomes are sought within a framework of ethically acceptable choices.

Framing ethical decisions separates outcomes from processes. This can create a sense of duality: personal self (behavior focused) and the work self (outcome focused) become separated (Spiegel, 2015; Tenbrunsel and Messick. 2004).

Last, there is an issue of ethical "laziness." You may recall that in Chapter 4, *The Philosophy of Ethics*,

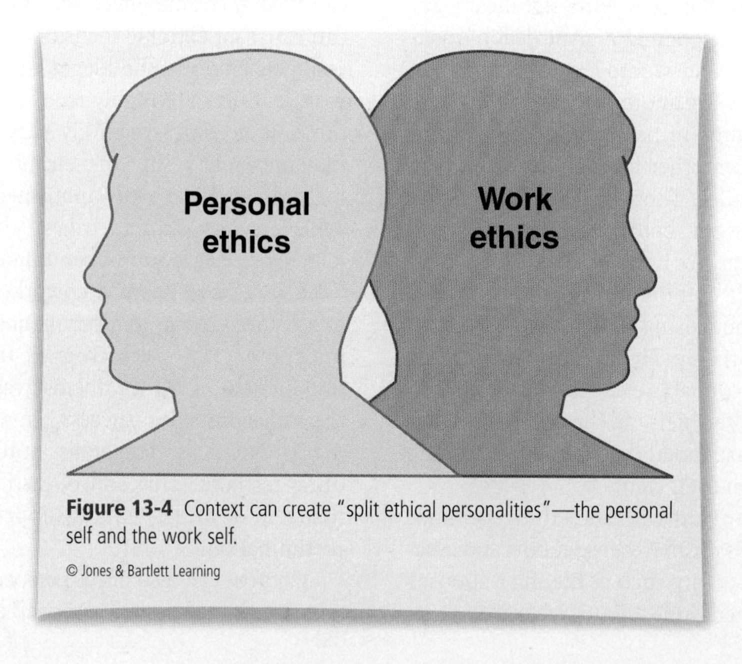

Figure 13-4 Context can create "split ethical personalities"—the personal self and the work self.

Aristotle's version of virtue ethics was introduced. A key concept in that ethical system is Aristotle's belief that being virtuous was the result of habituation. He felt that attention to acting ethically and thinking ethically eventually made virtues become second nature. For many, the habit of ethical and moral thinking has never been ingrained. Poor behavior is undertaken absent consideration of the outcome, intent, or consistency with personal values. Just as virtuous behavior can become a habit, so too can ignoring moral instincts (MacDonald, 2015). Fire fighters may act without consideration of consequences due to distraction or simply out of habit. Unethical behavior can be self-reinforcing: Poor ethics often leads to future unethical conduct.

Moral Disengagement

To this point, this chapter has explored how "good people" commit bad actions by failing to recognize ethical responsibility, or via contextual influences that alter moral attitudes.

It seems likely that most people who engage in bad behavior ostensibly realize what they are doing. Assuming that people are aware of the consequences of their actions, and assuming that they wish to maintain self-image through the avoidance of cognitive and moral dissonance, certain psychological processes must be employed. These processes are collectively known as moral disengagement, which can be defined as convincing yourself that ethical standards do not apply in a particular context (Bandura, 1999).

The Process of Moral Disengagement

To properly understand moral disengagement, it is necessary to think of it as a process. As normal individuals, we are constrained by our sense of right and wrong, our personal values, and our understanding of truth. When an individual rises on a given morning, there is likely no conscious decision to toss out moral restraint. Rather, the reshaping of ethical boundaries is an ongoing and gradual procedure. Getting around our conscience requires self-deception and rationalization, and the process becomes easier with each escalating unethical act (Tsang, 2002).

The act of moral disengagement may include one of several processes, but in most cases, it begins with a singular evaluation. Whether consciously or not, there is a judgment of the relative importance of perceived need or desired outcome against ethical values. If an action is considered inconsistent with personal values, it must be abandoned, or moral dissonance occurs. If the motivation to pursue an action is strong enough that conscience is overwritten, then the evaluation process must be altered if moral dissonance is to be avoided. To overcome that dissonance, one or more mental processes are employed supporting a course of action with questionable ethical implications. By focusing intently enough on the desired outcome, a natural tendency to rationalize its achievement is engaged. Notions that the ends justify the means or that the inappropriate behaviors constitute some form of justice are introduced into the thought process. Again, for the questionable action to occur, moral restraint must be overcome. Those restraints are undermined by one or more several processes that follow.

Cognitive Disengagement

Cognitive disengagement is a suspension of rational understanding through an exercise in self-deception. It can be an important first step in the process of moral disengagement in that it can create a false narrative in support of an otherwise unethical action Figure 13-5 (Moskowitz and Grant, 2009). An individual who self-identifies as a devout Christian should rationally understand the moral teachings of the faith. The person has likely heard them at church and in numerous biblical readings. The same person may identify him- or herself as a patriot and, as such, likely embraces the freedoms guaranteed by the Constitution. Cognitively, the person understands that freedom of religion is a fundamental right in this country.

For self-identified Christian patriots to firebomb a synagogue or mosque, they must not only rationalize a behavior that is inconsistent with their moral compass but also suspend their cognitive

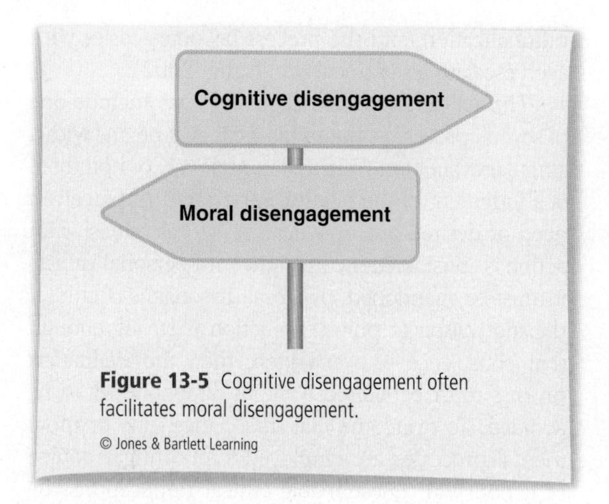

Figure 13-5 Cognitive disengagement often facilitates moral disengagement.

understanding of the precepts of their religion and the U.S. Constitution. The emotional urge to indulge in violence directed at other faiths simply does not sit squarely with the U.S. Constitution or the precepts of Christianity. The resulting incompatibility between hatred of other religions and understanding of legal and religious doctrine is an example of *cognitive dissonance.*

To overcome cognitive dissonance, individuals may employ several tactics that allow the creation of false logic or *cognitive disengagement* (Moskowitz and Grant, 2009). These may include:

- **Ignoring facts that are inconvenient to an opinion.** Individuals may ignore facts which are inconsistent with opinions while then simultaneously focusing on facts that support the desired opinion and resulting actions.
- **Reshaping facts.** For instance, a fire fighter may justify excluding minority hires from the department by quoting "scientific studies" on crime rates or employment aptitude out of context. A similar tactic is reshaping facts by inference. For example, someone may deny global warming based on the observation that there's been an unusually cold winter.
- **Questioning the source of an established fact.** Many people find it easy to dismiss information simply because it does not come from a source they support. In today's à la carte news environment, individuals can self-select news sources based on their predisposition of belief. Thus, belief is reinforced as fact. For example, the xenophobe who only visits ultranationalist websites has his or her beliefs constantly echoed until they become apocryphal "truth."
- **Cognitively justifying an action by simply drawing conclusions that are not supported by fact.** As an example, for over a century anti-Semites have justified violence against Jews based on the persistent yet thoroughly debunked myth of *The Protocols of the Meetings of the Learned Elders of Zion*. The inconvenient evidence of the myth's falsehoods are ignored by those seeking justification for hatred.

Creating False Moral Justifications

A second process by which moral disengagement can occur is by creation of false moral justifications. For clarification consider two examples. A fire fighter falsely turns in a request for overtime for work he or she did not do. The fire fighter may argue that he or she is entitled to the hours because, over the past year, he or she has worked several hours that were uncompensated. Even if true, the rationale does not support falsifying a document. Also assume that a fire chief decides that deliberately underestimating the cost of a new fire apparatus will likely ensure its approval in the budget, knowing that later the city council will approve the "unanticipated cost overruns." He or she may rationalize that the apparatus is needed and misinforming the city council is the only path that will lead to its acquisition. He or she convinces him- or herself that "the ends justify the means." The tactic is a false justification based on several principles. The chief is paid by the city to advise and represent the municipal administration. In deceiving them, he or she is failing to act as a faithful agent to the city council. He or she is also assuming powers not bestowed on the position. By undermining the council's ability to make an informed decision, the chief is subverting its ethical responsibilities.

In both examples, false moral justifications are created. Both the fire fighter and fire chief have displaced ethical restraint by internally justifying the actions. In the case of the fire fighter, there is a false argument that being deceitful attains justice. In the case of the fire chief, there is a belief that deceit attains a greater good. In both cases, personal values are set aside, as are professional responsibilities. The creation of false moral justifications does not represent a change in ethical values, for in each case the participants realized that their actions were deceitful. Rather, they seek to excuse inappropriate behavior to maintain a positive self-image.

Normalizing the Behavior

Another process in moral disengagement is the normalization of behavior. The individual undermines personal moral restraint through the false justification that group consensus can alter moral standards. Common phrases consistent with behavior normalization include: "Others do it more than me," "I don't make up the rules," "They started it," and "It's just the way that it is."

Fire fighters engaged in harassment disguised as hazing are likely to excuse the behavior as, "It's a tradition; they did it to me when I was a newcomer." The fact that others are doing an unethical behavior does not justify continuing the behavior Figure 13-6. It does, however, soothe the internal feeling of moral dissonance. Behavior normalization is particularly effective in displacing ethical restraint. History has demonstrated that entire communities will engage in looting, vandalism, and violence even though, as individuals, members of the crowd would likely never participate in such behavior (Bandura, 1999; Tsang, 2002).

Consider a fire fighter who normally would never participate in racist or sexist banter. Placed in the station house where such behavior is commonplace, peer participation and event frequency can eventually erode the fire fighter's personal inhibition toward the behavior. As personal restraint diminishes, the behavior becomes more likely to be repeated.

Involvement in an unethical event is made easier if it is endorsed by others in the group. It is a matter of perception. We are willing to suspend our own moral compass if given permission by community standards. In reality, group standards of behavior are irrelevant to the individual's values. In many cases, the others participating in the event are likely having the same morally dissonant reaction.

The Minimization of Consequences

An important tactic in moral disengagement is the minimizing of consequence. Consider the Robin Hood story. The modern interpretation of the myth has Robin Hood as a character who steals from the rich to give to the poor. Robin is regarded as a folk hero, although in strict terms he is a thief. Rationalization of his behavior is based on two premises. First, in stealing from the rich, there is the assumption that they can afford the loss. The act of stealing is somehow minimized because of the perception that the victim of theft suffers no significant injury. A common everyday example of this occurs when an individual steals a piece of candy out of someone's candy dish. If the dish is not full, the individual may likely refrain from taking candy if it is the last piece. The assumption is that the last piece is somehow more valuable, and the loss will be felt more, rather than if the dish has many pieces.

The second rationalization of the Robin Hood story is that Robin's intention offsets the consequences of thefts. By donating a portion of the stolen goods to "more deserving people," Robin is making a false argument that he is entitled to steal from those he has judged as "less deserving." You should recognize this is another iteration of the justification that "the ends justify the means." This justification only works if one accepts that Robin has the moral authority to judge other people's relative worth (Bandura, 1999).

Another form of consequence minimization is the rationalization that an unethical action is an isolated incident. The argument is that in only doing an action once, it is somehow more palatable than a repeated transgression. The argument asserts that stealing twice is more unethical than stealing once. By extension, a single theft is more easily justified. Although the logic is flawed, it does have the effect of easing moral dissonance.

Figure 13-6 Training is for training, not your amusement. Keep it real!

The problem with consequence minimization is that it is exacerbated by the argument of the slippery slope. Unethical actions tend to pave the way for further unethical actions. Once it is accepted that ends justify means or that consequences can be warranted depending on who is affected, then an ever-escalating series of unethical actions can likewise be justified.

Dehumanizing the Victim

Ethics is intrinsically tied to relationships with other individuals. One of the fundamental principles of ethics is that an action is ethical, or not, based on its impact on others or their humanity. An essential element of the moral disengagement process is the dehumanization of the victim. History is full of examples of atrocities excused by the dehumanizing of victims. For individuals who possess a conscience, a sense of right and wrong, and healthy self-image, it is nearly impossible to condone mistreatment of others viewed as equals. And so perpetrators of immoral acts (including those who witness but do not intervene) rationalize that the victims are unworthy or culpable.

Phrases such as, "They are not like us," "They don't belong here," "They have done worse," and "They brought it on themselves" are all hallmarks of victim dehumanization. Making others "less than" or culpable in their own fate is imperative to the process of moral disengagement. All other forms of rationalization sit upon this single important suspension of moral restriction. "Normal" people simply cannot justify hurting others without reformulating their opinion of the other's worthiness of treatment.

For example, Jane Smith tests for a job with the Anytown Fire Department. She scores well in all facets of the testing and has the top score of all the candidates. The fire commissioners are considering hiring her and ask the fire chief's opinion. The fire chief has a problem. He sees himself as a fair person, free from prejudice, and inherently ethical. He has heard all the arguments in favor of diversity, and he understands that passing over qualified candidates for a job is unfair and unethical. The problem is that he doesn't think women should be fire fighters and he does not want one in his department.

Given that Jane Smith is a top candidate, he cannot logically rationalize passing her over. His personal values would make him uncomfortable treating her unfairly. As a result, he is faced with a quandary. What he wants to do and what he knows he should do are not in agreement with each other.

The first step in moral disengagement will be to minimize her as a candidate. He may suggest to the commissioners that she likely does not really want to be a fire fighter, she is just trying to make a point, or her presence will likely disturb the working relationship within the department. Another excuse may be that hiring a woman may be the politically correct thing to do, but she will never be the fire fighter that a male would be, and the job is just too important to play politics.

In each case, the fire chief is attempting to diminish Jane Smith to a stereotype, consistent with his preconceived notions. By being a woman, she must be disruptive to team unity, or she is a woman making a feminist point. Further, by the simple benefit of her gender, she is less qualified. Because of his self-image as a fair person, the chief cannot emotionally justify denying a qualified candidate a position. It would be unfair and unethical. So he changes the conditions of the choice by diminishing Jane Smith Figure 13-7.

Figure 13-7 What to do about Jane Smith?

Avoiding Moral Disengagement

At its heart, moral disengagement is an exercise in self-deception. It is an unintended consequence of our very human need to maintain self-image when indulging in actions that are inconsistent with personal values. Each of the processes related to moral disengagement discussed previously has a common element; each requires some degree of moral rationalization. Recognizing **moral rationalization** in ourselves and others is a significant first step in maintaining high ethical standards.

Rationalization can take many forms and can intrude on our moral standards without being recognized Figure 13-8. The following are some of the more common rationalizations employed by those attempting to convince themselves and others that an action is appropriate (Ethics Alarms, 2014).

- **Inevitability.** "If I don't do it someone else will." This is, of course, a false premise. One person's actions or inactions do not change the likelihood of someone else participating in unacceptable behavior. Even if true, the notion that someone else's unethical behavior

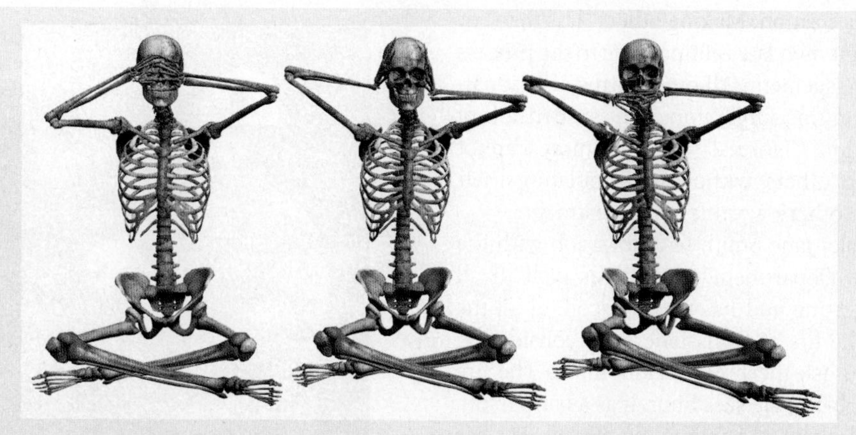

Figure 13-8 Moral rationalization: Favoring a particular conclusion as a result of some factor (such as self-interest) that is of little justificatory logical relevance.

justifies the same action on your part is false logic. The action remains ethical or unethical to the same degree regardless of whether anyone else participates or not.

- **Personal debt.** "I am owed this." Like so many other rationalizations, the logic of this assertion is that external factors can change the elemental morality of an action. This is a false premise. Any action undertaken by a fire fighter is either ethical or unethical of its own accord. Precipitating factors do not change the consequence of the behavior or its intent.
- **Nonresponsibility.** "It's not my problem." This rationalization is rooted in the debate over obligation to intervene on behalf of others. Egoism would argue that we are not responsible for others. However, deontology, virtue ethics, utilitarianism, and social contract theory would all argue that every individual has an obligation to help others. The moral codes of all established Western and Eastern religions endorse this concept. "You are your brother's keeper."
- **Humanity defense.** "I am only human" or "Nobody's perfect." The enduring popularity of this rationalization is rooted in the fact that it is fundamentally true. Humans do make mistakes, and no one's perfect. The fact that people make mistakes is not, however, a justification for indulging in unethical behavior. As a sentient being, each person has an obligation to aspire to ethical behavior as best as he or she can. As an example, if a person carelessly breaks a window, the action could be excused as an accident. However, the accident does not justify continually repeating the same careless act. This particular rationalization is also used by people who indulge in premeditated unethical acts and try to pass them off as honest mistakes.
- **Downplaying the scope.** "Yes, it was a bad choice, but I could've done worse." This is a common rationalization that attempts to excuse behavior by comparing it to a perceived worse offense. Consider this example; "Yes, I stole half your money, but I could've stolen all of it." This flaw of the logic is apparent; both actions are inherently wrong. Choosing the lesser of two evils does not mitigate the behavior.
- **The fake conundrum**. "I really had no choice." This form of rationalization seeks to mitigate responsibility for an immoral action by pleading that the circumstances insisted on it. There are certainly legitimate ethical dilemmas. By definition they require difficult choices that challenge ethical concepts. But it is important to realize

that choices exist. We all have the choice to do right and wrong. Not liking available choices is not the same as having no choice.

- **The rationalization of extraordinary circumstances**. "We can't afford to be the good guys right now." This rationalization assumes that doing the right thing will bring negative consequences. Again, this is an attempt at creating a false dilemma. Contrary to popular belief, wrong actions rarely yield good consequences. There are times when you must choose between doing the right thing that yields negative consequences or doing the wrong thing that appears to yield better consequences. The question becomes, Is right and wrong only determined by consequence? For example, it may be ethical to lie in order to save a life (consequentialism). It is not, however, ethically justified to lie in order to avoid punishment. The danger of consequential ethics is the temptation to rationalize those outcomes that are merely self-serving as "good."
- **Caveat emptor.** "It's not my fault that they were stupid enough to fall for it." This rationalization is a form of the argument that "might makes right." The assumption is that people who are unable to defend themselves are somehow undeserving of consideration, but ethics requires just the opposite. Those who are weakest deserve the greatest consideration. A person who suffers negative consequences due to another's actions may or may not be complicit in the outcome. However, his or her complicity does not diminish the ethical responsibility of the person doing the act.
- **Normalization.** "Everybody does it." Normalization is perhaps one of the most common rationalizations employed to excuse unethical conduct. It falsely assumes that the ethical nature of an act is somehow elevated if numerous people engage in it. The more people who engage in an unethical act, the greater the harm. Each must be held accountable for her or his shared responsibility in causing harm.

Avoiding Rationalization

Avoiding the natural tendency to rationalize behavior may be more challenging than you assume. The following are some simple questions to ask as an effective guide to determine whether or not intended actions are disguised rationalizations of inappropriate behavior.

- Under normal circumstances, would I consider this action to be right or wrong?
- Am I hearing or using language that is downplaying what's really happening?
- Am I excusing harm to others by placing blame?
- Am I responsible for the outcome of the action I am about to take?
- Am I about to harm someone?

Moral Engagement

Perhaps the best assurance of ethical behavior is to be morally engaged. **Moral engagement** is essentially the opposite of moral disengagement. It is a focus on ethical issues and an awareness of consequence (Malley-Morrison, 2010).

Moral engagement is not easy. It requires strength of character and commitment to behaving responsibly even in the face of significant contrary social pressures. Unfortunately, life is full of incentives to participate in or passively cooperate with behaviors that are harmful to others. Being a moral agent is an active process and, as Aristotle suggested, must be practiced until it becomes a habit (Malley-Morrison, 2010).

Being morally engaged requires:

- **Accepting responsibility for your own behavior.** A first step in ethical development is an acknowledgment of responsibility. The hallmark of unethical behavior is undertaking actions in self-interest followed by an attempt to shift responsibility for those actions.
- **Exercising empathy.** As has been repeated often in this text, ethics is essentially the evaluation of behavior as it affects other people and relationships. A fundamental principle of ethics

is the necessity to understand the needs, desires, and motivations of others. In Chapter 12, *Ethical Decision Making,* we discussed the concept of emotional intelligence. Being a genuinely ethical person is nearly impossible absent the ability to understand others' emotional responses.

- **Exercising sympathy.** Empathy allows us to understand the feelings and thoughts of others. Sympathy is the engagement of emotional attachment to those feelings and thoughts. Empathy allows us to know how others are feeling, whereas sympathy makes it possible to care how others are feeling. Absent an emotional connection to others, ethics is reduced to an exercise in calculation. Human connection is an absolute requirement for morality and ethics.
- **Respecting the rights of others.** Immanuel Kant's categorical imperative (Chapter 4, *The Philosophy of Ethics*) stated that each human being is a moral agent. He further asserted that failing to respect the rights associated with humanity was inherently unethical. An important element in maintaining moral engagement is to constantly be aware that others have rights, and even enemies are human and worthy of consideration.

Figure 13-9 Misplaced loyalty or obligation can lead to poor decision making.

Getting Sucked In

Until now this chapter has focused on how individuals suspend personal values in order to rationalize improper behavior. There is, however, another mechanism by which "good" people find themselves in ethical quicksand (Figure 13-9). It is not too unusual for well-intended individuals to find themselves involved in the misdeeds of others. This form of unethical activity is called **enabling behavior**. Enabling behavior is defined as actively or passively supporting the actions of others. The enabler's involvement may be as simple as looking the other way or may be as involved as participation in a conspiracy. In either case, cohorts in unethical behavior contribute to or enable others' unethical behavior.

The process of an ethical engagement among enablers is significantly different than that of an initiator of bad behavior. First, they tend to rationalize their participation through the intent of helping others. Second, their involvement typically starts off relatively innocently but escalates to the level of unethical behavior.

For example, at a fire scene fire fighter Smith is using a spare portable radio because the battery on his unit is dead. Somehow, he manages to lose the radio. The radio in question is assigned to the apparatus and not to Smith personally. Because it is not the first time he has lost equipment, he is afraid to report it. As luck would have it, the next two shifts fail to notice the missing spare radio during inventory. On Smith's next shift he writes the radio up as missing. Fire fighter Jones sees the missing radio report and remembers that Smith had the radio at the last fire, and asked him about it. Smith confesses that he lost it and begs Jones to keep quiet because he does not want to draw a suspension. Jones agrees to help his friend. Unfortunately, the chief is upset at the loss of the thousand-dollar radio and initiates an investigation as to what happened to it. To cover their tracks, the members of the other two shifts swear the radio was present during their inventory. No one has an idea what happened to it. During the investigation, Jones lies to his superior officer to protect Smith and assures the investigator that the radio was still present at the end of their last shift. One week later the owner of the fire building brings the radio to the chief's office saying he found it in the fire debris. Jones, like everyone else involved, is brought up on charges for falsely testifying in an investigation.

Jones obviously acted unethically in lying to the investigator. However, unlike Smith and the other fire fighters who lied to protect themselves, Jones's actions were altruistic. He lied to protect his friend. His involvement also escalated. Jones likely rationalized his choice to keep silent about the lost radio as a "small sin." However, when a formal investigation was launched, he escalated his behavior from a lie of omission to a lie of deception because he felt he was now complicit. What started out as simple silence became active deceit because of a desire to help someone in need.

Certainly, Jones's actions are unethical, but they do differ from the other characters in the example. Jones placed himself in a compromised position by following misguided good intentions. This pathway to unethical behavior is not uncommon. Unethical behavior is sometimes triggered by a desire to help others or by a sense of loyalty.

The individual's sense of right and wrong can be skewed by both perspective and proximity. Perspective is an individual's view of the behavior. Feelings of sympathy for another in a difficult position may exist even if the behavior that caused the trouble is condemned. This is certainly true for individuals with which we have an emotional connection. We recognize that their actions are inappropriate, but we feel sympathy for the likely punishment they are trying to avoid.

In the case of Smith and Jones, Jones likely can relate to fear of suspension and may well be rationalizing that Smith losing the radio was not an intentional act, and so not unethical. Jones's focus is displaced, as it is on Smith's misfortune rather than Smith's inappropriate failure to report the loss. This is because Smith is a sympathetic character, and the emotional response to Smith's plight is highly relatable, and so garners attention. If asked out of context about the scenario, Jones would recognize the responsibility to report the equipment loss. However, his perspective of the situation is obscured by his sympathy (emotion) for Smith, which displaces his sense of obligation (cognitive understanding) to the department and its rules (Spiegel, 2015).

The issue of Smith and Jones also suggests a problem of proximity. As stated previously, Smith is a coworker, an individual with whom Jones has a relationship and emotional attachment. That sense of loyalty is a value that should not be underestimated when weighing the ethical virtues of Jones's duty to report. Jones's responsibility to report is to the department; departmental policy is a conceptual understanding. Inventory control and expenses are removed from Jones's day-to-day thoughts. Smith, however, is a consistent part of Jones's daily existence. Therefore, Smith is more real to Jones than the concept of department policy.

The difficulty for Jones is that it is his humanity that is misleading his interpretation of his ethical responsibilities. Normally it is our "better nature" that controls our behavior. In the case of enablers, it is their values and conscience that are errantly pushing them toward an unethical act. If moral dissonance occurs within Jones, he is likely to describe the issue inaccurately as a moral dilemma because he is dealing with two competing values: a responsibility to act honestly versus his loyalty to a coworker. In reality, there is no ethical dilemma, and there is no ethical justification for misleading an investigation or for covering up a lie. Loyalty is normally a virtue, but a virtue is only a virtue if it necessitates positive action. Deceit is not virtuous.

Jones's unethical actions are brought on by his reluctance to see his friend suffer from his own mistake. This form of self-deception may be the most difficult for an individual to recognize. For many, emotional responses of sympathy, loyalty, and empathy may be more compelling than even that of self-preservation.

WRAP UP

Chapter Summary

- This chapter explored how fire fighters with a clear understanding of right from wrong ignore those sensibilities and undertake actions that later prove to be unethical.
- The internal processes by which individuals avoid cognitive and moral dissonance were reviewed.
- A popular concept known as attribution theory was introduced as the relationship between a person's need to relate cause to effect. Individuals judging others' behavior tend to attribute that behavior to internal causes, such as flaws in character. In judging their own behavior, individuals tend to attribute cause to external factors as mitigating circumstances.
- The attribution of external factors allows for the implementation of a rationalization process by which moral dissonance can be avoided. One such process is cognitive disengagement. This is a suspension of simple logic by ignoring or minimizing the relative value of facts in support of preconceived notions.
- Moral apathy is a condition where the ethical relevance of a situation is either not appreciated, or it is ignored.
- Moral apathy can be brought on by several contextual elements. An important contributing element is that of work environments in which peers and leaders model unethical behavior; this has a powerful contributory effect on reshaping individual behavior standards.
- Another environmental condition that can change moral standards is that of expectation. When individuals are *assumed* to be unethical, they are more likely to engage subsequently in unethical behavior.
- Perspective also contributes to moral apathy. Individuals in high authority positions are often granted institutional freedom. Because special rules apply to them, it is not uncommon for powerful people to assume that they are above all rules.
- People with significant responsibilities tend to believe that they operate in unique circumstances that are not understood or appreciated by "lesser" people.
- Assuming a fire fighter recognizes that a particular action may breach ethical standards, a process known as moral disengagement must occur if the action is to be pursued absent negative self-image consequences or absent moral dissonance.
- Moral dissonance is defined as internal stress that is felt when actions do not coincide with personal values.
- Moral disengagement is a process by which personal values are suspended, modified, or judged to be nonrelevant so as to accommodate the normally unaccepted behavior.
- The process of moral disengagement can be very complex and involve any number of coping mechanisms, including cognitive disengagement, the creation of false moral justifications (rationalization), the normalization of behavior through repeated practice, and the minimization of consequences.
- Moral disengagement requires rationalization. Methodologies were presented for recognizing and avoiding rationalization.
- An important strategy in avoiding rationalization is active moral engagement. Moral engagement includes the processes of accepting personal responsibility, exercising empathy, developing sympathy, and recognizing and respecting human rights.
- The chapter closed with a discussion of how otherwise ethical individuals can inadvertently entangle themselves in the misdeeds of others. Perspective of a problem's context and proximity of consequences can skew an individual's assessment of his or her ethical responsibilities.

Key Terms

attribution theory a theory explaining the discrepancy between judging your own behavior versus others' behavior.

cognitive disengagement a suspension of rational understanding, or an exercise in self-deception.

cognitive dissonance an internal stress that presents itself when we act irrationally.

enabling behavior actively or passively supporting the actions of others.

framing a theory describing how our responses to situations, including our ethical judgments, are affected by how those situations are posed or viewed.

moral apathy a lack of moral engagement springing from a lack of awareness or cognitive disengagement.

moral disengagement convincing yourself that ethical standards do not apply in a particular context.

moral dissonance an internal stress that presents itself when we act in a way that is inconsistent with our self-image.

moral engagement A focus on ethical issues and an awareness of consequence.

moral rationalization favoring a particular conclusion as a result of some factor (such as self-interest) that is of little justificatory logical relevance.

Challenging Questions

To check your understanding of this chapter's material, answer the following questions. It is highly recommended that you discuss your viewpoints with fellow students, peers, coworkers, and friends to discover their opinions as well.

- Based on this chapter, what steps can the fire fighter take to avoid rationalizing unethical behavior?
- Your text discussed a lack of moral engagement. What steps can the department take to encourage fire fighters to consider ethical consequences?
- In what ways are attribution theory and the theory of locus of control similar, and how might they affect each other?
- Why do people engage in rationalization?
- As a nation, the United States has been grappling with moral dissonance related to the conflicting interests of homeland security versus human rights. Restricting immigration based on religious beliefs and the use of enhanced interrogation methods are both inconsistent with established national values. Conversely, many argue that there is a compelling need to depart from moral standards to serve the greater good of protecting citizens from acts of terror. Analyze and critique the various ethical arguments of both positions. In your opinion, which has the greater weight and why? Be sure to reference the material within this chapter in your answers.
- Can a good intention justify a bad action? Articulate your answer referencing the principles found within this text.

WRAP-UP

Case Study Conclusion

Revisit the case study at the beginning of the chapter. Spend a few minutes considering the questions posed at the end of the case study. In light of the information shared in this chapter, have any of your original observations changed?

The commandeering of the alcohol and beer sign will certainly have negative consequences for the department. The owner, who has just lost everything, may well be put off by the sight of fire fighters turned vultures picking through what used to be his business. Taken out of context, a video of fire fighters packing away alcohol and beer signs from the fire certainly does not look good. Smith and Jones did not act wisely. But is taking the dumpster-bound material unethical?

The rationale employed by Smith and Jones that the material being "salvaged" was headed for the dumpster may make sense within the department, but the reality is that it is not a valid excuse. Ask yourself the following questions: Do Smith and Jones own the liquor and the sign? Is taking someone else's stuff ethical? Does the fact that the material has little or no value change the underlying issue of taking items that do not belong to them? Smith and Jones are working for the department and are being paid for that work. Were they acting in the department's interest when they took the goods? If they truly thought it was okay for them to have the material, why did they wait until no one was looking to load it? Have their actions caused problems for other people? Are other fire fighters suffering embarrassment because of their actions? Have their actions damaged their department? Do they have the ethical right to hurt others' reputation? Are their actions consistent with the best traditions of the fire service? Does permission from Lieutenant Franklin change any of the answers to these questions?

Is fire fighter Franklin culpable for the actions of Smith and Jones? Was he representing the best interest of the department in allowing Smith and Jones to act as they did?

The actions of Smith and Jones are problematic to say the least. In considering how to respond, the fire chief should keep the following in mind. Smith and Jones acted in their own self-interest and took property that did not belong to them. While they may have rationalized that the property had no value, it does not mitigate the fact that they took something that does not belong to them. They had the opportunity to seek the permission of the owner if they thought it was appropriate to salvage the material. The consent of their superior officer does not mitigate their actions. Further, Franklin is every bit as guilty as Smith and Jones. Franklin's actions may be worse in that he is entrusted to make sure things like this do not happen, and an officer is expected to have better judgment.

If the chief accepts the rationalization that the property "was headed for the dumpster" and fair game, he tacitly approves of his fire fighters having the authority to judge when and if citizens' property rights are valid. This sets an incredibly dangerous precedent.

In summary, the chief has three fire fighters who, regardless of the rationale, took property that was not theirs. He should react accordingly.

Chapter Review Questions

1. Define moral disengagement.
2. What is attribution theory?
3. In ethics, what is meant by normalization of behavior?
4. Explaine cognitive disengagement.
5. What effect do the ethical expectations of others have on an individual's likelihood of behaving unethically?

References

Bandura, A. 1999. "Moral Disengagement in the Perpetration of Inhumanities." *Personality and Social Psychology Review* 3, 193–209.

Ethics Alarms. 2014. "Unethical Rationalizations and Misconceptions." Accessed June 22, 2018. https://ethicsalarms.com/rule-book/unethical-rationalizations-and-misconceptions.

Kaptein, M. 2012. "Why Do Good People Sometimes Do Bad Things?" Accessed June 22, 2018. https://papers.ssrn.com/sol3/papers.cfm?abstract_id=2117396.

Kashtan, M. 2013. "Moral Dissonance." *Psychology Today*. Accessed June 22, 2018. https://www.psychologytoday.com/blog/acquired-spontaneity/201307/moral-dissonance.

MacDonald, C. 2015. "If You Think You Can't Be Ethically Compromised at Work, You're Wrong." *Canadian Business*. Accessed June 22, 2018. http://www.canadianbusiness.com/blogs-and-comment/if-you-think-you-cant-be-ethically-compromised-at-work-youre-wrong.

Malle, B. F. 2011. *Attribution Theories: How People Make Sense of Behavior*. Accessed June 22, 2018. https://pdfs.semanticscholar.org/dbc6/ca9548099b6f2b84d1cd81f3eb13c07cde7f.pdf.

Malley-Morrison, K. 2010. "Moral Engagement: Introduction." Accessed June 22, 2018. http://engagingpeace.com/?p=391.

McLeod, S. 2015. "Cognitive Dissonance." Accessed June 22, 2018. https://www.simplypsychology.org/cognitive.html.

Moskowitz, G. B, and H. Grant. 2009. *The Psychology of Goals*. New York: Guilford Press.

Pessah, J. 2015. *The Game*. Boston: Little, Brown and Company.

Spiegel, A. 2015. "Why People Do Bad Things." *Planet Money*. Accessed June 22, 2018. https://www.npr.org/sections/money/2015/07/03/419543470/episode-363-why-people-do-bad-things.

Tenbrunsel, A. E., and D. M. Messick. 2004. "Ethical F: The Role of Self-deception in Unethical Behavior." *Social Justice Research* 17, no. 2, 223–236.

Tsang, J. A. 2002. "Moral Rationalization and the Integration of Situational Factors and Psychological Processes in Immoral Behavior." *Review of General Psychology* 6, no. 1, 25–50.

Tversky, A., and D. Kahneman. 1981. "The Framing of Decisions and the Psychology of Choice." *Science* 211, no. 4481, 453–458.

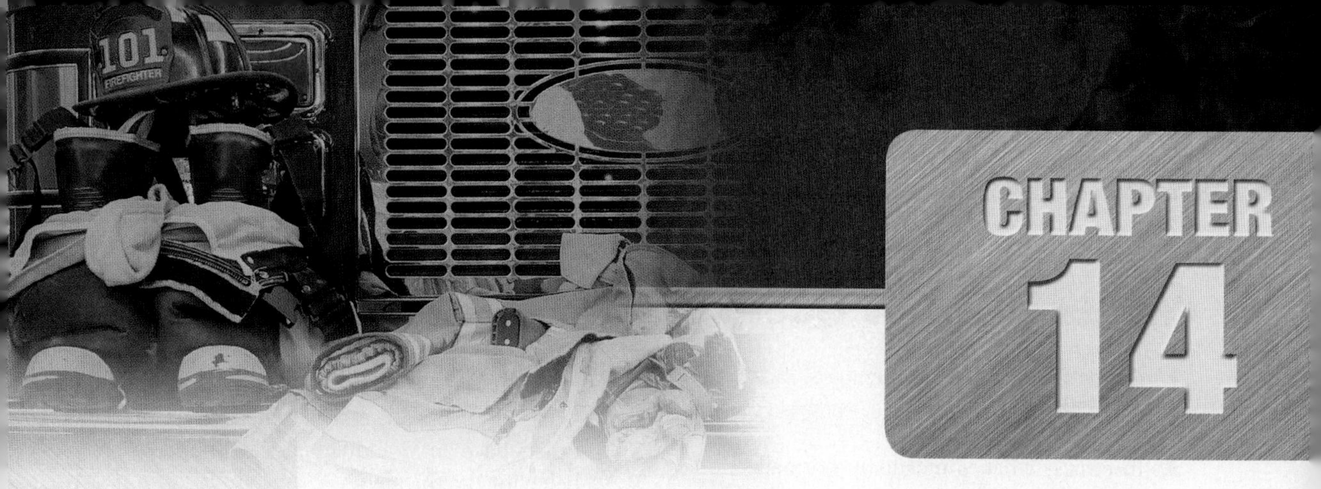

Contemporary Issues

"In a civilized life, law floats in a sea of ethics."

—Earl Warren

OBJECTIVES

After studying this chapter, you should be able to:

- Consider the privacy issues in society as well as the fire service.
- Understand the elements of HIPPA.
- Outline the ethics concerning the use of body cameras.
- Critique the impact of mass communications and social media within the fire service.
- Depict the challenges of off-duty activities.
- Analyze the social issues such as politics, religion, harassment, and hazing in the workplace.
- Discern the challenges with on-the-job relationships.
- Consider the impact of staffing and safety issues on fire fighters.
- Characterize the ethical concerns of alternative funding.

Case Study

For more than 15 years, fire fighter Smith has been the unofficial photographer of the Anytown Fire Department. It was not unusual to see Smith at calls and department functions with his camera. He especially liked to take fire ground pictures; he did so both off-duty and on-duty. Smith often shared the photos with fellow fire fighters, and some were even hung on fire station walls. In 2007, Smith moved from photographs to video. In some instances, his videos were used in departmental post-incident critiques and training sessions.

In 2012, Smith began his own website and a blog called *Anytown on Fire*. The website featured unedited film of fire ground operations and commentary. The site became a local favorite, and soon after, Smith started selling advertising on the site. The site generated several thousand dollars a month. Because of Smith's affiliation with the Anytown Fire Department, he had access to fire ground operations not enjoyed by local media. As a result, in 2015, Smith reached a financial agreement with the local TV station, which allowed the station to utilize his footage and granted him a byline as a reporter.

In 2016, the Anytown Fire Department responded to an apartment fire. The building received heavy fire and smoke damage, and there were two civilian casualties as well as one significant fire fighter injury. Smith was on-duty and was wearing a "body cam" on his helmet. Smith was convinced that the officer in charge of the fire scene made some tactical mistakes that may have contributed to the extensive damage to the building. He also had footage of the incident commander and a fire captain having a heated debate about hose placement.

He placed this footage on his website and wrote an article questioning the tactics of the incident commander. The local media outlet, by virtue of the agreement, used the footage and his commentary as the lead story the next day. The story led to the nearly immediate filing of lawsuits by the building owner and the fire victims' families.

Smith was brought up on charges for violation of department policy regarding the dissemination of unapproved information regarding fire service operations.

- What, if any, ethical issues are relevant to Smith's actions?
- Did Smith have a right to make money based on his affiliation with the Anytown Fire Department?
- Were there any issues in the reporting of the apartment fire that justified disciplinary action that were not present in Smith's previous website reports?

Introduction

The previous chapters have addressed ethical theory, the application of ethics, and the role of ethics within the fire service organization. These topics have been addressed generally. This chapter will focus on specific issues faced by fire fighters and explore the ethical considerations relevant to each. Some of the problems discussed in this chapter have been previously introduced as case studies in support of a concept or to help clarify a particular point. Here, the reverse is intended—we will use ethical concepts to define common problems that regularly cause ethical debate or confusion in the fire service.

The issues explored in this chapter are complex, hence the existence of debate and confusion. It is important to note that the value of this chapter is the journey rather than the destination. That is, the consideration of ethical implications has value beyond whether or not you completely agree with the assertions made here. By engaging in critical thinking regarding these issues, you can develop consistency in your approach. Each of the problems is unique, but there is a commonality in many of them. Each can be explored from common frames of reference including responsibility, social obligation, personal values, and virtues.

The issues discussed in this chapter can and should be reviewed from the viewpoint of professional ethics. As such, professional responsibility and obligation are the primary considerations. Not surprisingly many of the issues discussed in this chapter also have personal ethical implications. It is rare when personal issues are not intertwined with workplace behavior. As you have learned in previous chapters, personal values, needs, and wants often shape professional judgment. Private life and work life are not easily separated, and personal values and professional responsibilities sometimes clash.

Fire fighters have difficult jobs, performed in difficult circumstances. One of the great benefits of being a fire fighter is that every day will be different. Ironically, one major challenge of the profession is that every day brings something different. Not surprisingly, moral and ethical challenges faced by the fire service are diverse.

Figure 14-1 Privacy is becoming a societal priority.

Privacy Issues

The issue of privacy is becoming paramount in today's society. Information technology has revolutionized business, communications, and recreation. Information is power and, increasingly, it is currency. Privacy issues are a growing concern because the concept of privacy has fundamentally changed.

Members of contemporary society rarely have complete privacy. To function in modern society, individuals must make daily decisions regarding the distribution of personal information. Your Internet usage, shopping habits, banking habits, entertainment choices, and travel are all categorized and sold.

Identity theft is among the fastest growing crimes worldwide. Unfortunately, your personal and financial safety are directly tied to your ability to control the availability of your personal data. As information becomes increasingly valuable, privacy issues are becoming societal priorities. To function in society, we must routinely share personal information, and we must trust in others to protect our privacy. We trust that our bank will not share our financial information; we hope that our pharmacy's website is not hacked and that our medical information remains safe Figure 14-1. We assume that our smartphones and tablets are not leaking sensitive information.

What does fire service ethics have to do with the growing issue of privacy? Few other organizations are implicitly trusted as much as the fire service. As fire fighters, we are members of an organization dedicated to the welfare of others. We exist explicitly to help those in need. Vulnerable and even desperate people call us for help. Because of the nature of the services we promise and the privileges granted to us, we have an obligation to protect the privacy of those we serve. We have free access to people's homes, their businesses, and even their person. In most cases, we are unexpected guests. As a result, we see how people actually live. We see people at their worst, and we often see them at their most foolish. Any fire fighter with more than a few years on the job can tell story after story about the amazing things that people do.

Despite embarrassment, people call us because they feel they have no choice. They need help, and they trust and hope that the fire fighters will act professionally, treat them with dignity, and respect their privacy.

Medical Information

Maintaining the confidentiality of specific medical information regarding patients' conditions is not only an ethical requirement, it is enforced by the **Health Insurance Portability and Accountability Act (HIPAA).** However, HIPAA does have limitations and some elements of patient interaction are not necessarily covered.

Clarification

HIPAA was passed by Congress in 1996 to do the following:

- Provide the ability to transfer and continue health insurance coverage for millions of American workers and their families when they change or lose their jobs.
- Reduce health care fraud and abuse.
- Mandate industry-wide standards for health care information on electronic billing and other processes.
- Require the protection and confidential handling of protected health information (California Department of Health Care Services. n.d.).

Although certain information may not be legally protected, fire fighters are still ethically bound to protect patient privacy. Because of the nature of the relationship between the public and the fire service, the public has a right to expect that fire fighters respond with the intent of providing assistance. Access to property, person, and personal information are often necessary to facilitate the provision of emergency services. Violation of personal privacy, whether legally protected or not, is inherently unfair and likely harmful. As such, indiscrete and unofficial sharing of private information is unethical.

General Gossip

Consider the following scenario. Fire fighters respond to a call for "smoke in the house." On arrival, they find themselves at the home of a well-known local TV weatherman. The call ends up being routine except for the fact that the homeowner is a "hoarder." A few days later fire fighter Smith is having supper with several friends at a local bar and grill. Smith entertains his friends with a humorous and slightly exaggerated account of how the local celebrity lives in squalor. Not surprisingly, fire fighter Smith's friends share the story with their friends, and much to the horror of the weatherman, the story ends up being posted on social media. The department now has a victim who called for help from the fire service and is subsequently personally humiliated. Further, the story could be damaging to his career and very likely may affect his standing in the community. Whether intentionally or unintentionally, fire fighter Smith caused personal harm to an individual who called him for help.

Fire fighters quickly learn that their friends and family find what they do for a living fascinating. They want details about their calls and hang on every word of their stories. Unfortunately, fire fighters are not immune to gossip. It usually seems harmless enough. After all, who doesn't like telling a good story to a receptive audience?

The problem is that spreading gossip about "customers" is a betrayal of trust. People do not *usually* call a fire department because they want attention. They call because they need help. They are vulnerable and often desperate. To cause them further harm by damaging their reputation, or simply embarrassing them, is a breach of faith. Even if they are not aware that they are the focus of the fire fighter's story, it is still demeaning to exploit someone's misfortune for our entertainment **Figure 14-2**.

Two primary concepts of ethics are first, to do no harm, and second, to act in good faith. The careless disregard of victims' privacy violates both these principles. Further, the betrayal of trust is inconsistent with virtues associated with ethical behavior.

The precepts of professional ethics also tend to reproach fire fighter gossip. For the fire service to do its job effectively, the public must trust that the department will not harm them in some way. Public confidence is therefore essential to the fire department's efficiency. Further, public trust leads to public support, which is also vital to the fire department's

Figure 14-2 Trust is the fire fighter's most important asset.

© Comstock/Stockbyte/Getty

operations. Actions taken by fire crews that diminish public confidence are counterproductive to the well-being of the fire department. Fire fighters are paid by the agency to promote the department's mission. Actions that are counterproductive to that purpose are ethically suspect.

Privacy rights should be extended to businesses as well as individuals. As a result of both emergency response and inspections, fire fighters are often granted access to a business's "backstage" areas. They often witness production techniques, storage areas, and working conditions. Their level of access also gives insight into the relative economic health of the business, and even trade secrets. As an example, many fire fighters have been surprised by the appearance of the kitchen at a favorite local restaurant and, fairly or unfairly, questioned its cleanliness and the safety of eating there to their friends. Theoretically, their comments can have catastrophic effects on a restaurant's business.

Certainly, if health and safety concerns are witnessed, fire fighters should report those conditions to the proper authorities. Beyond that, there is a responsibility to maintain confidentiality as to what they see and hear. There is an ethical responsibility to maintain professionalism. Fire fighters must be cognizant of the harm they can cause to business owners and employees through careless comments.

The Ethics of Body Cameras

The last 20 years have seen a revolution in technology. One of the greatest changes is the proliferation of cameras. Cameras constantly surround us; almost every individual carries a cell phone, which has both a still and video camera. Surveillance cameras have become cheap and small and, as a result, their usage has skyrocketed. Because of the proliferation of cameras, a person's actions are being filmed almost daily, and the ethical ramifications to individual privacy are stunning. Frankly, the old rules regarding consent and usage are outdated and irrelevant.

In addition, individual attitudes toward camera usage have changed. Digital imagery has made photography and videography cheap and convenient. An entire generation has become accustomed to documenting anything and everything. Individual attitudes toward privacy are changing as a result.

A significant new development in videography is the **body camera**. Until recently the act of shooting video was relatively unchanged. True, the prevalence of digital cameras and phones made the number of recording units much greater, but the act of actually shooting video was much the same. The photographer stopped what he or she was doing, took out a camera, aimed it, and pressed record. The body camera, however, has made it possible for fire fighters to record personal activity even while working. Because it is essentially unattended, it is indiscriminate—it sees what the fire fighter sees. Because body cameras are small and unobtrusive, rugged and essentially hands-free, many fire fighters have bought them for personal use while on duty. The body camera has allowed for the extensive documentation and proliferation of emergency scenes and emergency

Figure 14-3 Fire fighter or reporter?

operations Figure 14-3. Unfortunately, the same cameras candidly capture people and property without explicit permission.

Fire fighters have made several arguments in favor of the use of body cameras (Snyder, 2017). Their ability to document fire ground operations in real time is seen as a potential benefit for training programs. There's also an argument that fire ground documentation is beneficial for cause and origin investigations and as potential evidence in civil proceedings. Still others argue that there is a benefit in transparency and that the public has a right to know what fire fighters do. As a result, the argument goes, the filming of firefighting operations should be in the public domain (Rucke, 2013).

While each of these arguments has merit, it would be naïve to discount the personal advantages of body camera usage. First among these is that fire fighters enjoy having "home movies" of their calls. There are also more problematic incentives for capturing fire scenes on video—many fire fighters have personal websites and blogs, and they use the video on those sites. Other fire fighters have sold their videos to local news outlets.

Many, including fire chiefs and some civil libertarians, have grave concerns about the proliferation of video cameras at emergency scenes. There is a growing belief that filming victims without their consent is a privacy concern. Some of the naysayers point out that in most cases patient information is protected, and in all cases, victims have a reasonable expectation of privacy and should be respected by first responders. While some fire chiefs acknowledge that there may be value in video, many are not convinced potential benefits outweigh privacy concerns.

The ethical questions associated with body cameras are numerous. Who owns the video? Do victims have a *legal* expectation of privacy? Do they have a *reasonable* expectation of privacy? Does a fire department have a right to control the release of a fire fighter's personal video? Is an on-duty fire fighter's personal video actually personal? Is it ethical for a fire fighter to publicly release, or even sell, the video taken on duty (Dennis, 2015)?

On-duty fire fighters (including volunteers) are employed to carry out emergency operations. Fire fighters have no reasonable right to expect facilitation of private pursuits while working for the fire department. The fire fighter is working at a scene legally controlled by the authority having jurisdiction; their presence is a condition of employment. Any video shot by fire fighters while on duty theoretically belongs to the fire department. As such, it would be unethical to distribute video without the department's permission. Logically, it follows that the department has a right to control the release of that video (subject to FOIA rules; see Chapter 11, *Ethics and the Law*).

The legality of privacy rights is complicated, although the pertinent ethics related to privacy is relatively straightforward. Fire fighters are employees of the community. There is an assumption that fire fighters responding to a scene are acting in the best interest of the community in general, and the victim in particular. While a reasonable argument may be made for capturing video for purposes of training may be justifiable, however there is no such argument for the use of candid video for either personal entertainment or public dissemination. The public has a right to expect assistance from the fire department. Violation of public privacy is counterintuitive to the terms of that arrangement. Some have suggested that the public does not have an expectation of privacy, as fire department responses are in the public domain and are covered by news agencies. Ethically, the comparison is a false one. The media covers fire department responses for the specific purposes of disseminating information, both as a public service and for profit. There is no expectation that they are there to help the victim. As a result, their access to victims and their property are restricted by emergency personnel for safety reasons *and to protect victim privacy*. Conversely, the public can reasonably assume that fire department personnel are present for the sole purpose of rendering aid. As a result they grant access to personal property and personal space. Capitalizing on that access for personal gain is unethical.

In all privacy issues related to emergency response, the reasonable expectation of privacy is a legal argument to determine whether an individual's actions are lawful. Ethically speaking, whether or not someone has a legal expectation of privacy is irrelevant. In calling a fire department, victims have an absolute expectation of assistance. We are by our nature and founding principles a helping organization—fire fighters are ethically bound to do no harm to the people who call for our help. It is reasonable to assume that the public release of a candid video will likely cause embarrassment and a sense of violated trust.

Mass Communication

The emergence of mass communication has changed the social landscape. Prior to this century very few individuals have had the capability of communicating directly with an entire community, much less a global community. The advent of the World Wide Web, and in particular social media, has given each citizen the ability to reach a global audience. The social and ethical ramifications of that power are significant. Just a few decades ago we assumed that everything we saw on TV was accurate, and we also assumed that news was in fact newsworthy.

Professionals vetted information for distribution, reviewing both the accuracy and relevance. The advent of new media technology has given each individual the potential of a global audience. Unlike traditional media, social media also has the element of anonymity. Individuals can upload pictures, videos, and written messages behind the veil of hidden identity and, as such, are often able to escape accountability for content. More now than ever, individuals are responsible for self-censorship. What we choose to "broadcast" is entirely up to individual discretion, honesty, and intention. Our use of mass communication can have a significant impact on the lives of others, and therefore it has ethical implications.

Social Media

Usage of social media has become a hot button topic in today's fire service **Figure 14-4**. The last decade has seen numerous disciplinary actions based on the indiscreet use of social media. Rules and policies regarding social media are common topics in fire service magazines and websites. Many fire chiefs will tell you that the inappropriate use of social media is one of their foremost ethics concerns. Their concern is rooted in three different social media qualities. The first is that social media is hard to control, as fire fighters have individual access to it and oversight is nearly impossible. Second, social media is pervasive. Most fire fighters in the country have one or more social media accounts. Third, unlike other forms of communication, social media has an unlimited shelf life. Once posted, comments, pictures, and videos never go away.

To understand the ethics of social media, it is important first to understand that social media is a "thing." As an inanimate object, social media is neither ethical nor unethical in and of itself; the ethics regarding social media are entirely based on content and intent. As a result, the use of social media does not present any new ethical issues but rather a new perspective on old ones. What social media has changed is the scope of ethical infringement and the consequences. Consider this example. In 1985, fire fighter Smith is called for a medical emergency related to alcohol intoxication. He finds a 35-year-old woman passed out in a state of undress. Smith uses a Polaroid camera to take a picture of the unconscious woman and passes it around to his firehouse buddies.

Figure 14-4 Social media has become a hot button issue in firehouse policy.

© Jones & Bartlett Learning

Fast-forward a few decades. In 2017, fire fighter Jones responds to the same type of call and uses his cell phone to snap a picture of an unconscious female and sends it to his buddies via Snapchat. One of his buddies captures the picture and posts it Instagram.

In both cases, the ethics of the situation are the same. Fire fighters Smith and Jones have both violated the patient's privacy, violated trust, and potentially damaged the reputation of their entire department **Figure 14-5**. The difference in the two examples is that technology allowed the 2017 picture to be distributed to a wider audience, and so the likelihood of being caught is higher.

Whether one person or a thousand see the picture, the act is equally unethical.

The point of the comparison is this: The underlying problem is not one of social media usage; rather, it is about the inappropriate behavior of the

LOCAL NEWS

Firefighter Under Investigation for Pictures

thebridge

Figure 14-5 Social media is not the problem, the underlying ethics of its use is.

Courtesy of Scott Walker.

fire fighters involved. A policy on social media will not correct the underlying issue. A social media policy may exacerbate the problem by appearing to focus the department's objection to sharing the picture rather than taking the picture in the first place.

The abuse of social media is a symptom; it is not the cause. The cause of social media scandals can almost always be traced to an individual, ethical, or organizational culture issue. Restricting the use of social media is a Band-Aid at best. Fire administrators must address the underlying ethics issues that facilitate abuse of social media. This can only be accomplished through training, education, and engaged leadership.

The ethics regarding social media content is relatively straightforward. The communication of content that is misleading, deceptive, unfair, harmful to others, or an invasion of individual privacy is likely unethical. Communications that may be inherently truthful, but made with malice or with intent to harm, are likely unethical. It is also unethical for fire fighters to misrepresent themselves as spokespersons for their department if not officially empowered to do so.

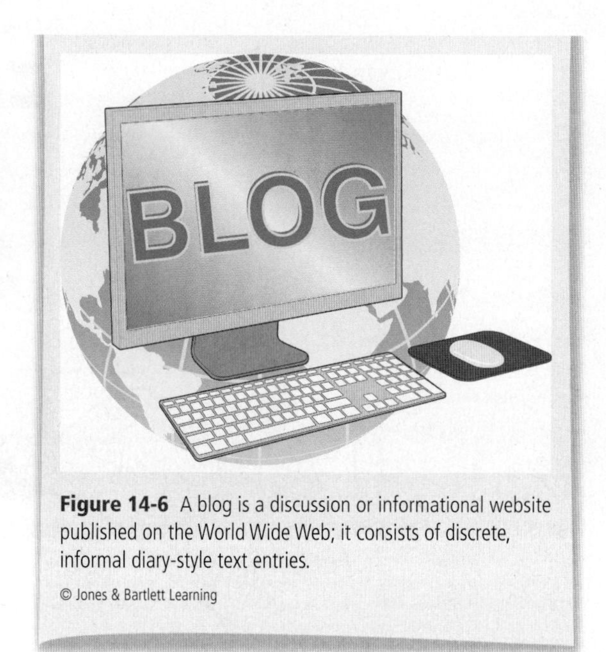

Figure 14-6 A blog is a discussion or informational website published on the World Wide Web; it consists of discrete, informal diary-style text entries.

Blogs

In general, **blogs** may be one of the most productive innovations in modern communication technology. The ability of individuals to share their opinions, interests, and wisdom with the world does much to advance understanding and learning Figure 14-6. Unfortunately, many bloggers do not properly distinguish between their First Amendment rights to free speech and the concept of ethical responsibility. *Having legal protection to say something does not mean that an expression is ethical.* The issues of integrity, honesty, and the consequences of expression on others are ethical issues—not legal ones. Words can be incredibly powerful: they can be uplifting or destructive; they can be educational or deceptive. A basic principle of ethics is that each of us is responsible for the choices we make, both in action and word. Fire fighters who write blogs or editorials are ethically responsible for the content and for the ramifications of that content. Untrue statements are of course ethically suspect, yet the condition of truthfulness does not necessarily mitigate ethical responsibility. A truthful statement may still be damaging and, as such, the individual is ethically responsible for the consequences of that statement. As in social media, blogs are neither ethical nor unethical. Fire fighters must realize that if they write blogs, they are responsible for the results of those writings. It is the content and intent of the material that ultimately defines the relative ethics of the blogger. As such, the ethical guidelines for blogging are similar to those of social media.

Off-Duty Activities

In most professions, the line between an employee's private life and professional life is evident. Behavior away from the workplace has little or no effect on the employer or the individual's professional obligations. The case of the fire fighter is a bit different; the profession is as much a lifestyle as employment. Being a fire fighter makes demands on your personal life, and as a high-profile public employee, off-duty behavior matters. There is a long history of union grievances and legal battles regarding sanctions of off-duty conduct. At face value, it seems fair to draw the line between the fire fighter's personal life and his or her professional life. However, whether fire fighters like it or not, whether they think it is fair or not, fire fighters are public employees who are held to a higher standard of behavior than most. The nature of our profession blurs the lines between off duty and on duty. A fire fighter is always a fire fighter whether on the clock or not. As a result, there are several common ethics issues relative to fire fighters' off-duty activities.

Employment

A common source of friction between fire chiefs and fire fighters is off-duty employment. Over the years, fire officials have sought to restrict fire fighters' off-duty employment for numerous reasons. For public relations reasons, a few cities have sought to restrict fire fighters from licensed trades such as plumbers and electricians. Some cities have attempted to restrict fire fighters from engaging in business considered controversial such as tattoo parlors and gentlemen's clubs. These restrictions are usually based on the maintenance of some perceived community moral standard.

Commonly, many cities have sought to restrict fire fighters from working for more than one fire department or even volunteering for fire departments. These restrictions are often based on insurance and pension issues. Many fire departments have asserted that off-duty employment must not interfere with emergency calls or mandatory off-duty training.

All the cases mentioned previously have some sort of practical intention, but what are the ethical issues with off-duty employment? If the reader accepts the following precepts, then some ethical conclusions can be drawn regarding off-duty employment.

- By joining the fire department, fire fighters accept the responsibilities associated with the profession, including providing critical health and life safety services to others.
- Because of the important nature of the fire fighter's work, there is an obligation for competency and efficiency.
- Because of the complexity of a fire fighter's duties, competency depends on continued professional development and training.
- Because of the dangerous nature of the fire fighter's duty, robust medical insurance benefits are typically extended to fire fighters.

Based on these premises, the following ethical conclusions can be made. For instance, any off-duty employment that interferes with a fire fighter's physical or mental ability to perform his or her duties is a breach of responsibility and thus unethical. Fire fighters have a responsibility to be physically, mentally, and emotionally capable of performing their duties because people's lives may depend on it. Because a fire fighter's work literally deals with life-and-death responsibilities, off-duty employment should not interfere with requisite training and continued education. Because fire fighters are typically given the benefit of the doubt regarding orthopedic injuries and various medical conditions, it is unethical for fire fighters to avail themselves of fire service benefits for compensation of injury or disease related to off-duty employment. In particular, heart and lung presumptive benefits granted fire fighters bring into question the ethics of fire fighters engaging in off-duty activities that pose high risks to respiratory and cardiac health.

What Do You Think?

Many departments limit their fire fighters from working for or volunteering with another fire department. The rationale for the prohibition is often rooted in workmen's compensation liabilities. For instance, if a fire fighter were to develop cancer or have a heart attack, which department would be held responsible for medical and pension obligations? *Are these restrictions fair?*

Other off-duty employment issues are more complicated. Is there an ethical issue with fire fighters who prioritize off-duty employment requirements ahead of fire department needs? Do fire fighters have an ethical obligation to show up for off-duty training? Do fire fighters have a duty to respond to emergency and nonemergency callbacks? Collective-bargaining agreements tend to set "objective obligations" regarding these issues. However, there is a difference between contractual obligation and ethical responsibility. Professional ethics would indicate that fire fighters have a duty to maintain competency. For instance, if a fire fighter agrees to join the hazmat team, there is certainly an obligation that he or she maintain proficiency by participation in special training sessions. Another example may be an aspiring fire officer. If he or she were to accept additional responsibilities, then there is a professional obligation to build competency even if that requires attending off-duty seminars and training sessions. These are professional responsibilities based on voluntarily accepting additional responsibilities.

The fire service speaks often and passionately about teamwork and loyalty. An unofficial motto of the fire service is, "You go, we go." Inherent in that worthy motto is that we accept responsibility for being our brother's and sister's keeper. If we accept responsibility for the welfare of our fellow workers, do we not have an obligation to come to their aid when an emergency callback occurs? Do we not have a duty to step into the breach when there are emergency staffing issues? The fire service celebrates its image as the American hero. We clearly tell the public that we are here to protect their lives and property. Should the desire for extra income outweigh that pledge? The tenets of virtue ethics, contract theory, and care ethics all strongly imply an ethical responsibility to uphold that promise.

The obligation to provide for our families and our future is a legitimate pursuit. However, by voluntarily joining the fire service we accept obligations for the welfare of others. While there must be balance in the fire fighter's life, it is incumbent on fire fighters to maintain perspective regarding employment priorities.

Off-Duty Response

A common ethics question within emergency services regards the ethical obligation to render aid off duty. Is there an ethical requirement to stop at a car wreck to see if assistance is needed? While on vacation, is there an ethical obligation to see if a local fire crew needs any assistance with the structure fire you just happened upon? Are fire fighters and medics really ever off duty Figure 14-7? Professional requirements to render aid may likely vary from state to state, and even from municipality to municipality. However, ethical obligations tend to be universal. Primary ethical principles include: do no harm, minimize harm whenever possible, contribute to the common good, and honor responsibility. Assuming the truth in these notions, there is an ethical responsibility to act if assistance is needed, you have the resources and ability to provide help, and your assistance will minimize harm to others. Conversely, there is no *ethical* requirement to help if assistance is not needed, or you are incapable of providing assistance beyond what is already available. Further, there is no ethical duty where your intervention will have no meaningful positive impact. For example, you are in a restaurant and an individual chokes. As a trained medic, you are familiar with the Heimlich maneuver. Its application may likely be successful, and no other qualified individual seems to be coming to the person's aid. You then have an ethical obligation to intervene. Conversely, assume an off-duty fire fighter comes across a house fire. An adequate crew is fighting the fire, and the individual has no

Figure 14-7 Is a fire fighter ever really off duty?

bunker gear with him or her. In this case, the intervention will likely make no positive impact as the fire fighter is limited to what he or she can contribute. As a result, the off-duty fire fighter has no ethical obligation to assist.

Conduct Unbecoming

The discipline of fire fighters for off-duty behavior has long been a point of contention between fire chiefs and unions. Incidents are varied and have included fire fighters suspended for off-duty employment as exotic dancers, posing for explicit pictures and calendars, the participation in controversial protests, and questionable business practices.

The primary questions involved with off-duty behavior are twofold. Does a department have a right to impose standards for off-duty conduct? And are there ethical obligations for fire fighters to refrain from controversial activities even if they are legal? The motives for fire departments to impose punishment for off-duty behavior can generally be classified into one of two categories. The first is reputation management; second is the imposition of perceived community standards for their own sake. For instance, a fire fighter was recently disciplined for expressing controversial opinions about race. The statements were made in public but were not related to the fire fighter's official capacity with the department. The reason for the discipline was simply that the mayor and city council were offended by the behavior.

Most fire chiefs recognize the importance of maintaining a positive public image for the department. Under the assumption that off-duty behavior negatively affects the fire service as much as on-duty behavior, fire administrators often seek to impose disciplinary standards for off-duty behavior called **morality clauses**. Determining the legality of imposing penalties for off-duty behavior rests in collective-bargaining agreements and the courts. While these clauses may be contentious and subject to interpretation, they remain relatively straightforward.

The ethics of a department controlling off-duty behavior become somewhat more complicated. Certainly, egregious behavior by fire fighters can damage the department's reputation, which, in turn, can undermine public support. Lack of public support can have a negative impact on the department's financial and operational well-being. As such, the department may have ethical standing to impose restrictions on behavior that may pose a risk to the department's well-being. Additionally, beyond the personal ethics of the questionable conduct, the fire fighter may have a professional ethical obligation to refrain from behavior that can harm the department and, by extension, his or her fellow fire fighters.

When a fire fighter joins a department, he or she should know that there are behavior expectations beyond that of the average citizen Figure 14-8. It is simply part of the job. It is also incumbent on the department to articulate clearly its behavior expectations.

The imposition of behavior standards apart from reputation management are problematic. The temptation to control others' behavior so that it conforms to our expectations of right and wrong is a powerful one. There is, however, no actual ethical imperative to restrict off-duty behavior if no harm can be demonstrated from it. It is also the obligation of the department to balance individual rights and freedoms against the likelihood of harm and the extent of harm. Just because the

Figure 14-8 Like it or not, when a fire fighter joins a department, there are behavior expectations beyond that of the average citizen.

© Radius Images/Getty Images Plus/Getty

fire chief does not like a behavior does not necessarily empower the chief to restrict the behavior. While fire fighters have a tacit obligation to protect the reputation of their department and fellow fire fighters, they also have personal rights. Generally, those activities that do not directly or indirectly harm others or negatively affect departmental operations are likely ethical.

It should be noted that the previous statement applies to the issue of ethics, not morality. As an example, a fire fighter having an extramarital affair may be deemed to be acting immorally by a significant portion of the population. However, it may not necessarily be *professionally* unethical, assuming that it does not involve or have a negative impact on the workplace.

Substance Abuse

The term "substance abuse" typically evokes images of illegal drugs, but the use and abuse of alcohol, tobacco, and prescription drugs are far more widespread. In the discussion of off-duty activities earlier in this chapter, we introduced several basic premises of ethical obligation.

Those same obligations are relevant to the issue of substance abuse. A fire fighter's job is physically demanding, more so than most other professions. Fire fighters shoulder tremendous responsibility for the welfare of their community as well as the welfare of their fellow fire fighters. The abuse of any substance that physically or mentally impairs their ability to perform those duties is a breach of those responsibilities Figure 14-9. To be paid for doing a job and not doing it to the best of your ability is inherently unethical. To make a promise to others that you will assist and protect them, and not do it to the best of your ability, is likewise unethical. Long-term abuse of chemical substances, including tobacco and alcohol, has been demonstrated to have significant health consequences. Those consequences are usually borne by the employer. Moreover, the same negative health consequences can have a disastrous impact on a family's financial security. As a matter of ethics, addicted fire fighters have an obligation to seek treatment, and nonaddicted fire fighters are required to protect their ability to perform. Others depend on it.

Figure 14-9 Firefighting is a physically demanding profession with immense responsibilities. As a result, fire fighters have an ethical obligation to maintain their physical abilities.

Social Issues

There are any number of issues in the world that bring people together and conversely pull people apart. Most of the great social debates can be categorized as either political or religious in origin. There is general agreement that politics and religion are topics best left out of the workplace, yet they almost always are present nonetheless. While there is near universal agreement that everyone is entitled to his or her opinions about religion and politics, differences of opinion in these areas frequently cause conflict. We mostly agree that politics and religion should not be discussed, yet we discuss both topics anyway, and while we honor the right of others to disagree with us, we get angry when they do. While the wisdom of avoiding the topics of religion and politics seems self-apparent, the ethical implications of politics and religion are nearly unavoidable.

Politics

At the time this book was written, the political climates of North America and Western Europe have never been more divisive. Political acrimony has led to increasing political passion, and political discourse often seems to be dominated by extremists. As a result, the issue of politics is becoming increasingly poignant in the workplace, especially in the close quarters of the firehouse. More and more fire fighters have become politically vocal, and more are politically active. As a result, the ethics of politics in the workplace becomes increasingly relevant.

With frustration and anger comes the urge to vent. With political passion comes the urge to climb on a soapbox and preach. At best, political outbursts are annoying to others; at their worst they can make a firehouse atmosphere completely toxic. Social scientists warn us that intolerance is on the rise. At some point over the last 25 years, many have come to believe that it is perfectly acceptable to insult and belittle individuals who have different political views. Political expression can sometimes cross over into vehemence and blatant hostility; this can create a hostile work environment, demoralize others, and potentially interfere with departmental operations. Verbal attacks, personal attacks, and political hostility are unethical in the workplace. Even absent insult or blatant hostility, the continued unwelcome expression of political views in the workplace can be tantamount to harassment. This behavior is also unethical as it causes harm. It is the responsibility of company officers to maintain a nonhostile work environment. They should not confuse an individual's right to free speech with the apparent belief that a fire fighter has a right to browbeat other employees.

It should go without saying that as public employees, fire fighters have an ethical obligation to act in the public interest. Political beliefs must not ever interfere with the fair and unbiased provision of public safety services. All citizens have a right to equal access to government services. The law guarantees equal access to government services, and as a matter of ethics, equal access is a question of justice and fairness. If fire trucks lead a political parade, they have an ethical responsibility to lead all political parades. If a fire department supports a religious function, it should support all religious functions.

As political fervor grows, increasing numbers of fire fighters are becoming interested in participating in the political process. This can range from the simple act of placing bumper stickers on vehicles to actually running for office. The reality of politics is that there are winners and losers. Most fire chiefs appreciate the political hazards of backing the wrong candidate. As a result, many fire chiefs endeavor to prohibit political activity by their fire fighters. As an example, several years ago a fire chief issued a policy restricting fire fighters from placing any political bumper stickers on personal vehicles if those vehicles were to be parked on the fire department property. Many communities have also endeavored to prohibit fire fighters from running for public office. Several court cases have found that such restrictions are constitutionally prohibited.

However, there are some ethical restrictions placed on politically active fire fighters. It is unethical for a fire fighter to falsely assert that he or she represents the political views of the fire department or fellow fire fighters unless explicitly empowered to do so. Fire fighters certainly have the legal right to free speech, but this does not relieve them of personal responsibility for the harm they may cause to others as a result of their speech. As such, provocative or inflammatory political activity may be legal, yet harmful to your fellow fire fighters and department as a whole. The ethical ramifications of such activity must be carefully weighed before engaging in political activism. The fundamental question becomes, Are the actions of the individual serving a greater good?

Apart from the ethical requirements of all elected officials, there are unique ethical issues for fire fighters who attain elected office. It is unethical for fire fighters elected to office to use their authority in a way that conflicts with departmental obligations. This often requires fire fighters acting in the capacity of elected officials to recuse themselves from voting on issues that directly affect the department. To the best of their ability, fire fighters have an ethical obligation to compartmentalize the responsibilities of elected office and those of their career.

Finanly, it should be noted that as government officials, fire fighters are covered by political restrictions specified in the Hatch Act (see Chapter 11, *Ethics and the Law*).

Religion

Similar to politics, religion in the workplace can present ethics challenges. For many people, faith is a primary force in their personal lives. As such, it is quite natural to want to talk about their religious beliefs. It is also quite common for individuals to want others to embrace their religion. Finally, some people have a hard time distinguishing between the moral obligations of their faith and their professional ethical responsibilities. One of the primary functions of religion is to provide guidance in daily life, but sometimes professional obligations and personal faith may appear to be in conflict. In Chapter 6, *Fire Fighter Responsibilities*, we introduced you to a Kentucky county clerk named Kimberly Davis who had a religious objection to a court decision that required her to issue marriage licenses to same-sex couples (Payne, 2015). In 2015, a volunteer fire fighter suggested that a good Christian should not put out a fire in a mosque (Gershon, 2015). Personal beliefs, especially religious beliefs, can inflame passion, but conflicts between personal beliefs and professional ethical obligations are relatively straightforward. As a government employee, a fire fighter is ethically obligated to provide services without bias toward race, religion, nationality, or sexual orientation **Figure 14-10**. This obligation arises out of the right of all citizens to have equal access to government services. Certainly, fire fighters have a right to their religious beliefs. Consequentially, fire officials do not have the right to impose their religious beliefs on their fire fighters. Recently, a fire chief in Georgia was dismissed for publicly condemning homosexuality for religious reasons. While no one would argue that he is entitled to his views, whether based on religion or otherwise, he has an ethical obligation to refrain from creating a hostile work environment for any gay employees he may have, and by the benefit of his position, tacitly creating a policy inconsistent with stated municipal and state policies. Because each citizen, including fire fighters, has a right to religious freedom, government officials must refrain from religious-based policy.

Ethics issues concerning religion are not strictly limited to policy. Similar to politics, individuals have absolute rights to their religious beliefs. However, they do not have the right to impose those beliefs on others, nor do they have the right to create a hostile work environment by repeatedly interjecting religious doctrine into the workplace. While it is normal to want to share one's beliefs, it is an ethical imperative to respect the rights of others. In the United States, Christianity is by far the predominant religion.

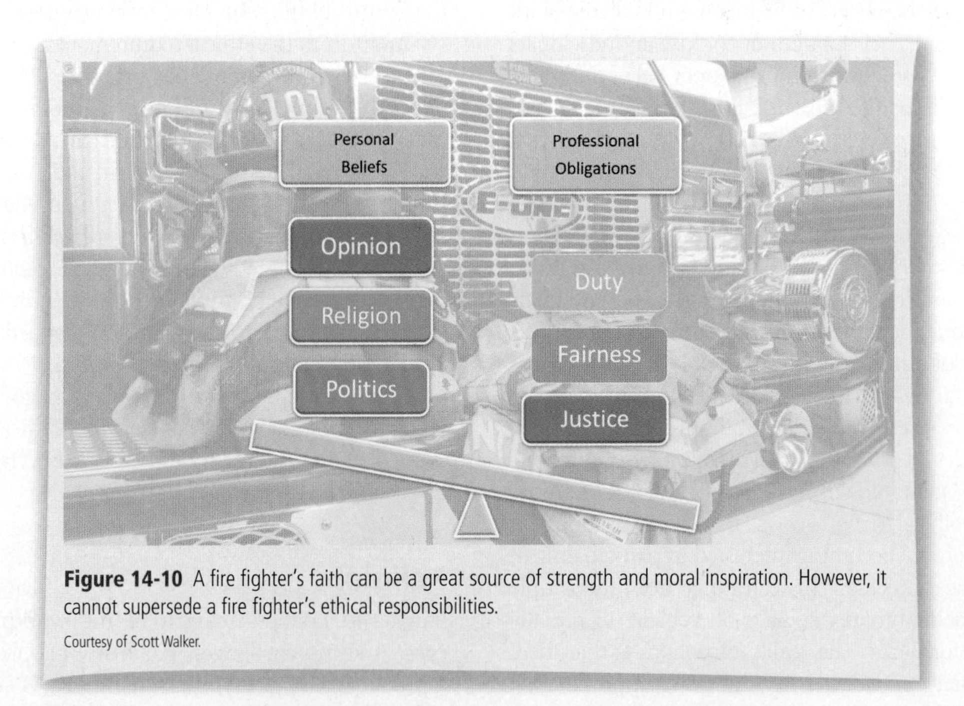

Figure 14-10 A fire fighter's faith can be a great source of strength and moral inspiration. However, it cannot supersede a fire fighter's ethical responsibilities.

Courtesy of Scott Walker.

It is sometimes easy to forget that religious freedoms were created to protect the minority from the influence of the majority. Given that most individuals are Christian does not excuse the insistence that others conform to the majority's beliefs and traditions. Separation of church and state exist to protect religious freedom, not limit it. Fire departments have an ethical obligation to respect those individual rights.

Harassment

In theory, harassment no longer exists in the fire service. In practice, it remains alive and well. The ethics concerning harassment is straightforward. Individuals have a right to expect a nonhostile workplace. Harassing behavior causes harm, it disrespects the humanity of the person being harassed, and it interferes with the primary responsibilities of the fire service and individual fire fighters. The harassment of others is deemed inappropriate by nearly every philosophical theory of ethics. Harassment can take on many forms. Specific focus has been given to sexual harassment over the last 25 years, as it is one of the most common variations of harassing behavior Figure 14-11. However, harassment can and does occur unrelated to gender. Fire fighters have been harassed for numerous reasons, including for physical appearance, speech idiosyncrasies, hygiene habits, nationality, religion, and ethnic background. Sometimes harassing behavior is not intended to be hurtful, and sometimes it is not malicious in intent. However, in all cases, harassment is unacceptable. Any behavior that belittles, humiliates, intimidates, or ostracizes is unethical and also destructive to the fire department's morale and efficiency.

Figure 14-11 Any behavior that intimidates or belittles another is unethical and destructive.

Courtesy of Scott Walker.

Note that lack of intent does not mitigate moral culpability regarding harassing behavior. Cultural sensitivity and empathy are reasonable requirements of fire fighters. Being insensitive and socially inept are not excuses for the tolerance of harassing behavior. Adults should know better.

Harassment can have dire consequences. In 2015, a fire fighter committed suicide as an apparent result of workplace harassment (Barton, 2016).

There have been numerous instances of fire fighters who have left the fire service because of hostile working environments. This is unacceptable. As a matter of ethical principle, each member of society has an obligation to avoid causing harm to others and to respect others' autonomy rights. Further, it is the ethical responsibility of fire officers to ensure that individual rights are respected.

Hazing

A unique form of harassment is **hazing**, which is defined as the ritualistic emotional or physical abuse of new group members by existing members. Nearly every fire department has a policy against hazing. Still, it occurs on a regular basis. Over the last few years, several notorious cases of hazing have led to personal injury, legal action, and unwanted national attention. Defenders of hazing argue that it is intended to be harmless fun that initiates members into the department. They further claim that hazing reinforces traditions and bonds new members to the group.

Tradition notwithstanding, hazing rituals are designed to embarrass and therefore are abusive Figure 14-12. They can and do cause harm. While the perpetrators will likely not admit it, hazing is about dominance. It is about extending control over another for entertainment purposes or perpetrator

Figure 14-12 Hazing is counterproductive to team building.

empowerment. In most cases, the tradition of hazing is perpetrated because those who were subjected to it seek to regain self-image by normalizing it, and so they participate in it as an aggressor when the opportunity arises (Total Life Counseling, 2018). By definition hazing ceremonies are intended to embarrass, humiliate, cause fear, or physically stress the victim. This is the very definition of unethical behavior.

Many departments have traditional ceremonies to commemorate individual achievement and to welcome new employees. Some involve humor and even practical jokes. So long as ceremonial rituals are not humiliating, dangerous, or culturally insensitive, they may have a team-building effect. The ethics of hazing does not lie in intent but rather in consequence.

On-the-Job Relationships

Romantic relationships within the workplace have been going on for as long as people have shared workspaces. Unfortunately, workplace romances have a long and notorious history of negative consequences. Ongoing relationships can be distracting and make fellow workers uncomfortable. When workplace relationships end, they can wreak havoc on productivity and on crew morale. Because of the difficulties that workplace romances can create for management, many businesses (including fire departments) have attempted to institute policies prohibiting or restricting workplace dating. Those policies are rarely successful.

Firehouse relationships can be particularly problematic. From a strictly professional point of view, in most cases dating fellow fire fighters is unwise. Fire departments are typically close-knit units; there are no secrets; and, similar to a family, everyone is affected by the actions of anyone. As such, romantic entanglements can have a negative impact on group culture. Unfortunately, relationships often end badly and the fallout can be difficult to manage. Because of the close quarters in which fire fighters work, separation after a breakup can be impractical. Common wisdom is that workplace relationships are ill-considered.

Wisdom aside, the ethics of workplace relationships vary greatly depending on circumstance. A consenting relationship between peers, kept separate from on-duty activities, is essentially ethical provided that the relationship does not interfere with a fire fighter's responsibilities. However, displays of affection in the workplace or even undue familiarity can make fellow workers uncomfortable. Participants in the relationship have an ethical responsibility to ensure that their behavior does not negatively affect the work environment or coworkers.

Conversely, a relationship between a superior officer and a subordinate is inherently unethical. Professionally, the relationship undermines the authority of the supervising officer. It also likely creates a negative impact on other fire fighters. The rest of the crew will likely assume (with cause) that the relationship will result in some form of favoritism extended to the subordinate employee. Even in the absence of favoritism, the supervising officer is inviting suspicion of motive in every decision and work assignment. The relationship may also have an adverse impact on the subordinate's relationship with his or her peers. The supervising officer has the responsibility to maintain a positive and productive work environment. Dating a subordinate interjects problematic issues that interfere with those responsibilities.

On a personal level, a relationship between a subordinate and a supervising officer is inherently unequal in power. Even though the relationship is consensual, the subordinate employee has less control, as his or her romantic partner has official authority over the subordinate's work life and career. Even if not exercised, the supervising officer has the ability to bully and coerce. Knowing that someone can abuse authority may be as threatening as the actual abuse of power.

The final ethical consideration applying to any workplace relationship is the potential harm that can come to the department via legal exposure. Relationships that end badly can be acrimonious. The problems created for management can include accommodating the involved employee's desire to be separated, allegations of harassing behavior, and even the loss of valued employees. All of these issues can have legal ramifications. Professionally, a fire fighter is obligated to competently perform his or her duties and support the mission of the department. You can make an argument that the pursuit of a relationship that undermines the department's efficiency, or potentially exposes the department to significant legal expense, is a breach of that responsibility.

Staffing and Safety

Perhaps one of the most important and challenging conversations being held in the fire service today regards the threat of understaffing to the safety of fire fighters, and what to do about it. Recent trends in tax policy, a diminishing middle class, and increasing operating costs have conspired to create a funding crisis affecting municipalities across the nation. For more than two decades there has been nearly constant pressure on fire department staffing levels. Fire fighters have been asked to "do more with less." As stewards of public money they have every obligation to be as efficient as possible. However, fire fighters perform dangerous and labor-intensive work; adequate staffing is an absolute requirement to maintain safety standards.

Figure 14-13 Go or no go? Diminished staffing levels are creating some tough decisions for incident commanders. At what point does "doing more with less" become irresponsible? How much risk is too much risk?

For many years, fire departments have struggled to maintain operational status quo despite decreasing staffing. Many in the fire service are questioning the wisdom of this practice. They argue that departments are risking the well-being of fire fighters by attempting to continue past practices with inadequate resources Figure 14-13. As two-person engine companies have become a norm, there have been suggestions that understaffed fire departments should consider discontinuation of interior attack and provide only those fire suppression activities that can be safely performed with available personnel. Others argue that the policy of "letting it burn" is contrary to the mission of the fire service, and risk is inherent in the profession.

The problem of understaffing and fire fighter safety is not just a logistical one; it is also a multifaceted ethical dilemma. On the one hand, fire fighters willingly accept the obligation to save lives and protect property, and they accept the associated risks inherent in that mission. Fire fighters are the last hope for our neighbors in dire need. Fire fighters make a promise to help and so there is an ethical obligation to live up to their responsibilities.

The public expects heroic effort from their fire department, even with reduced staffing. A significant amount of their tax dollars goes to public safety. Moreover, fire fighters want to perform their duties as expected. Witness the fact that some fire departments are routinely conducting interior fire operations with as few as four people on the fire ground.

On the other hand, fire departments have an ethical obligation to protect the health and safety of their fire fighters. Fighting fires should be considered a profession, not a cause, and no one should be expected to commit suicide to save a 1500-foot, ranch-style house. If a municipality cannot or will not provide adequate resources, then fire administration has an ethical responsibility to scale back services to protect their employees.

There is an old saying in the fire service: "Risk a little to save a little, and risk a lot to save a lot." Logically, the ethics of risk-benefit analysis seems clear. A fire officer's ethical responsibility to protect the lives of his or her fire fighters must supersede any mission obligation to protect property. The difficulty in performing a risk-benefit analysis is in determining acceptable risk. Exactly what constitutes a little risk? Even after decades of focus on fire fighter safety, fire departments are still routinely sending people into unduly dangerous situations. It is only after a serious injury that hard questions are asked. When a citizen's home and property are going up in flames, it is incredibly difficult to explain that having fire fighters attempt to save the property is too big a risk for too little reward. And so too often, understaffed departments make ill-considered fire attacks. The responsibility of an officer to be competent in fire tactics is paramount, and because fire fighters' lives depend on it, competence and caution are ethical imperatives.

The "risk a lot to save a lot" half of the risk-benefit equation is much more ethically problematic. The euphemistic term "a lot" usually refers to human life. Philosophically and logically, a fire fighter's life is worth as much as any other life. Therefore, the concept of "risking a lot" is ethically problematic. However, fire fighters voluntarily undertake the responsibility and risks associated with the profession, specifically to save lives Figure 14-14. Thus, the ethics of "risking a lot to save a lot" seems justifiable. The balancing of fire fighter lives against the imperative of protecting civilian lives is a significant ethical issue, complicated by high emotion.

Figure 14-14 The difficulty in assessing the ethics of risk analysis is determining just what constitutes "reasonable risk."

The public has the absolute expectation that fire fighters will intercede (even at great risk) to save a civilian. Further, fire fighters are deeply emotionally invested in that responsibility, even in the face of great personal risk. Again, it falls on the fire officer to make tactical decisions based on two separate realities: that fire fighters' lives are as valuable as those of the public, and any undertaking must be balanced against the likelihood of success versus injury. At play are two very distinct ethical priorities. First, do no harm. Second, when possible, minimize harm to others. The second ethical priority is not justification for ignoring the first.

In keeping with the first ethical principle, it is the responsibility of fire officers to avoid placing fire fighters in unreasonably dangerous conditions. Understaffing creates inherently and unreasonably dangerous circumstances. Hence, a fire officer is ethically justified in limiting response if, in his or her honest opinion, there are inadequate resources to reasonably assume the safety of fire crews or a successful outcome.

Alternative Funding

Pay for Spray

Pay for spray is a euphemism for the practice of contracting fire protection, usually on an annual basis, to unincorporated areas. Unfortunately, there have been several highly controversial and very public incidents involving fire departments selling fire service. The cases all seem to have a common thread—a local fire department or district sells service to an otherwise unprotected or under-protected area and is called to respond to a call for help from someone who has not bought protection.

From a strict business sense, the withholding of service to noncustomers is both logical and ethical. In business, a sales company has no obligation to provide a service that is not purchased, and it would be foolish to do so. However, fire departments are not businesses in the typical sense. First, we provide protective services of life and property. Therefore, withholding such services has significantly stronger ethical implications than does withholding other types of services. Second, the fire service has existed for over 200 years as a prototypical benefice organization. We are organized around the idea of helping others. The public's loyal support of the fire service is rooted in the image of the heroic protector of the community. Withholding services when a call for help is placed does not sit comfortably with the traditions or values of the fire service. Neither does it sit well with the public's expectations. Not surprisingly, when incidents of fire departments withholding service to nonpaying customers occurs, only a small group supports the department's decision as common sense **Figure 14-15**. A larger and more vocal group typically expresses outrage and disappointment.

As departments are feeling more and more financial urgency due to shrinking budgets, and as the

PALOS VERDES PENINSULA NEWS — SATURDAY, MARCH 13, 2004

LOCAL NEWS

Norris Center for the Performing Arts, 27570 Crossfield Drive in RHE, on Saturday, March 20 at 2 and 4 p.m. Tickets are $38 for adults and $19 for students for the evening performance and $30 for adults and $15 for students at the matinee performance. For reservations, call 544-0403.

• UPCOMING — The Palos Verdes Peninsula Unified School District and Friends of School Music host the 15th Palos Verdes Elementary Choral Festival on March 23, 24 and 25 at the Norris Center for the Performing Arts, 27570 Crossfield Drive in RHE. All shows begin at 7:30 p.m. For tickets, call the Norris box office at 544-0403.

• ONGOING — The Distinctive Edge, 29050 S. Western Ave., Suite 113 in RPV, continues "Third Time's a Charm," an exhibit of 3-D collages by artist Steve Jacobsen, through March 30. For gallery hours, call 833-3613.

• ONGOING — "Natural Treasures" exhibition contin-

Firefighters Watch House Burn to The Ground: No Money, No Help

Two years ago, U.S. Navy personnel and their families assigned to the Atsugi Navy base, home of the U.S.S. Kittyhawk, were treated to a rare experience when Terry Fleming and his local Irish/American band, Innisfree, traveled to the base to entertain them on St. Patrick's Day. Fleming and the other five members of Innisfree were delighted and honored to be able to go to Japan and lift the spirits, if only for a few hours, of the Navy personnel and their families.

For the third year in a row, Fleming — a local insurance broker in Rolling Hills by day and an entertainer by night — and the band travel to entertain the Navy men, women and families at various bases throughout Japan.

Fleming, the leader of the band on accordion and harmonica, actually is the only member of the band from Ireland. Other members include lead singer Julie Delaney, a civil engineer in Newport Beach; Terry Doyle, guitar, a news director with CBS news; Denis Doyle, Celtic harpist, a professor at Glendale College; Kevin Weed, keyboards and bagpipes, music teacher and assistant director of the Orange County Symphony; and Mike Tiffany, bass, a computer engineer. The band has been playing the length and breadth of California for the past 23 years. They have played at pubs, wakes, weddings, birthdays and on occasions where there was little excuse for throwing a party.

Fleming says it was by coincidence the band got the opportunity to travel to Japan. Another band was unable to travel at the last minute and so he and his band were offered the opportunity to go in their place.

With some trepidation they made their first trip and with the overwhelming response they received at Atsugi, any fears they had were quickly allayed. On a damp St. Patrick's Day, hundreds of families, clad in many shades of green, whooped it up, sang their hearts out and danced up a storm. As the evening wore on, many in the audience were emboldened to try their hand or foot at the Irish gig, with much encouragement from the band.

Even though far from home, the Atsugi base — situated a few hours south of Tokyo — felt like home away from home, with its lush green rolling landscape and its multitude of cherry blossom trees. "Yet," Fleming says, "one was struck by the commitment and dedication of our men and women in uniform as they played their part in protecting and serving in an ever challenging and hostile world."

A couple of the families even took time out from their busy schedule to host Terry and the band members. They treated them to a guided tour of the base and accompanied them on a few exciting trips off the base, visiting beautiful ancient temples, monuments and revered giant Buddas.

On a visit to downtown Tokyo, the band came across what they assumed was a very rare sight, a place called "Scruffy Murphys," an Irish pub located in the heart of a bristling downtown. Upon checking the establishment out, they discovered a real authentic Irish Pub with excellent Guinness and good pub grub. It also happened to be open mic night, so the band members took over the stage and entertained the locals for a few fun-filled hours. It turned out that it was just one of many establishments in the city.

A special bond developed between the band members and these families and already exchange visits have occurred when the same families were on leave in the United States.

For more information about the band, log on to [illegible]

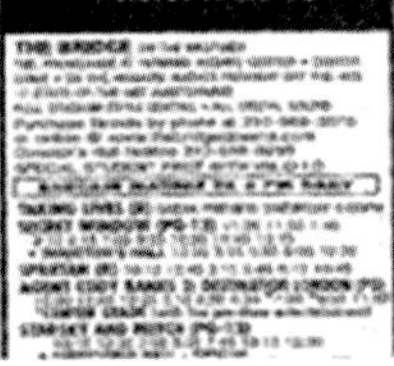

Figure 14-15 It is difficult to reduce emergency response to a business model.

Courtesy of Scott Walker.

prevailing political climate has called for government to run more like a business, fire departments are exploring entrepreneurial initiatives. Unfortunately, this can create a public relations nightmare when fire departments are forced to withhold service for nonpayment (MSNBC.com, 2010).

The question becomes: Are the ethics associated with a fire department withholding service for nonpayment any different than the same practice of any other business?

Hospitals may provide an answer to that question. Similar to fire departments, they provide life-saving services. Both ethical and legal principles require hospitals to provide emergency service regardless of ability to pay. Ethics assert that human life and well-being must supersede commercial consideration. As such, a fire department at the scene of a structure fire or in response to a life-and-death emergency has an ethical obligation to intercede to protect life if possible. Human life supersedes financial concerns.

For a fire department, the further issue of property protection must also be considered. In the private sector, businesses such as roofing and plumbing contractors have no ethical obligation to provide property protection services absent payment. However, many hold that fire departments are unique in that, first, they are public-funded organizations, and second, withholding service for financial consideration is inconsistent with the fire service's values.

Withholding services for financial consideration is questionable from a public relations standpoint, as it is not consistent with the stated values or image of the fire service. However, acting contrary to stated values is not necessarily unethical. If no direct promise of service is made absent payment, then a fire department is likely not ethically obligated to provide coverage outside its district. However, the conditions of this ethical argument are modified if a property owner has a reasonable expectation of service delivery absent a contract.

The department officer faced with this issue may do well to ask the following question: Finances aside, would my department respond to a request for mutual aid from another department? If so, how can I justify not extending the same courtesy to an individual citizen? The complication is that fire departments routinely provide services to nontaxpayers. Fire departments do not check the license plates or registration of a vehicle before they put the car fire out. Nor do they check for residency before administering emergency medical care. The fire service motto is *not* to save lives and protect property subject to payment. Given that the majority of the public is surprised and disappointed by a fire department withholding service indicates that they assume that fire departments exist to help, regardless of status. Realistically, that assumption is fostered by the fire service itself. The fire service does not market itself as a business. Whether we admit it or not, we embrace and encourage the image of life-saving heroes. Although open for debate, you can make the argument that ethical attachment to provide service is created by the fire service's promise to provide help in times of need. Promises create responsibility, which can create ethical obligation.

In the end, the ethics of withholding services for nonpayment rests in a gray area, but it is inconsistent with the fire service's values, is a public relations nightmare, and is likely bad for morale.

Entrepreneurial Initiatives

Beyond fire suppression services, many departments have engaged in creative entrepreneurial endeavors to augment budgets. Examples include: contracting EMS and fire apparatus at sporting events and festivals, providing OSHA-required standby teams for confined space entry, provision of mandated hazardous materials training, and nonemergency window boarding.

The ethical considerations relative to these endeavors are similar to those of normal business ethics. However, certain ethical issues unique to the fire service can arise. For instance, there may be a conflict of interest when for-fee services are provided that are required by law and are enforced by the local fire department. For instance, assume the local municipality requires an annual inspection of sprinkler systems with results reported to the local authority having jurisdiction. It is a conflict of interest for a fire department to provide that inspection on a fee-based arrangement if it also has the power to fine or in some other way punish noncompliance with the requirements for inspection.

Clarification

It is ethically questionable for a department to charge a fee for a service that is legally mandated and enforced by the department, and for which the department can penalize for noncompliance.

A common problem faced by fire departments providing entrepreneurial services is complaints by local businesses providing similar services. They may see the fire department activities as unfair competition. This is a political issue, not necessarily an ethics issue. While a local businessperson may not like the fire service providing services, there is nothing inherently unethical about the practice assuming local elected officials support it. Again, one must also assume that the department does not use its ministerial authority to create an unfair advantage in the marketplace.

A final common complication faced by fire departments providing entrepreneurial services are union objections. Historically, there has been pushback by fire fighters who have been assigned nontraditional work details for which the local fire department is charging service fees. Certainly, fire departments have an ethical obligation to comply with the terms of the fire fighters' collective bargaining agreement, as well as local and state labor-related law. However, there is no *ethical* obligation to refrain from work assignments that are inconsistent with the terms and spirit of the fire fighters' union contract.

As with the issue of contracted fire protection, the wisdom of entrepreneurial initiatives should be considered carefully before implementation. The likelihood of contentious reactions by fire fighters and competing local businesses is possible, as are allegations of unfair practices. The issue of fairness is subjective and will ultimately be determined by local government. But, fairness aside, entrepreneurial activity is not inherently unethical.

Hear No Evil, See No Evil

The issue of loyalty often presents ethical dilemmas. In many ways, the fire service is particularly affected by loyalty issues. The fire service is team oriented—mutual interdependence and trust are core values. At emergency scenes, we must depend on each other, and we trust that our fellow fire fighters will be there for us when we need them **Figure 14-16**. That same attitude rightfully extends to the engine house. Fire fighters work together, eat together, laugh together, and sometimes cry together. In many important ways, fire departments are like families. Loyalty and trust should be celebrated. However, loyalty and trust can be misplaced and abused. Hence, issues of loyalty are often at the root of ethical and moral dilemmas.

Loyalty and Silence

Perhaps the most common question raised in ethics classes is the proper response to the knowledge that a friend is behaving inappropriately. For example, you learn that one of your fellow fire fighters is abusing

Figure 14-16 Firefighters are often like family. Loyalty is important, but it can be misplaced.

© Ken Redding/Corbis/Getty Images

prescription drugs and has reported for duty with diminished capacity. Do you report him or her? Do you confront the person? Alternatively, do you simply look the other way? As a hypothetical ethics question, the issue is relatively straightforward. You have an obligation to report that person. His or her diminished capacity poses a significant threat to his or her ability to perform duties. Further, that person poses a safety threat to him- or herself and fellow fire fighters.

Unfortunately, when fire fighters are faced with this issue in reality, the choices seem somewhat more complicated. Our sense of loyalty and our sense of trust are very real emotions, and they often appear to outweigh some abstract ethical obligation. Earlier we explored the concept of **enabling behavior**, which can be described as assisting others in unethical behavior through one's own silence. Ethically, assisting someone in behaving unethically is equally unethical.

Loyalty and trust must be confined to the support of appropriate behavior and can even extend to assisting fellow fire fighters in dealing with the consequences of unacceptable behavior. For example, I knew of a battalion chief who was forced to "file charges" on a friend for falsely reporting overtime in order to pay overdue debts. The fire fighter was suspended, and the battalion chief personally loaned the fire fighter money to pay off the debts. The battalion chief's reporting of the issue was an act of loyalty to duty. The loaning of money was an act of loyalty to the fire fighter. Loyalty cannot constitute an ethical argument that supports "looking the other way." From an ethical standpoint, loyalty to duty is as real as loyalty to people. This applies in particular to supervisors, who not only have an ethical obligation to refrain from assisting in unethical behavior but also have a responsibility to report that behavior as an obligation of rank.

Union Activity

A particularly common ethical dilemma fire fighters (in particular fire officers) face are the demands associated with union leadership. As union officials, fire fighters have an ethical obligation to represent their members. As department paid members, they have an ethical obligation to act in a way that furthers the mission of the department and best serves the public's interests.

In most business settings employees with managerial and disciplinary responsibilities are typically not in the collective bargaining unit, but that is not the case with the fire service. Although individual departments vary, it is not uncommon for chief officers to be in collective bargaining units, and it is almost the norm that company officers are union members.

Fortunately, most of the time, union and fire officer obligations do not conflict; they are usually mutually beneficial. In general, what is good for fire fighters is good for the department. However, there are times when the demands of both positions seem to compete. This is particularly the case in disciplinary issues. A fire officer has an obligation to enforce policy and to report inappropriate behavior. As a union official, that same fire officer has a duty to present a best defense to an individual charged with inappropriate conduct.

In many cases, union officials are required to "walk a thin line" to meet both obligations, and they often do so successfully. However, there is a legitimate ethical question to be raised when the responsibilities of being a fire fighter or officer directly conflict with those of being a union official. If a person cannot serve "both masters," where should ethical priorities lie? Although I was a former union officer and recognize the valuable role of fire fighter unions in maintaining working conditions, participation in union leadership is voluntary and therefore secondary to the obligations of being a fire fighter or fire officer.

As repeated several times throughout this text, fire fighters' responsibilities are rooted in life safety and property protection. By joining a fire department, you accept an ethical obligation to conduct yourself in a way that supports the department's mission. Further, as public officials supported by tax money, fire fighters have a duty to act in the public interest. Acceptance of those managerial responsibilities supersedes union interests in the rare cases where the two are in conflict.

WRAP-UP

Chapter Summary

- Privacy is becoming an issue of grave concern for contemporary Western society.
- Technology has had a significant impact on the availability of information and the immediacy of access to it. While most people recognize the benefits of the information revolution, there are certainly ethical issues that arise.
- Because fire fighters routinely provide emergency medical services, the privacy requirements imposed by HIPPA were presented.
- Beyond legal obligations, fire fighters have the ethical obligation to maintain patient dignity and demonstrate respect for patient privacy.
- This chapter also considered the ethical implications of respecting victim privacy rights in all aspects of fire service response and honoring the trust placed in emergency responders by individuals in need of help.
- Those most desperate for help have little choice but to call the local fire department. It is inherently unfair to violate trust by those in desperate need and who have no choice but to confide in fire fighters.
- The harm done to personal reputations and self-image can be devastating when fire fighters betray the trust that is extended to them. Under the ethical principle of "do no harm," damaging reputations and self-image are inherently unethical.
- The growing practice of utilizing body cameras to record fire service operations was examined. While there was an acknowledgment that the use of body cameras could have value in training and operational critique, respect for victim privacy must be given priority.
- Rules and policies regarding social media usage are commonplace in the fire service. The distribution of images captured on duty for personal enjoyment or for profit over social media are of ethical concern.
- The ethics of using department time and equipment for personal profit were explored in depth, as was the ethical considerations of privacy violation.
- Ethical limitations of "unofficially" representing a local fire department without official permission to do so through social media was discussed. Any social media post or blog that tacitly or directly implies an author's editorial opinion as being "representative of the department's" was found unethical.
- Individuals who publicly identify themselves as members of the department are ethically responsible for the consequences of inflammatory content within social media and blog postings. Having a First Amendment right to make inflammatory statements does not necessarily absolve a person from the ethical responsibilities associated with damage done to others. Communications that are misleading, deceptive, or harmful to others are inherently unethical.
- The ethical consequences related to social media usage versus traditional media were explored. It was found that mass communication allows for greater consequences for unethical communication, but the underlying issue of ethics remains rooted in the content—not the medium.
- The use of social media is no more or less ethical than older technology. Only the consequences and likelihood of its abuse are different.
- Off-duty employment, off-duty response, substance abuse, and actions considered conduct unbecoming were discussed. It was found that there is a fine line between the fire fighter's right to conduct his or her life according to personal moral values and

WRAP-UP

interests, and the person's responsibilities to the department.

- It was found that behaviors that impair the fire fighter's ability to perform his or her duties were ethically problematic. Further, activities that bring disrepute or in some way damage a fire department's image are contrary to the ethical obligation a person has to his or her fellow fire fighters.
- As often discussed in this text, a fire department's reputation has value. Ethically, a person does not have the right to destroy something of value that belongs to another.
- Fire fighters have protected rights to engage in the political process and to practice religion as they deem appropriate. Fire fighters do not have the right to impose political or religious beliefs on others, nor do groups of fire fighters have the right to directly or indirectly compel conformance to majority beliefs.
- All fire fighters have a right to work in a nonhostile environment. Fire officers have the responsibility to ensure that political and religious beliefs do not become contentious within the workplace.
- Fire fighters and officers have an ethical obligation to respect the individual privacy and dignity of their fellow fire fighters. All fire fighters have a right to a nonhostile workplace. As such, harassment of any kind is inherently unethical.
- All ethical systems recognize the individual's rights to autonomy, and all condemn harming others for personal gain or satisfaction. Harassment—including hazing—is unacceptable, even in cases where harm is unintended.
- Fire fighters have a responsibility to be "emotionally intelligent." It was pointed out that cultural insensitivity does not excuse harmful speech or action.
- The ethics of relationships between employees is dependent on the relative status of the consenting participants. Relationships were found to be inherently ethical assuming that all parties are free to provide consent without coercion and are of equal empowerment (rank), and that the relationship is not causing disruption within the workplace.
- It was found that relationships between supervisors and employees were inherently unethical as they have an unequal distribution of "power" within the relationship, can compromise supervisor effectiveness, and expose the department to operational and legal risk.
- The ethical implications of limiting fire operations as a result of insufficient staffing were explored. Many fire departments struggle with the competing obligations of providing fire protection and protecting the lives of their fire fighters in the presence of understaffing.
- Fire fighter life and safety are at least equally valuable as the life and safety of civilians, and clearly more valuable than that of property. It was found that fire officers have an ethical obligation not to expose fire crews to unreasonable risk.
- The topic of entrepreneurial initiatives was discussed, including the ethical implications of pay for spray programs, and the withholding of fire protection from individuals who had not subscribed.
- From a business perspective, it was found that fire departments do have an ethical obligation to provide life-safety services even absent contractual obligation, but that ethical obligation does not extend to the protection of property. However, fire departments are not a traditional business and have a long history of providing mutual aid and emergency services to nontaxpayers.

- Withholding assistance from people in need is not consistent with fire department values or traditions, nor is it consistent with the image that the fire service embraces. Because of the nature of a fire department's mission and relationship with the public, traditional business ethics are insufficient.
- The chapter closed with a discussion of the ethical dilemmas fire fighters often face as a result of competing values of loyalty to comrades and the duty to report inappropriate behavior by fellow fire fighters.
- Fire officers have an obligation to enforce policy; it is fundamental to their job description and their responsibilities. Further, loyalty to the department is as significant as loyalty to individual fire fighters.
- While the value of loyalty and trust among fire fighters is recognized, supporting unethical or illegal behavior by remaining silent ignores ethical responsibility. Enabling unethical behavior through silence was found as problematic as the unethical act.

Key Terms

blog a discussion or informational website published on the World Wide Web and consisting of discrete, often informal diary-style text entries.

body camera also known as body worn video (BWV); a video-recording system that is typically utilized by law enforcement officers to record their interactions with the public or gather video evidence at crime scenes. It has been known to increase both officer and citizen accountability.

enabling behavior actively or passively supporting the actions of others.

hazing the ritualistic emotional or physical abuse of new group members by existing members.

Health Insurance Portability and Accountability Act (HIPAA) A medical privacy act passed in 1996 that mandates industry-wide standards for health care information on billing and other processes, and requires the protection and confidential handling of protected health information.

morals or morality clause a provision within a contract or department policy that curtails, restrains, or prohibits certain behavior of individuals as inappropriate.

pay for spray a euphemism for the practice of contracting fire protection, usually on an annual basis, to unincorporated areas.

Challenging Questions

To check your understanding of this chapter's material, answer the following questions. It is highly recommended that you discuss your viewpoints with fellow students, peers, coworkers, and friends to discover their opinions as well.

- Many ethical dilemmas exist as a result of conflicts between personal values and professional responsibilities. How would you prioritize these elements should they be in conflict?
- Your text asserts that personal rights do not necessarily justify the ethics of actions. Can you articulate an argument in support of that assertion?
- Many departments have "morality clauses" that limit off-duty behavior deemed inconsistent

with community values. Give two arguments in favor of and opposed to this practice.

- By either visiting fireethics.com (one of the major fire service web portals) or conducting a general Internet search, select a recent account of a fire official under investigation or sanctioned for inappropriate behavior. Identify the underlying ethical issues, explain the relevant ethical principles involved, and summarize your views regarding the correct outcome.
- Based on ethics theory, provide both supporting and opposing arguments regarding the cessation of interior firefighting operations in cases of understaffing. Critique both the pros and cons and summarize your opinion of the issue.

Case Study Conclusion

Revisit the case study at the beginning of the chapter. Spend a few minutes considering the questions posed at the end of the case study. In light of the information shared in this chapter, have any of your original observations changed?

Fire fighter Smith's photographic pursuits were initially undertaken as a hobby whose results were beneficial to his local department. He sought to make no money from them, and the distribution of the material was limited to the department. Assuming that Smith's actions did not violate the individual privacy of anyone in the photographs, his actions were essentially ethical. No deceit was involved nor was any personal gain sought at the expense of others. Further, his hobby did not conflict with professional responsibility, and no harm came to anyone, nor was it intended.

The scope of the issue changes dramatically when Smith begins to disseminate information on his public website. Even if he did not receive money for his efforts, he began using department time and the privileges given to him as a condition of employment for a reason other than the benefit of the department. Once he began making money from the website, he undertook an ethical liability for an apparent conflict of interest between his responsibilities as a fire fighter and his profit incentive.

By making a financial arrangement with the local media, Smith also created a potential conflict of interest. As he is paid by the fire department, his superiors and peers can reasonably assume his actions are consistent with an obligation to his professional responsibilities. As a paid agent of the news media, his employers there can also assume his actions are consistent with their needs. When the interests or the missions of the two agencies diverge, Smith will have an ethical conflict of responsibility.

The issue of discipline is complex. It is evident that for several years the department had permitted Smith to use images shot both on and off duty on his website. As such, it would appear that the department has been inconsistent in its reaction to his use of the material. When it was complimentary it was tolerated; when it had a negative outcome for the department, it was censured. From this inconsistency, an argument can be made that the department failed to recognize the ethical implications of privacy and conflict of interest. Smith's behavior was ethically questionable. However, the lack of department response earlier on probably limits the department's ability to do so from a legal point of view.

Chapter Review Questions

1. What does HIPPA legislation provide?
2. Do constitutional rights equate to ethical conformance?
3. What are morality clauses?
4. Define hazing.
5. Is it ethical for a career fire fighter to hold public office?
6. What is the definition of pay for spray?
7. Define enabling behavior.

References

Barton, M. A. 2016. "Did Fire Fighter Commit Suicide After Cyber Bullying? Fire Chief to Investigate." Fairfax Station Patch. Accessed June 23, 2018. https://patch.com/virginia/fairfaxstation/did-firefighter-commit-suicide-after-cyber-bullying-fire-chief-investigate.

California Department of Health Care Services. n.d. "Health Insurance Portability and Accountability Act." Accessed May 30, 2017. http://www.dhcs.ca.gov/formsandpubs/laws/hipaa/Pages/1.00WhatisHIPAA.aspx.

Dennis, L. 2015. "MPD Looks to Navigate Privacy and Ethical Concerns That Come with Body Cameras." WUWM. Accessed April 17, 2017. http://wuwm.com/post/mpd-looks-navigate-privacy-and-ethical-concerns-come-use-body-cameras#stream/0.

Gershon, D. H. 2014. "As Houston Islamic Center Burns, Fire Fighter Posts, 'Let It Burn . . . Block the Fire Hydrant." Daily Kos. Accessed June 23, 2018. http://www.dailykos.com/story/2015/2/14/1364404/-As-Houston-Islamic-Center-Burns-Firefighter-Posts-Let-it-burn-block-the-fire-hydrant.

MSNBC.com. 2010. "No Pay No Spray." Accessed May 30, 2016. http://www.nbcnews.com/id/39516346/ns/us_news-life/t/no-pay-no-spray-firefighters-let-home-burn/#.WQ9E29LyuUk.

Payne, E. 2015. "Who Is Kentucky Clerk Kim Davis?" CNN. Accessed June 23, 2018. http://www.cnn.com/2015/09/04/us/kentucky-clerk-kim-davis/.

Rucke, K. 2013. "Is a Ban on Fire Fighter Helmet Cameras Really About Protecting Privacy?" MintPress News. Accessed June 23, 2013. http://www.mintpressnews.com/is-a-ban-on-firefighter-helmet-cameras-really-about-protecting-privacy/167445/.

Snyder P. 2017. "Facing Hostility Fire Fighters in Rural Oregon Raising Money for Body Cameras." *Seattle Times*. Accessed June 23, 2018. http://www.seattletimes.com/seattle-news/crime/facing-hostility-firefighters-in-rural-oregon-raising-money-for-body-cameras/.

Total Life Counseling. 2018. "5 Reasons Why Hazing Still Happens and How to Stop Hazing." Accessed June 23, 2018. http://www.totallifecounseling.com/5-reasons-why-hazing-still-happens-how-to-stop-hazing-robert-champion-bria-shante-hunter/.

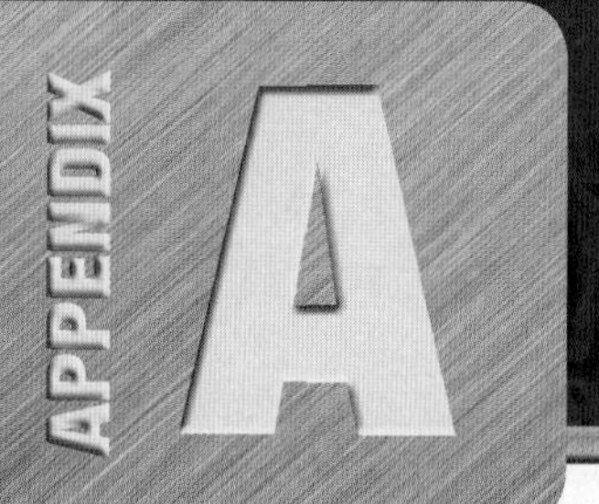

Fire and Emergency Services Higher Education (FESHE) Correlation Guide

In June 2017, the National Fire Academy in Emmetsburg, Maryland, hosted the Fire and Emergency Services Higher Education Professional Development Symposium. An important element in the symposium is the meetings of the workgroups who oversee the review of the FESHE standardized curriculum.

At the 2017 meeting the baccalaureate workgroup took a significant step forward in support of fire service professional development by adding a new (seventh) course to the core curriculum dedicated to fire service ethics. Course objectives and a model course outline were submitted to the FESHE body and were nearly unanimously approved. It was the consensus of the group that ethics is an important element in fire service education, and that its inclusion should be a required course in fire programs within the academic setting.

I participated in those FESHE meetings and played a significant role in the curriculum's development. This text was created specifically to support the FESHE ethics curriculum.

FESHE Content Area Matrix

The table that follows correlates the learning outcomes and model course outlines to this book's chapters.

Fire Service Ethics		
Course Description	This course examines the basic principles of ethics as related to fire service operations and management, with special attention given to current issues in the fire service.	
FESHE Outcomes	**Description**	**Chapter Reference**
1	Identify what the term "ethics" means, and why it is important to the fire service.	1
2	Distinguish between social norms, morality, ethics, and the law.	2
3	Compare and contrast the concepts of values, beliefs, and attitudes.	3
4	Explore how the concepts of accountability, obligation, and responsibility define ethical behavior.	10
5	Contrast modern and classical philosophy of ethical study as they relate to the fire service.	4
6	Contrast and compare fire service ethics standards and guidelines.	5, 6
7	Review a fire fighter's professional obligations and responsibilities.	5, 6
8	Appraise the ethical responsibilities associated with leadership.	7
9	Evaluate current issues in fire service ethics.	14
10	Identify the values of a diverse workplace.	8
11	Identify benefits, hindrances, and tactics related to achieving diversity.	8

FESHE Outcomes	Description	Chapter Reference
12	Compare and contrast internal versus external ethical control systems.	10
13	Review the principles of integrity-based management programs.	9
14	Apply relevant state and federal ethics laws to fire service administration.	11
15	Appraise various influences on ethical decision making.	12, 13
16	Evaluate tactics for implementing an ethical culture.	9
17	Explore best practices in building an ethical culture.	9

Fire and Emergency Services Higher Education (FESHE) Model Outline

FESHE Outline	Section 1: Foundations	Chapter Reference
I	Introduction a. What is ethics? b. Why study ethics? c. Why be ethical? d. Why is ethics of value to the fire service?	1
II	Understanding Ethics a. Social Norms i. Norms ii. Folkways iii. Mores iv. Taboos b. Morality i. What is it? ii. Where does it come from? c. Ethics i. How do they differ from morals? ii. How are ethics determined? iii. Professional versus personal ethics d. Law and Ethics i. *Malum in se* and *malum prohibitum* ii. Law and ethics iii. Law and morality e. The role of religion i. Religion and morality ii. Religion and free will	2
III	Behavior Influences a. Values b. Beliefs c. Attitudes d. Distinguishing needs and wants e. The role of responsibility i. Subjective responsibility ii. Objective responsibility f. Integration of behavior	3

IV	Ethics and Philosophy a. Validity b. Normative ethics i. Consequentialism ii. Deontology iii. Virtue ethics c. Constructivism i. Benefice ethics ii. Egoism d. Social contracts e. Subjectivism f. Metaethics	4
Section 2: Ethics in the Fire House		
V	Professional Ethics within the Fire Service a. Are we a profession or a job? b. Professional standards c. Fire fighter's code of ethics d. The importance of character	5
VI	Fire Fighter's Responsibility a. Fire service values i. Duty ii. Compassion iii. Honesty iv. Team work v. Competency vi. Bravery vii. Loyalty—thin red line b. Objective responsibilities i. Department policy ii. Competency iii. Safety c. Honoring trust i. Privacy ii. HIPPA iii. Gossip d. Subjective responsibility i. Personal values ii. Career ambition iii. Tradition iv. Balancing home life with work life e. Benefice ethics and contact theory applied to emergency response	6

VII	Ethics and Leadership a. Ethics and company officers i. Enforcing policy ii. Modeling behavior b. Duty i. Duty to teach ii. Duty to superiors iii. Duty to subordinates c. Balancing leadership i. Unions and management d. Servant leadership	7
VIII	Diversity a. What is diversity? i. Value in diversity ii. Organizational adaptation iii. The ethics of workplace fairness b. Recruitment and hiring i. Legal issues ii. The ethics of affirmative action iii. The ethics of quotas c. Promotions d. Women's issues in the fire service e. Hostility in the workplace i. Harassment (sexual and otherwise) ii. Hazing	8
Section 3: Administrative Ethics		
IX	Managing Ethics a. The relevance of responsibility i. Obligation and accountability b. Objective responsibility i. Public safety ii. Fire fighter safety iii. Administrative responsibility c. Subjective responsibility i. Traditions ii. Morals iii. Department growth iv. Quality of life d. Conflicts of responsibility i. Prioritizing obligations ii. Conflict of role e. Conflicts of authority i. Corruption ii. Conflicting priority f. Conflicts of authority	10

X	Leading an Ethical Culture a. What is an ethical culture? i. Why culture management is critical b. Unbounded ethics c. Compliance-based ethics i. Elements of compliance-based systems ii. Analysis of effectiveness d. Building an ethical culture i. The role of value-based leadership ii. Testing for ethics iii. Change management e. Maintaining an ethical culture i. Modeling ethics ii. Training in ethics iii. Ethics within training iv. Continuity of values through chain of command f. Barriers to creating an ethical culture i. Negative culture influences	9
XI	Fire Service Ethics and the Law a. Ethical conflicts i. Conflict of interest 1. Gifts 2. Quid pro quo 3. Financial reporting requirements ii. Outside employment iii. Abuse of position b. Financial ethical controls i. Accounting practices ii. Bidding and purchasing iii. Appropriation of funds c. Ethics and labor management i. Commonality of task ii. Contract conformance iii. Bargaining iv. Grievances v. Sexual harassment reporting vi. Discrimination restrictions d. Ethics and transparency i. Freedom of information ii. Open meetings acts iii. Reporting of economic interest requirements iv. Record keeping and disposal 1. E-mail and correspondence v. Whistle-blowers	11

Section 4: Applied Ethics		
XII	Ethical Decision Making a. Levels of ethical reflection i. Morals rules level ii. Ethical analysis iii. Post-ethical analysis iv. Descriptive models b. Conscience or obligation i. Moral imperatives ii. Fiduciary responsibility c. Outcome versus process d. Loci of control	12
XIII	Moral Disengagement a. Moral disengagement i. Rationalization ii. Context iii. Priority confusion iv. Cognitive/moral dissonance b. Normalizing bad behavior i. Culture effect on behavior ii. Focus on outcome rather than process iii. Attribution theory c. Stages of moral/ethical development	13
XIV	Contemporary Issues a. Privacy b. Off-duty activity c. Pay per spray d. Social media/blogging/body cameras e. Politics f. On-the-job relationships g. Harassment h. Politics and religion/separating beliefs from obligations	14

Flames: Drx/Dreamstime.com; Steel texture: © Sharpshot/Dreamstime.com

GLOSSARY

accountability an element of responsibility associated directly with another individual or group.

affective component the emotions, such as anxiety, sorrow, or excitement, that a person has regarding the object.

affirmative action a diversity effort intended to increase diversity representation through recruitment and employment from within a qualified pool of eligible minorities.

agent-based virtue ethics a system of ethics that is in the idea that each of us has an internal sense of what is right and wrong.

ALIR model a guideline for assessing ethical dilemmas rooted in the application of fundamental public responsibilities to public accountability, respect for law, personal integrity, and responsiveness to public need.

antisocial behavior aggressive, impulsive, and often violent actions that violate protective rules, conventions, and codes of a society.

applied ethics a system of ethics that seeks to identify a morally correct path or action relative to issues in everyday life.

argumentum ad baculum a Latin phrase meaning "appeal to the stick." It attempts to justify the use of coercion to bring about a desired outcome.

attitudes a collection of feelings, beliefs, and behavior tendencies directed toward specific people, groups, and ideas.

attribution theory a theory explaining the discrepancy between judging your own behavior versus others' behavior.

behavioral component the way attitude influences how we act or behave.

beliefs those things that we believe to be true absent proof.

beneficiary (aka principal) the group or individual for whom a fiduciary is responsible.

blog a discussion or informational website published on the World Wide Web and consisting of discrete, often informal diary-style text entries.

body camera also known as body worn video (BWV); a video-recording system that is typically utilized by law enforcement officers to record their interactions with the public or gather video evidence at crime scenes. It has been known to increase both officer and citizen accountability.

bright lines clearly identifiable boundaries of behavior.

care ethics a system of ethics based on the importance of being a positive force within a social group.

Carl Friedrich a theorist who asserted that ethical responsibilities within public agencies could best be assured by internal controls.

categorical imperative according to Immanuel Kant, an action that is required out of duty, whether it benefits the actor or not.

certification the assertion by a certifying body that an individual has been exposed to training and tested competency.

cognitive component a person's belief or knowledge about an attitude or object.

cognitive disengagement a suspension of rational understanding, or an exercise in self-deception.

cognitive dissonance the feeling of discomfort we feel when we perceive an inconsistency among values, beliefs, and attitudes.

community-based initiatives programs in support of improving quality of life within the community.

compliance-based diversity strategy diversity strategy targeted at complying with equal opportunity employment numerical objectives.

compliance-based ethics an ethics system utilizing specific policies and enforcement of imposed standards.

compliance creep the gradual overdependence on rules to define ethical behavior.

conditional egoism assertion that acting in self-interest is ethically acceptable only if it leads to moral outcomes.

conflict of interest a situation in which a person is in a position to derive personal benefit from actions or decisions made in his or her official capacity.

conflicts of loyalty ethical dilemma springing from inconsistency of needs and expectations of stakeholders with legitimate expectations of loyalty or authority.

conflicts of obligation ethical dilemma springing from competing obligations.

consequence analysis the prediction of likely outcomes and their ramifications.

consequential ethical assessment ethical choices are made based on an anticipated result of doing the most good.

consequentialism the theory that the ethics of an action should be judged by its outcome.

contextual standard of behavior a situational condition that suspends normal behavior standards, replacing them with new situationally relative standards.

conventional moral reasoning Kohlberg's second level of moral development.

cultural inertia the tendency of groups of people to maintain the status quo.

cultural relativism the belief that actions are moral or immoral relative to the beliefs and traditions of a culture.

decision ethics a genre of applied ethics in which various normative theories are applied to the decision-making process.

demographics categorizing the population based on specific identifying markers.

deontology-based ethical assessment decision making based on compliance to Immanuel Kant's categorical imperatives.

deontology the ethics of an action should be judged relative to its compliance with a code of conduct based on certain categorical imperatives.

descriptive belief a truth as we see it.

descriptive ethics a study to understand and catalog the attitudes of individuals or groups of individuals.

disparate impact programs or policies that, at face value, treat all groups equally but have a disproportionate effect on one group over another.

disparate treatment programs and policies that facilitate an unequal treatment of a minority group.

divine command theory assertion that universal moral truth came from God's will.

dual consequentialism consequences can be measured both morally and objectively.

due process the legal requirement that an authority must respect all legal rights that are owed to a person.

egoism theory asserting individuals ought to promote self-interest above any other value.

emotional intelligence (EI) the ability to understand and manage emotional aspects of human interaction.

empathy the action of understanding, being aware of, being sensitive to, and vicariously experiencing the feelings, thoughts, and experiences of others.

enabling behavior actively or passively supporting the actions of others.

equal opportunity a diversity initiative focusing on the creation of a "level playing field" by legally restricting unfair hiring and promotional practices that facilitate the disparate treatment of workers.

ethical altruism consequential doctrine that holds that each of us has a moral obligation to help serve and benefit others.

ethical attachment a condition in which an action has ethical implications. A condition of attachment is that an action has an impact on the well-being of others.

ethical dilemma a complex situation where values or moral imperatives conflict with each other.

ethical fading the displacement of personal values and judgment as a result of overdependence on rules.

ethical intelligence an awareness of moral principles and their contextual application.

ethical leadership the act of leading individuals using ethical methods and seeking ethical outcomes.

ethically bounded a condition in which right and wrong are defined strictly through conformance to policy.

ethical restraint expressions such as etiquette, politeness, and self-restraint that guide us in our interpersonal relationships.

ethical work culture a culture where doing the right thing is normal and easy, and doing the wrong thing is unusual, difficult, and unrewarded.

ethics the development, evaluation, and study of behavior boundaries and expectations within personal and professional interactions with others.

eudemonism a virtue ethic theory that centers on either moral or intellectual traits that cause the individual to "flourish."

evaluative belief a belief that a particular thing is good (has value) or is bad arising from descriptive beliefs.

external controls a form of accountability relying on the supervision of public agencies by elected officials, policy, and law.

externalized values values focused on responsibilities and obligations.

external locus of control attributing consequences to external forces such as luck, fate, or powerful individuals.

external motivations influences that are rooted in a sense of responsibility to someone or something

fairness judgment or equal distribution without bias, or regard for an individual's needs or feelings.

Federal Freedom of Information Act (FOIA) enacted in 1966, it was the first law that gave Americans the right to access the records of federal agencies.

fiduciary a person who holds either an imperative ethical or legal responsibility for the welfare of another person (referred to as beneficiary or principal) or a person's assets.

fiduciary responsibility a mandate for highest standard of care under the law.

fire administrators people responsible for taxpayer money, signing contracts, disciplining employees, and writing policies affecting stakeholders both in the fire service and the community at large.

Fire Fighter Code of Ethics code of ethics developed by the Cumberland Valley Volunteer Fireman's Association and adopted by the U.S. Fire Administration.

fire fighters' bill of rights legislation intended to apply commonsense principles of fairness and professionalism to the process of investigating and disciplining first responders.

folkways the unofficial practices and rituals that define routine and everyday behaviors.

framing a theory describing how our responses to situations, including our ethical judgments, are affected by how those situations are posed or viewed.

functional imperatives guidelines for assessing ethical dilemmas rooted in the concepts of morality, stakeholder legitimacy, duty, and primacy of responsibilities.

game theory as applied to ethics, the belief that an otherwise unethical action is justifiable if there is no expectation of fairness.

grievance procedure a formalized process to clarify contract wording and intent.

Hatch Act a federal law passed in 1939 that limits certain political activities by governmental officials.

Hawthorne effect scientific theory that suggests that the observation of an action tends to modify it.

hazing the ritualistic emotional or physical abuse of new group members by existing members.

Health Insurance Portability and Accountability Act (HIPAA) A medical privacy act passed in 1996 that mandates industry-wide standards for health care information on billing and other processes, and requires the protection and confidential handling of protected health information.

Herman Finer a social scientist who asserted that ethical standards of behavior among government officials could only be ensured by a clearly defined policy body, accompanied by significant oversight.

human need something that a human must have in order to live a recognizably human life.

individual relativism moral values are created by individuals, relative to their place, time, and situation.

innate existing in, belonging to, or determined by factors present in an individual from birth.

integrity-based ethics the individual self-regulates based on a shared understanding of behavior standards, and with the intention of contributing to the common good.

intellectual virtues virtues that are inherent in an individual's nature.

interest condition that is directly related to a need or the acquisition thereof.

internal controls a form of accountability relying on individual ethical behavior reinforced by ethics education, the maintenance of ethical cultures, and the removal of disincentives for ethical misbehavior.

internalized values values focused on needs and wants.

internal locus of control attributing consequences to choices.

inward motivations influences that seek to meet perceived needs or wants.

justice-based ethical assessment ethical choices made on the perceived basis of fairness or a more advanced concept of social justice.

justice the quality of being equitable, compliant with law, or acting with respect to individual rights.

justitia socialis a body of law that seeks to maintain justice and fairness within society

Kohlberg's stages of moral development a scale depicting three stages of moral reasoning in an attempt to illustrate how people assess and justify their actions: preconventional moral reasoning, conventional moral reasoning, and post-conventional moral reasoning.

Labor Management Relations Act also known as the Taft-Hartley Act, it delineated unfair labor practices imposed on unions.

laws a collection of policies enforced by government to regulate the behavior of the governed.

leadership in ethics leadership methods intended to raise the ethical standards of followers.

licensure permission by a recognized authority to practice a profession or trade.

locus of control the extent to which people believe they have power over the events in their lives.

macro ethics a system of ethics focusing on global issues such as the geopolitical ramifications of wealth distribution, technology emergence, international business practices, and environmental issues.

malum in se a Latin term meaning "wrong" or "evil in itself." This is a fundamental concept in understanding morality.

malum prohibitum a thing is illegal because the law prohibits it.

Maslow's hierarchy of needs a theory that suggests, as humans, we have needs that can be categorized as physiological, safety, social, esteem, and self-actualization. These needs are listed in their order of primacy.

mentoring programs as applied to diversity, programs that identify, recruit, and prepare minority candidates for eventual employment.

metaethics a system of ethics that explores the meanings of concepts, words, and values associated with ethics.

modeling a theory suggesting behavior cues from trusted individuals influence levels of ethical behavior within organizations.

moral apathy a lack of moral engagement springing from a lack of awareness or cognitive disengagement.

moral disengagement the process of convincing the self that ethical standards do not apply to yourself in a particular context.

moral dissonance an internal conflict between an individual's actions and his or her personal values.

moral engagement A focus on ethical issues and an awareness of consequence.

morality a doctrine or system of moral conduct.

moral rationalization favoring a particular conclusion as a result of some factor (such as self-interest) that is of little justificatory logical relevance.

moral relativism a concept that right and wrong are predicated on context.

morals or morality clause a provision within a contract or department policy that curtails, restrains, or prohibits certain behavior of individuals as inappropriate.

morals universal principles of goodness usually attributed to religion or some other higher authority.

moral virtues virtues based on character and acquired through learning.

mores behavior expectations associated with the concepts of right and wrong. Terms like "decency" and "morality" are often connected to social mores.

motivation the act or process (such as a need or desire) that causes a person to act.

National Labor Relations Act also known as the Wagner Act, it granted employees the rights of self-organization and collective bargaining.

need an object or condition for which negative consequences will occur in its absence.

normalizing behavior patterns reassertion of the status quo by encouraging group minorities to adopt the behaviors and patterns of the group while simultaneously marginalizing nonconformists.

normative egoism acting in our own self-interest.

normative ethics a system of ethics that investigates questions regarding how an individual should morally act.

Nuremberg trials a series of military tribunals judging Nazi officials accused of war crimes. They were conducted from 1945 to 1949 in Nuremberg, Germany.

objectively in a way that is not influenced by personal feelings or opinions.

objective responsibility a responsibility rooted in obligation to task that is externally imposed.

objectivism moral values are universal, existing outside human conventions.

obligation a responsibility to a thing, task, or ideal.

official values those values expressed within the department's mission and value statements.

organization development a field of study dedicated to expanding the knowledge and effectiveness of people to accomplish more successful performances.

pay for spray a euphemism for the practice of contracting fire protection, usually on an annual basis, to unincorporated areas.

personal ethics code a formal declaration of values and priorities.

personal ethics the basic principles and values that govern interactions among individuals.

personal inertia the tendency of individuals to be resistant to change.

pluralistic deontology theory of morality with imperatives stating that individuals have the duty to keep promises, a duty to pursue justice, a duty to improve the conditions of others, a duty to self-improvement, and most importantly a duty not to injure others.

positional abuse using official status to acquire benefits to yourself or your immediate family.

positional authority authority bestowed as a condition of rank.

postconventional moral reasoning Kohlberg's third level of moral development.

preconventional moral reasoning Kohlberg's first level of moral development.

prescriptive belief an expression of justifiable outcome and is based on either descriptive or, more likely, evaluative beliefs.

prime theory of obligation a description of the base ethical obligation of the fire service; it is a touchstone for all other responsibilities and obligations associated with fire department activities.

professional code of ethics a formalized statement of behavior expectations created by a recognized authority.

professional ethics professionally accepted standards of personal and business behavior, values, and guiding principles.

professionalism the skill, good judgment, and polite behavior that is expected from a person who is trained to do a job well.

proximity of interest the relative immediacy of an action as related to a need.

proximity the relative value of a consequence either in time or scale.

psychological egoism suggestion of a universal truth that individuals will always act out of self-interest.

quid pro quo translates to "something for something," and generally describes the conditions for bribery.

quota systems mandated hiring targets of underrepresented minorities based on demographics.

rational egoism assertion that placing someone's interests above your own was contrary to rational thought and arose only from emotions of guilt.

referent authority authority related to respect and trust.

revolving door provisions typically restrict government employees from accepting employment with agencies that deal directly with their former employer for a specific period of time.

rule consequentialism rules that must have best-intended consequences.

rule of utility an assertion that good is that which brings the greatest happiness to the greatest number of people

Sarbanes-Oxley Act mandates agencies with fiduciary responsibilities to adopt a code of ethics; many affect municipal agencies.

self-awareness an awareness of your personality or individuality.

self-control restraint exercised over your impulses, emotions, or desires.

servant leadership leadership theory developed by Robert Greenleaf oriented to putting the well-being of people and communities first as a management priority.

situational ethics ethical theory in which right and wrong are judged by context and intent.

social norms pattern of behavior in a particular group, community, or culture that is accepted as normal and to which an individual is expected to conform.

social role a form of personal identity associated with particular responsibilities and obligations.

social skills the tools that enable people to communicate, learn, ask for help, and get needs met in appropriate ways.

subjective characteristic of or belonging to reality as perceived rather than as independent of mind.

subjective responsibility a responsibility that springs from internal values, beliefs, and priorities.

subjectivism moral values are developed by people out of necessity or social convention.

sunshine laws a body of ethical requirements intended to promote governmental transparency.

taboos a strict prohibition of behavior. Violation of taboo will likely elicit social censure.

tokenism a general term applied to the practice of selecting minorities simply for aesthetics or out of political correctness.

tolerance sympathy or indulgence for beliefs or practices differing from or conflicting with one's own.

transformational approach to diversity the promotion of diversity through the making of a business case for social responsibility and the economic advantages of having a diverse workforce.

transformational leadership leadership theory by James M. Burns that stresses employee empowerment and development.

unbounded having no limit; all encompassing.

unofficial values values held by fire fighters that are a reflection of the department's culture and traditions.

utilitarianism the ethics of an action can most effectively be judged by its positive consequence to a majority of stakeholders.

values a collection of virtues, principles, standards, or qualities that an individual or group holds in high regard.

value statement a formal declaration of an organization's priorities and core beliefs.

viral post a blog or social media message that is continuously shared by everyone who receives it.

virtue a habit or an acquired character quality that is deemed universally good.

virtue-based ethical assessment basing ethical choices on perceived consistency with character.

virtue ethics the belief that ethics is rooted within compliance to particular virtues.

want a desire for an object or condition for which there is no interest.

whistle-blower a person who exposes any kind of information or activity by an organization that is deemed illegal or unethical.

whistle-blower protections prohibitions against authorities taking (or threatening to take) retaliatory personnel action against any employee or applicant because of disclosure of information by that employee or applicant.

workplace diversity differences among employees in terms of age, cultural background, physical abilities and disabilities, race, religion, sex, and sexual orientation.

INDEX

Note: Page numbers followed by "*b*" denote boxes; "*f*" denote figures; and "*t*" denote tables.

A

B

C

D

J

K

L

Q

R

S